CLINICAL CHEMISTRY

IN

DIAGNOSIS AND TREATMENT

OTHER PUBLISHED WORK
OF RELATED INTEREST

Multiple-Choice Questions
on
"Clinical Chemistry in Diagnosis & Treatment"
by P. R. Fleming MD FRCP and P. H. Sanderson MB BChir FRCP

Clinical Chemistry

in

Diagnosis and Treatment

JOAN F. ZILVA

M.D., B.Sc., F.R.C.P., F.R.C.Path., D.C.C. (Biochem.)

*Reader in Chemical Pathology, Westminster
Medical School, London; Honorary Consultant in Chemical
Pathology, Westminster Hospital, London*

P. R. PANNALL

M.B., B.Ch. (Witwatersrand), F.F.Path. (S.A.),
M.R.C.Path., F.A.A.C.B.

*Senior Consultant Clinical Chemist,
The Queen Elizabeth Hospital, Adelaide*

Third Edition
(Reprint)

LLOYD-LUKE (MEDICAL BOOKS) LTD

49 NEWMAN STREET

LONDON

1981

FIRST EDITION	1971
Reprinted	1972
Reprinted	1973
SECOND EDITION	1975
Reprinted	1978
Italian translation	1978
Turkish translation	1978
THIRD EDITION	1979
Spanish translation	1979
Serbocroat translation	1979
Reprinted	1981

PRINTED AND BOUND IN ENGLAND BY
HAZELL WATSON AND VINEY LTD
AYLESBURY, BUCKS
ISBN 0 85324 139 2

FOREWORD

THIS book aims at giving, within a single cover, all the relevant bio-chemical and pathological facts and theories necessary to the intelligent interpretation of the analyses usually performed in departments of Clinical Chemistry or Chemical Pathology. The approach is firmly based on general principles and the authors have gone to great trouble to ensure that the development of ideas is logical and easy to follow. A basic knowledge of medicine and of elementary biochemistry is assumed but, given this, the reader should be able to understand even the more involved metabolic inter-relationships without undue difficulty and should thereafter be in a strong position to apply this knowledge in medical practice.

I have appreciated very much the opportunity of reading this book during its preparation and feel that it represents a distinctly novel approach to the interpretation of biochemical data. In my opinion it could be read with advantage by all the categories of reader mentioned in the Preface, and moreover I believe that they will enjoy the experience.

N. F. MACLAGAN

January, 1971

PREFACE TO THE THIRD EDITION

WE have again made several major modifications in this edition, some because of new advances in knowledge, some as a result of useful comments from our readers, and some because our own view of the topic has evolved. Gonadal function now merits a separate chapter, and we have radically modified that on hydrogen ion homeostasis and part of that on carbohydrate metabolism. Newer advances in prenatal monitoring have been included. After many of the chapters we have introduced sections in which we give the clinical indications for the investigations that have been discussed in that chapter. Table XXXV, which quotes approximate mean "normal" values, now also includes some indication of the expected imprecision due to *analytical factors*, although, as we stress, other sources of variation often outweigh these.

Many reviewers have suggested that, because of their importance at the ward/laboratory interface, the chapters entitled "The Clinician's Contribution to Valid Results" and "Requesting Tests and Interpreting Results" should come first rather than last. Because the book is aimed primarily at undergraduates, we feel that we cannot usefully discuss these topics before outlining the clinical significance of results in the rest of the book. However, we agree with the underlying sentiment, and regret that not everyone seems to have the staying power to read (or, at least, to remember) these chapters. This may be due to a mistaken impression that questions based on them will not be asked in examinations: students (at both undergraduate and postgraduate level) should be warned that this is increasingly untrue, for the good reason that a failure to understand the topics may have very serious clinical consequences.

The list of those to whom we are indebted for help grows with each edition, and, if complete, would almost take up a book on its own. We would particularly like to thank Dr. Arnold Fertig, Mr. Donald Mc-Lauchlan, Dr. Frances Murray and Dr. Philip Nicholson for each providing lively and useful comments (sometimes under pressure of time) on almost every chapter. Some of the others who have been equally helpful in their comments on one or more chapters are Dr. C. Beng, Professor F. V. Flynn, Professor S. C. Frazer, Dr. P. G. Frost, Professor J. R. Hobbs, Dr. Jane Maher, Professor D. M. Matthews, Mr. D. E. Perry, Dr. P. R. Raggatt, Dr. Pamela Riches, Professor M. C. Scrutton, Dr. Judith Scurr, Dr. J. Tendall, Dr. G. Walters, Dr. M. Wellby, and Dr. J. T. Whicher. The impact of Dr. Paul Carter on the second edition is still apparent in the third. Mrs. Joanne Payne and Miss Manuella Artini have skilfully typed the script and drafts, and Mr. David Gibbons, of the Department of Medical Photography and Illustration of the Westminster Medical School, has again prepared the illustrations. We thank the publishers for their continuing co-operation.

JFZ
PRP

January, 1979

PREFACE TO THE FIRST EDITION

THIS book is intended primarily for medical students and junior hospital staff. It is based on many years practical experience of undergraduate and postgraduate teaching, and of the problems of a routine chemical pathology department. It is written for those who learn best if they understand what they are learning. Wherever possible, explanations of the facts are given: if the explanation is a working hypothesis (like, for instance, that for the ectopic production of hormones) this is stressed, and where no explanation is known, this is stated. Our experience of teaching and our discussions with students have led us to believe that many of them are willing to read a slightly longer book if it gives them a better understanding than a shorter one. Electrolyte and acid-base balance have been discussed in some detail because, in our experience, these are the most common problems of chemical pathology met with by junior clinicians and ones in which there are often dangerous misunderstandings. Some subjects, such as the porphyrias and conditions of iron overload, are discussed in greater detail than is necessary for undergraduate examinations: however, the incidence of these is high in some areas of the world, and an elementary source of reference seemed to be needed.

We have tried to stress the clinical importance of an understanding of the subject: by including in the chapter appendices some details of treatment, difficult to find together in other books, we hope that this one may appeal to clinicians as well as to students. Two chapters are included on the best use of a laboratory, including precautions which should be taken in collecting specimens and interpreting results. As we have stressed, pathologists and clinicians should work as a team, and full consultation between the two should be the rule. The diagnostic tests suggested are those which we have found, by experience, to be the most valuable. For instance, in the differential diagnosis of hypercalcaemia we find the steroid suppression test very helpful, phosphate excretion indices fallible and tedious to perform and estimation of urinary calcium useless.

Our own junior staff have found the drafts helpful in preparing for the Part 1 examination for Membership of the Royal College of Physicians, as well as for the primary examination for Membership of the Royal College of Pathologists, and our Senior Technicians are using it to study for the Special Examination for Fellowship in Chemical Pathology of the Institute of Medical Laboratory Technology. We feel that those studying for the primary examination for the Fellowship of the Royal College of Surgeons might also make use of it. It could provide a groundwork for study for the final examination for the Membership of the Royal College of Pathologists in Chemical Pathology, and for the Mastership in Clinical Biochemistry.

Initially each chapter should be worked through from beginning to end. For revision purposes there are lists and tables, and summaries of the contents of each chapter. The short sub-indices in the Table of Contents should facilitate the use of the book for reference.

Appendix A lists some analogous facts which we hope may help understanding and learning. The list must be far from complete, and the student should seek other examples for himself.

We wish to thank Professor N. F. Maclagan for his unfailing encouragement and his helpful advice and criticisms. We are also indebted to a great many other people, foremost amongst whom we should mention Dr. J. P. Nicholson who has read and commented on the whole book, and Professor M. D. Milne, Professor D. M. Matthews, Mr. K. B. Cooke and Dr. B. W. Gilliver for helpful advice and criticism on individual chapters. Many registrars and senior house officers in the Department, in particular Dr. Krystyna Rowland, Dr. Elizabeth Small, Dr. Nalini Naik and Dr. Noel Walmsley, have been closely involved in the preparation of the book and have provided invaluable suggestions and criticisms. Students, too, have read individual chapters for comprehensibility, and we would particularly like to thank Mr. B. P. Heather, Mr. J. Muir and Miss H. M. Merriman for helpful comments. Mr. C. P. Butler of the Westminster Hospital Pharmacy was most helpful during the preparation of the sections on therapy. Mrs. Valerie Moorsom and Mrs. Marie-Lise Pannall, with the help of Mrs. Brenda Sarasin and Miss Barbara Bridges, have borne with us during the typing of the drafts and the final transcript. The illustrations were prepared by Mr. David Gibbons of the Department of Medical Photography and Illustration of the Westminster Medical School.

Finally, we would like to thank the publishers for their co-operation and understanding during the preparation of this book.

JFZ
PRP

May, 1971

CONTENTS

UNITS IN CHEMICAL PATHOLOGY

RESULTS in chemical pathology have been expressed in a variety of units. For instance, electrolytes were usually quoted in mEq/l, protein in g/100 ml and cholesterol in mg/100 ml. The units used might vary from laboratory to laboratory: calcium might be expressed as mg/100 ml or mEq/l; in Britain, urea results were expressed as mg/100 ml of *urea*, whereas in the United States it is usual to report mg/100 ml of *urea nitrogen*. This situation can be confusing and with patients moving, not only from one hospital to another, but from one country to another, dangerous misunderstandings could arise.

Système International d'Unités (SI Units)

International standardisation is obviously desirable; such standardisation has already been introduced into many branches of science and technology.

The main recommendations for chemical pathology are as follows:

1. Where the molecular weight (MW) of the substance being measured is known, the unit of quantity should be the *mole* or submultiple of a mole.

$$\text{Number of moles (mol)} = \frac{\text{wt. in g}}{\text{MW}}$$

In chemical pathology millimoles (mmol), micromoles (μmol) and nanomoles (nmol) are the most common units.

2. The unit of volume should be the *litre*. Units of concentration are therefore mmol/l, μmol/l or nmol/l.

Examples

1. *Results previously expressed as mEq/l*

$$\text{Number of equivalents (Eq)} = \frac{\text{wt. in g}}{\text{Equivalent wt.}}$$

$$= \frac{\text{wt. in g} \times \text{valency}}{\text{MW}}$$

(a) In the case of univalent ions, such as sodium and potassium, the units will be numerically the same. A sodium of 140 mEq/l becomes 140 mmol/l.

(b) For polyvalent ions, such as calcium or magnesium (both divalent), the old units are numerically divided by the valency. For instance, a magnesium of 2·0 mEq/l becomes 1·0 mmol/l.

2. *Results previously expressed as mg/100 ml*

If results were previously expressed in mg/100 ml the method of conversion to mmol/l is to divide by the molecular weight (to convert from mg to mmol), and to multiply by 10 (to convert from 100 ml to a litre). Thus effectively the previous units are divided by 1/10 of the molecular weight. For instance, the molecular weight of urea is 60, and of glucose 180. A urea value of 60 mg/ 100 ml and a glucose value of 180 mg/100 ml are both equivalent to 10 mmol/l.

The factor of 10 is, of course, only used for concentrations. The total amount of urea excreted in 24 hours in mg is numerically 60 times that in mmol.

Exceptions

1. *Units of pressure* (e.g. mm Hg) are expressed as pascals (or kilopascals— kPa). 1 kPa = 7·5 mm Hg, so that a Po_2 of 75 mm Hg is 10 kPa. Pascals are SI units.

2. *Proteins.* Body fluids contain a complex mixture of proteins of varying molecular weights. It is therefore recommended that the gram (g) be retained, but that the unit of volume be the litre. Thus a total protein of 7·0 g/100 ml becomes 70 g/l.

3. The expression 100 ml is to be expressed as *decilitre* (dl).

4. *Enzyme units* are not to be changed yet. Note that *the definition of international units for enzymes does not state the temperature of the reaction* (p. 337).

5. Some constituents, such as IgE and some hormones, are still expressed in "international" or other special units.

At the moment different laboratories are at different stages of implementation. In this book we have adopted the following policy.

1. Where the old and new units are numerically the same, we have given only the new units (e.g. for sodium and potassium).

2. For proteins we give only g/l.

3. Where it is generally accepted that the new units be adopted we have given these, with the equivalent old units in brackets.

A conversion table for some of the commoner results is given opposite.

Note that:

1 mol	= 1 000 mmol
1 mmol (10^{-3} mol)	= 1 000 μmol
1 μmol (10^{-6} mol)	= 1 000 nmol (nanomoles)
1 nmol (10^{-9} mol)	= 1 000 pmol (picomoles).

FURTHER READING

BARON, D. N., BROUGHTON, P. M. G., COHEN, M., LANSLEY, T. S., LEWIS, S. M., and SHINTON, N. K. (1974). *J. clin. Path.*, **27**, 590.

BARON, D. N. (1974). *Brit. med. J.*, **4**, 509.

BOLD, A. M., and WILDING, P. (1975). *Clinical Chemistry Conversion Scales for SI units with Adult Normal (Reference) Values.* Oxford: Blackwell Scientific Publications.

Some Approximate Conversion Factors for SI Units

	From SI Units		To SI Units	
Bilirubin	$\mu mol/l \times 0\cdot058$	$= mg/dl$	$mg/dl \div 0\cdot058$	$= \mu mol/l$
Calcium				
Plasma	$mmol/l \times 4$	$= mg/dl$	$mg/dl \div 4$	$= mmol/l$
Urine	$mmol/24\,h \times 40$	$= mg/24\,h$	$mg/24\,h \div 40$	$= mmol/24\,h$
Cholesterol	$mmol/l \times 39$	$= mg/dl$	$mg/dl \div 39$	$= mmol/l$
Cortisol				
Plasma	$nmol/l \times 0\cdot036$	$= \mu g/dl$	$\mu g/dl \div 0\cdot036$	$= nmol/l$
Urine	$nmol/24\,h \times 0\cdot36$	$= \mu g/24\,h$	$\mu g/24\,h \div 0\cdot36$	$= nmol/24\,h$
Creatinine				
Plasma	$\mu mol/l \times 0\cdot011$	$= mg/dl$	$mg/dl \div 0\cdot011$	$= \mu mol/l$
Urine	$\mu mol/24\,h \times 0\cdot11$	$= mg/24\,h$	$mg/24\,h \div 0\cdot11$	$= \mu mol/24\,h$
Gases				
Po_2 Pco_2	$kPa \times 7\cdot5$	$= mm\,Hg$	$mm\,Hg \div 7\cdot5$	$= kPa$
Glucose	$mmol/l \times 18$	$= mg/dl$	$mg/dl \div 18$	$= mmol/l$
Iron TIBC	$\mu mol/l \times 5\cdot6$	$= \mu g/dl$	$\mu g/dl \div 5\cdot6$	$= \mu mol/l$
Phosphorus	$mmol/l \times 3$	$= mg/dl$	$mg/dl \div 3$	$= mmol/l$
Proteins				
Serum				
Total Albumin Immuno-globulins	$g/l \div 10$	$= g/dl$	$g/dl \times 10$	$= g/l$
Urine				
Concentration	$g/l \times 100$	$= mg/dl$	$mg/dl \div 100$	$= g/l$
Daily Output	$g/24\,h$		No change	
Urate	$mmol/l \times 17$	$= mg/dl$	$mg/dl \div 17$	$= mmol/l$
Urea	$mmol/l \times 6$	$= mg/dl$	$mg/dl \div 6$	$= mmol/l$
5-HIAA HMMA	$\mu mol/24\,h \times 0\cdot2$	$= mg/24\,h$	$mg/24\,h \div 0\cdot2$	$= \mu mol/24\,h$
Oestriol 17 Oxosteroids Total 17-Oxogenic steroids	$\mu mol/24\,h \times 0\cdot3$	$= mg/24\,h$	$mg/24\,h \div 0\cdot3$	$= \mu mol/24\,h$
Faecal "Fat"	$mmol/24\,h \times 0\cdot3$	$= g/24\,h$	$g/24\,h \div 0\cdot3$	$= mmol/24\,h$

Abbreviations Used in the Book or in Common Use

ACP	Acid Phosphatase
ACTH	Adrenocorticotrophic Hormone (corticotrophin)
ADH	Antidiuretic Hormone ("Pitressin": Vasopressin)
ALA	5-Aminolaevulinate
ALP	Alkaline Phosphatase
ALS	Aldolase
ALT	Alanine Transaminase (=SGPT)
AMS	α-Amylase
APRT	Adenine Phosphoribosyl Transferase
APUD	Amine-Precursor Uptake and Decarboxylation
AST	Aspartate Transaminase (=SGOT)
BJP	Bence Jones Protein
BMR	Basal Metabolic Rate
BSP	Bromsulphthalein
BUN	Blood Urea Nitrogen $\left(\text{in mg/dl} = \frac{28}{60} \times \text{blood urea in mg/dl}\right.$
	$\left.\text{or } 2\cdot8 \times \text{blood urea in mmol/l}\right)$
CBG	Cortisol-Binding Globulin (Transcortin)
CC	Cholecalciferol
CK	Creatine Kinase (=CPK)
CoA	Coenzyme A
CPK	Creatine Phosphokinase (=CK)
CRF	Corticotrophin Releasing Factor
CSF	Cerebrospinal Fluid
1,25-DHCC	1,25-Dihydroxycholecalciferol
DIT	Di-iodotyrosine
DNA	Deoxyribonucleic Acid
DOC	Deoxycorticosterone
DOPA	Dihydroxyphenylalanine
DOPamine	Dihydroxyphenylethylamine
ECF	Extracellular Fluid
EDTA	Ethylene Diamine Tetra-acetate (Sequestrene)
EM Pathway	Embden-Meyerhof Pathway (Glycolytic Pathway)
ESR	Erythrocyte Sedimentation Rate
FAD	Flavin Adenine Dinucleotide
FBS	Fasting Blood Sugar
FFA	Free Fatty Acids (=NEFA)
FMN	Flavin Mononucleotide
FSH	Follicle-Stimulating Hormone (follitropin)
FTI	Free Thyroxine Index
GFR	Glomerular Filtration Rate
GGT	γ-Glutamyltransferase (γ-Glutamyltranspeptidase)
GH	Growth Hormone (somatotropin)

GMD	Glutamate Dehydrogenase
GOT	Glutamate Oxaloacetate Transaminase (=AST)
G-6-P	Glucose-6-Phosphate
G6PD	Glucose-6-Phosphate Dehydrogenase
GPT	Glutamate Pyruvate Transaminase (=ALT)
GTT	Glucose Tolerance Test
HAA	Hepatitis Associated Antigen (=Australia Antigen; HBsAg; hepatitis B antigen)
HBD	Hydroxybutyrate Dehydrogenase
HBsAg	Hepatitis B Antigen (HAA)
25-HCC	25-Hydroxycholecalciferol
HCG	Human Chorionic Gonadotrophin
HCS	Human Chorionic Somatomammotrophin (=HPL)
HDL	High-Density Lipoprotein
HGH	Human Growth Hormone
HGPRT	Hypoxanthine Guanine Phosphoribosyl Transferase
5HIAA	5-Hydroxyindole Acetic Acid
HMMA	4-Hydroxy-3-Methoxymandelic Acid (=VMA)
HPL	Human Placental Lactogen (=HCS)
5HT	5-Hydroxytryptamine (=Serotonin)
5HTP	5-Hydroxytryptophan
ICD	Isocitrate Dehydrogenase
ICF	Intracellular Fluid
ICSH	Interstitial Cell-Stimulating Hormone (=LH)
Ig	Immunoglobulin
LATS	Long-Acting Thyroid Stimulator
LCAT	Lecithin Cholesterol Acyl Transferase
LD	Lactate Dehydrogenase (=LDH)
LDH	Lactate Dehydrogenase (=LD)
LDL	Low-Density Lipoprotein
LH	Luteinising Hormone (=lutropin; ICSH)
LH-RH	LH-Releasing Hormone
MEA	Multiple Endocrine Adenopathy (=pluriglandular syndrome)
MIT	Mono-iodotyrosine
MSH	Melanocyte-Stimulating Hormone (melanotropin)
NAD	Nicotinamide Adenine Dinucleotide (=DPN)
NADP	Nicotinamide Adenine Dinucleotide Phosphate (=TPN)
NEFA	Non-Esterified Fatty Acids (=FFA)
5'NT	5'-Nucleotidase (=NTP)
NTP	5'-Nucleotidase (=5'NT)
17-OGS	17-Oxogenic Steroids (T 17-OGS)
11-OHCS	11-Hydroxycorticosteroids ("Cortisol")
17-OHCS	17-Hydroxycorticosteroids (17-oxogenic steroids)
OP	Osmotic Pressure
PBG	Porphobilinogen
PBI	Protein-Bound Iodine
PHLA	Post-Heparin Lipolytic Activity

PIF	Prolactin-Release Inhibiting Factor (prolactostatin)
PP factor	Pellagra Preventive Factor (nicotinamide: niacin)
PRPP	Phosphoribosyl Pyrophosphate
PTH	Parathyroid Hormone ("Parathormone")
RF	Releasing Factor (=RH; liberin)
RH	Releasing Hormone (=RF; liberin)
RNA	Ribonucleic Acid
RU	Resin Uptake (of T_3 or T_4)
SG	Specific Gravity
SGOT	Serum Glutamate Oxaloacetate Transaminase (=AST)
SGPT	Serum Glutamate Pyruvate Transaminase (=ALT)
SHBD	Serum Hydroxybutyrate Dehydrogenase (=HBD)
T_3	Tri-iodothyronine
T_4	Thyroxine (Tetra-iodothyronine)
TBG	Thyroxine Binding Globulin
TBPA	Thyroxine-Binding Prealbumin
TBW	Total Body Water
TCA cycle	Tricarboxylic Acid Cycle (=Krebs' Cycle or Citric Acid Cycle)
TIBC	Total Iron-Binding Capacity (usually measure of transferrin (siderophilin))
T-17-OGS	Total 17-Oxogenic Steroids (17-OGS)
TP	Total Protein
TPN	Triphosphopyridine Nucleotide (=NADP)
TPP	Thiamine Pyrophosphate
TRF	Thyrotrophin-Releasing Factor
TRH	Thyrotrophin-Releasing Hormone
TSH	Thyroid-Stimulating Hormone (Thyrotrophin)
UDP	Uridine Diphosphate
UTP	Uridine Triphosphate
VLDL	Very Low-Density Lipoprotein
VMA	Vanillyl Mandelic Acid (=HMMA)
WDHA	Watery Diarrhoea, Hypokalaemia and Achlorhydria (=Werner-Morrison syndrome)
Z–E syndrome	Zollinger-Ellison Syndrome

Chapter I

THE KIDNEYS: RENAL CALCULI

THE KIDNEYS

THE kidneys excrete waste products of metabolism and are the most important organs in the maintenance of normal body homeostasis. The renal tubules reabsorb some metabolically important substances, such as glucose, from the glomerular filtrate, secrete other substances into it, and exchange ions across the cell wall (for instance potassium and hydrogen in exchange for sodium ions). It should be remembered that although the renal tubular cells are of the greatest importance from the point of view of homeostasis, many of these processes occur in other cells of the body, including those of the intestinal mucosa. The normal functioning of these cells in the kidney depends on:

an adequate volume of glomerular filtrate with which the exchanges can occur (and therefore on normal glomerular function);

concentrations of ions in the tubular cells representative of those in the body as a whole (for example, potassium and hydrogen ions);

the ability of the endocrine glands to secrete such hormones as anti-diuretic hormone (ADH) and aldosterone;

the integrity of the feed-back mechanisms controlling the hormones;

functionally intact renal cells.

If all these factors are normal the kidney retains just as much of each constituent as the body requires.

Other functions of the kidney, which will not be dealt with further in this chapter, are:

the production of erythropoietin—a hormone stimulating erythro-poiesis. The student is referred to textbooks of haematology for further details;

the production of renin (p. 38);

the conversion of 25-hydroxycholecalciferol to the active 1,25 di-hydroxycholecalciferol (p. 235).

NORMAL RENAL FUNCTION

Glomerular Filtration

Glomerular filtration is a passive process. Protein and protein-bound plasma constituents are filtered in negligible amounts by the normal glomerulus and most of the small amount of protein that is filtered is probably reabsorbed. Diffusible plasma constituents such as sodium, potassium and *ionised* calcium are present at concentrations almost identical with those in extracellular fluid. Changes in the blood supply to the glomerulus, or reduction of the permeability of the membrane, affect the *volume, but not the composition*, of the filtrate. Increased permeability of the glomerulus may lead to proteinuria.

Tubular Function

The glomerular filtrate is modified in two ways. Many substances are dealt with actively by the tubular cells, while others are reabsorbed passively.

Passive transport can be explained by the presence of physicochemical gradients. For instance, a hydrostatic pressure gradient accounts for glomerular filtration and in the tubules water will move passively from an area of relatively low to relatively high osmotic pressure, or solute will move in the opposite direction: ions can also move along an electro-chemical gradient produced by reabsorption of charged ions (for instance, active reabsorption of cation may be accompanied by passive reabsorption of anion). For such movements to take place the cell wall must be permeable to the substances concerned, and selective permeability may explain preferential absorption of one ion or another. Passive reabsorption requires no metabolic energy, but cell death may affect it by altering the permeability of cell walls.

Active transport can occur against physicochemical gradients and this process requires energy, usually supplied by adenosine triphosphate (ATP). It is affected directly by cell death, enzyme poisons, and by hypoxia which impairs ATP production by oxidative phosphorylation. Such types of exchange occur in all cells in the body (for instance the "sodium pump" p. 66): in absorptive cells, such as those of the renal tubule and the intestinal mucosa, the process is one of passing substances through the cell from the lumen into the blood stream; in other cells transport is in the same direction on all sides of the cell, and substances pass in or out of the cell rather than through it.

Many substances, such as urea and hydrogen ion, can reach concentrations in the urine well above those in the blood. Whether this differential concentration is the result of reabsorption of water without solute (urea) or of secretion by the tubular cell (hydrogen ion), its maintenance depends on the relative impermeability of that part of the tubule distal to the site of concentration. If this breaks down high concentrations may not be reached.

We will now consider some of the more important urinary constituents dealt with by the renal tubular cells.

Glucose and amino acids are normally reabsorbed from the proximal renal tubule, and fluid entering the descending limb of the loop of Henle usually contains very low concentrations of these constituents, which can then be reutilised in body metabolism. The higher the concentration in the glomerular filtrate the more escapes reabsorption; at glucose concentrations above 10 mmol/l (180 mg/dl) enough glucose is usually present in the urine to be detectable by routine tests (but see p. 184). Impairment of tubular function may cause detectable glycosuria at lower blood levels.

Phosphate is incompletely reabsorbed by the proximal tubule: the presence of phosphate in the urine is important for buffering (p. 88). Reabsorption is inhibited by parathyroid hormone (PTH), and this action accounts for some of the changes in the plasma when PTH is circulating in excess (p. 233). Vitamin D also has some inhibitory effect.

Calcium and magnesium.—The reabsorption of calcium and magnesium is poorly understood. It is probably increased by parathyroid hormone, and these ions may compete with sodium.

Urate is probably normally completely reabsorbed in the proximal tubule. Normal urinary urate is derived from active tubular secretion.

Sodium.—Sodium can be reabsorbed by three mechanisms:

Isosmotic reabsorption.—About 70 per cent of the sodium in the glomerular filtrate is reabsorbed by an active process in the proximal tubule. This reabsorption is limited by the availability of chloride (see below).

Exchange with hydrogen ion.—Sodium reabsorption in exchange for hydrogen ion is linked with bicarbonate reabsorption, and is dependent on the enzyme carbonate dehydratase (carbonic anhydrase), present in cells throughout the renal tubule (see p. 85 for more detailed discussion).

Exchange with potassium ion.—Sodium is reabsorbed in exchange for potassium (and hydrogen) ions in the distal tubule. This exchange is stimulated by aldosterone.

Chloride.—Although an active mechanism for chloride reabsorption has been suggested, probably most of the chloride in the glomerular filtrate is reabsorbed passively in the proximal tubule along the electrochemical gradient created by sodium reabsorption. Sodium cannot be reabsorbed isosmotically without an accompanying anion, because the resulting electrochemical gradient would soon halt the process. Reabsorption of sodium is limited by the availability of chloride, the most abundant anion in the glomerular filtrate (p. 103).

Potassium.—Urinary potassium can be affected in two ways:

Active reabsorption.—Potassium is almost completely reabsorbed by an active process in the proximal tubule, and fluid entering the distal tubule contains little potassium.

Exchange with sodium ion.—In the distal tubule potassium is secreted in exchange for sodium, and this process is stimulated by aldosterone. Hydrogen and potassium ions compete for this exchange.

Hydrogen ion.—Hydrogen ion is secreted throughout the tubule in exchange for sodium. Final adjustment takes place in the distal tubule. As potassium competes for this exchange, disturbances of hydrogen ion homeostasis can be initiated by abnormalities of potassium balance (p. 66).

Bicarbonate.—Bicarbonate is recovered from the glomerular filtrate although the tubular cells are impermeable to it. It is converted to carbon dioxide when hydrogen ion is secreted into the urine, and the carbon dioxide diffuses passively into the tubular cell where it is reconverted to bicarbonate in the presence of carbonate dehydratase and returned to the blood stream (p. 85).

Urea.—Urea diffuses passively into the blood from the proximal tubule as water reabsorption increases its concentration in the filtrate. However, this diffusion is slow. Urinary urea can reach concentrations well above those in blood.

Creatinine.—Creatinine, in man, is not reabsorbed but is secreted by the renal tubule in very small amounts.

Water Reabsorption: Urinary Concentration and Dilution

Water is always reabsorbed *passively* along an osmotic gradient. However, *active* solute transport is necessary to produce this gradient. There are two main processes involved in water reabsorption:

isosmotic reabsorption of water in the proximal tubule;

differential reabsorption of water and solute in the loop of Henle, distal tubule and collecting ducts.

Isosmotic reabsorption of water in the proximal tubule.—Approximately 200 litres of water is filtered daily by the glomeruli, and yet only about 2 litres of urine enters the bladder. Therefore the nephron as a whole reabsorbs 99 per cent of the filtered water, about 70 to 80 per cent (that is 140 to 160 litres a day) being returned to the body by the proximal tubules.

The proximal tubules pass through the renal cortex and their walls are freely permeable to water. Active reabsorption of solute such as sodium and glucose from the glomerular filtrate is accompanied by passive reabsorption of an osmotically equivalent amount of water. Blood flow is brisk in this area and solute and water are removed rapidly. As water and solute reabsorption are almost concurrent, fluid entering the loop of Henle, though much reduced in volume, is still almost isosmotic and this process cannot adjust extracellular osmolality; it merely reclaims the bulk of filtered water and solute.

Differential reabsorption of water and solute in the loop of Henle, distal tubule and collecting duct.—Normally between 40 and 60 litres of water a day enters the loops of Henle. Not only is this volume further reduced to about 2 litres, but, if changes in extracellular osmolality are to be corrected, the proportion of water reabsorbed must be varied according to the body's needs. At extremes of water intake urinary osmolality can vary from about 40 to about 1400 mmol/kg. (These figures should be compared with the normal for plasma, and therefore for glomerular filtrate, of about 290 mmol/kg.) It will be seen that the proportion of solute to water reabsorption must be capable of varying by a factor of approximately 35. As the proximal tubule cannot dissociate water and solute reabsorption, this must occur between the end of the proximal tubule and the end of the collecting duct.

It is generally agreed that two mechanisms are involved, although it must be stressed that there are differences of opinion as to the exact details.

Countercurrent multiplication, an *active* process occurring probably in the *loop of Henle*, whereby high medullary osmolality is created, and urinary osmolality is reduced. This acts by itself in the absence of ADH, and a dilute (hypo-osmolal) urine is produced.

Countercurrent exchange, a *passive* process, only occurring in the *presence of ADH*, whereby water without solute is reabsorbed from the *distal tubules and collecting ducts* into the *ascending vasa recta* along the osmotic gradient created by multiplication; by this means the urine is concentrated and the plasma diluted.

Countercurrent multiplication.—The most generally held theory considers that this occurs in the loops of Henle, sodium (or perhaps chloride) being actively pumped from the ascending to the descending limb while fluid is flowing through the loop.

Fluid entering the descending limb from the proximal tubule is almost isosmolal—that is, is of the same osmolality as that in the general circulation. This is normally a little under 300 mmol/kg, and for ease of discussion we will use the figure 300 mmol/kg.

Suppose that the loop has been filled, no pumping has taken place, and the fluid in the loop is stationary. Osmolality throughout the loop and the adjacent medullary tissue will be about 300 mmol/kg.

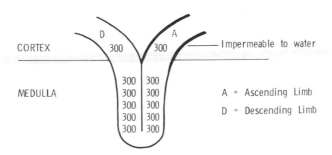

Suppose 1 mmol of solute per kg is pumped from the ascending limb (A) into the descending limb (D), the fluid column remaining stationary.

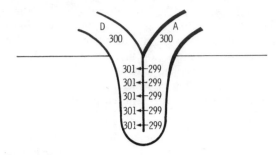

If this pumping were continued and there were no flow, limb D would become very hyperosmolal and limb A equally hypo-osmolal.

Let us now suppose that the fluid flows so that each figure "moves two places".

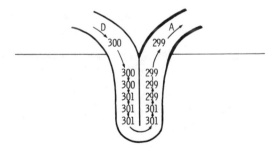

As this happens more solute is pumped from limb A to limb D.

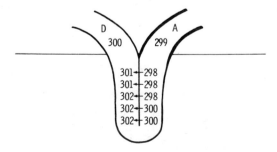

If the fluid again flows "two places", then the situation will be:

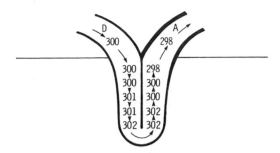

If these steps occur simultaneously and continuously, the wall of the *distal part* of limb A being *impermeable to water*, the consequences would be:

Increasing osmolality in the tips of the loops. As the *walls of the loops are permeable* to water and solute, osmotic equilibrium would be reached with all the surrounding tissues and the deeper layers of the medulla including the blood in the vasa recta, which will also be of increasing osmolality;

hypo-osmolal fluid leaving the ascending limb.

The final result might be:

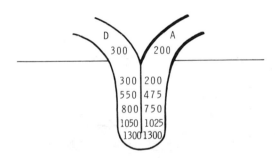

In the absence of ADH the walls of the distal tubules and collecting ducts are impermeable to water, no further change in osmolality occurs, and hypo-osmolal urine would be passed.

Countercurrent exchange is essential, *together with multiplication*, for *concentration of urine*. It can only occur in the presence of ADH, and depends on the "random" apposition of collecting ducts and ascending vasa recta, a result of the close anatomical relations of *all* medullary constituents (Fig. 1) (apposition to descending vasa recta will also occur, but this will have little effect on urinary osmolality). The action of ADH makes the walls of the distal part of the tubule and the collecting ducts permeable to water, which then passes along the osmotic gradient created by multiplication; urine is thus concentrated as the

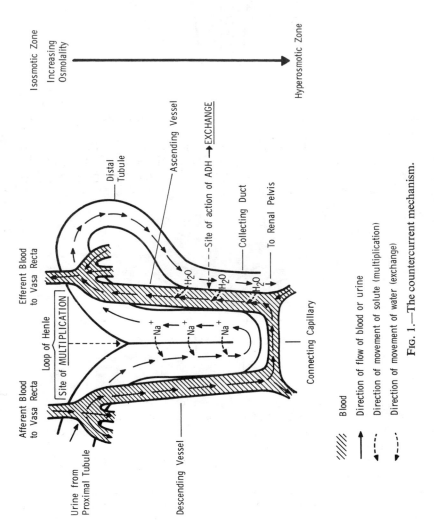

Fig. 1.—The countercurrent mechanism.

collecting ducts pass into the increasingly hyperosmolal medulla. The increasing concentration of the fluid as it passes down the ducts would reduce the osmotic gradient if it did not meet even more concentrated blood flowing in the opposite (countercurrent) direction. The gradient is thus maintained, and water can continue to be reabsorbed until urine reaches the osmolality of the deepest layers (four or five times that of plasma). The diluted blood is carried towards the cortex and soon enters the general circulation, thus tending to reduce plasma osmolality.

Whether this is the exact mechanism or not, both concentration and dilution of urine depend on active processes, which may be deranged if tubules are damaged.

Let us now look at the process in more detail as it works at extremes of water intake, bearing in mind that the tendency to change in *plasma* osmolality is normally rapidly corrected.

Water load.—A high water intake dilutes the extracellular fluid and the fall in osmolality *cuts off ADH* production (p. 40). As the walls of the collecting ducts are then impermeable to water, *countercurrent multiplication* is acting *alone*, and as we have seen, a dilute urine will be produced. However, plasma osmolality would not be corrected unless the hyperosmolality created by multiplication could be carried into the general circulation.

It has been found that during a maximal water diuresis the osmolality at the tips of the papillae may only reach about 600 mmol/kg, rather than the maximum of about 1400 mmol/kg. Since increasing the circulating volume increases renal blood flow, the *more rapid flow in the vasa recta* may tend to "wash out" medullary hyperosmolality, returning some of the solute, without extra water, into the circulation. Thus, not only is more water than normal lost in the urine, but more solute is also "reclaimed" into the general circulation.

Water restriction.—Water restriction, by increasing plasma osmolality, *leads to ADH production* and allows countercurrent exchange. Reduced circulating volume results in *sluggish flow in the vasa recta*, allowing build up of the medullary hyperosmolality produced by multiplication and therefore increasing this exchange.

Osmotic diuresis.—Under physiological circumstances by far the largest contribution to the osmolality of the glomerular filtrate comes from sodium salts, and active removal of sodium in the proximal tubule is followed by passive reabsorption of water. Suppose that another osmotically active substance is circulating in significant amounts, that this substance is freely filterable at the glomerulus, but that it cannot be reabsorbed either actively or passively to any significant extent in the proximal tubule. *Mannitol*, which cannot diffuse significantly through cell walls, is such a substance. As water reabsorption takes place with that of sodium the mannitol will become more concentrated, and its osmotic effect will inhibit further water diffusion. Since less water is reabsorbed a larger volume than usual will enter the loop of Henle and flow through the rest of the tubular system. The rate of water reabsorption in the distal tubule and collecting duct is limited, and water diuresis will result. Note that urine leaving the proximal tubule is still isosmotic with blood, but that the contribution to the osmotic pressure of the tubular fluid from sodium is less than that in blood; the difference is made up by mannitol (in the absence of osmotic diuretics the sodium concentrations are the same in the proximal tubular lumen and in the blood). *Urea* can diffuse back to a limited extent in the proximal tubule, and most *glucose* is actively reabsorbed. However, if these substances are filtered at very high concentrations more is unabsorbed and they then act as osmotic diuretics (see p. 41).

Summarising the Functions of the Kidney

1. Excretion of most substances depends initially on normal permeability of the *glomerulus* (cortex), and on a hydrostatic pressure gradient between the blood flowing through it and the lumen of the glomerulus. If either of these factors is significantly reduced the tubules may not receive adequate amounts of material for "fine adjustment" to the body's needs.

2. The *proximal tubule* (cortex), almost completely reabsorbs many substances which can be reutilised by the body (e.g. glucose and amino acids). About 70 per cent of the sodium and water filtered at the glomerulus is reabsorbed at this site, the active process being reabsorption of sodium.

3. The *loop of Henle* (medulla) by means of countercurrent multiplication, creates the osmotic gradient which, in the presence of a normally functioning ADH feedback mechanism, enables water reabsorption to be increased or reduced according to the body's need, with formation of urine of higher or lower osmolality than that of the extracellular fluid.

4. The *distal tubule* (cortex) makes the final adjustment of sodium, potassium and hydrogen ions by means of exchange mechanisms.

5. Final adjustment of water excretion takes place in the *distal tubule* and the *collecting ducts* (medulla) under the influence of ADH, and is dependent on normal functioning of the loop of Henle.

CHEMICAL PATHOLOGY OF KIDNEY DISEASE

Although some diseases affect primarily the glomeruli and others the renal tubules, it is rare that disturbances of either of these are completely isolated from one another: the parts of the nephron are so closely associated anatomically, as well as being dependent on a common blood supply, that the end result of almost all renal disease is disturbance of the nephron as a whole. In most cases not all nephrons are equally affected, and while some may be completely non-functional, others may be quite normal, and yet others may have some functions disturbed to a greater extent than others. Some of the effects of chronic renal failure can be explained by this patchy distribution of disease (compare the effects of patchy pulmonary disease, p. 105).

However, in the initial stages some diseases affect either tubules or glomeruli predominantly and can occasionally be arrested or even reversed at this stage. It is probably easier to understand renal failure if the effects of these two types of lesion are first discussed separately.

Glomerular Dysfunction

Consequences of a reduced glomerular filtration rate.—If the GFR is reduced an abnormally small volume of filtrate is produced: its composition remains that of a plasma ultrafiltrate. Abnormally small amounts of all constituents of the filtrate are presented to the tubular cells and, as the flow is relatively sluggish, stay in contact with them for longer than normal.

1. The reduced volume of filtrate may, of itself, result in *oliguria* (less than 400 ml per day): this will occur only if the GFR is reduced to less than about 30 per cent of its normal value (i.e. to less than about 30 ml/minute). If the low GFR is due to factors that also cause maximal ADH secretion, water reabsorption in the collecting ducts will also be high, and *urine of high osmolality and specific gravity* will be passed. These findings are due to the presence of solute not normally reabsorbed significantly by the tubular cells: the concentrations of urinary urea and creatinine, for instance, will be high.

2. The total daily amount of *urea and creatinine* lost in the urine depends mainly on the amount filtered at the glomerulus during the same period, since tubular action on them is quantitatively insignificant in this context. The amount filtered at the glomerulus depends on plasma concentration and the GFR. If the rate of excretion falls below the rate of production the plasma levels will rise: as a higher plasma level results in a higher rate of excretion the concentration may equilibrate temporarily at this raised level. If, however, the rise in concentration cannot compensate for the fall in filtered volume, levels continue to rise.

Phosphate and urate, like urea, are products of cell breakdown. They can be reabsorbed in the proximal tubule, and reabsorption may be complete if the amount filtered is low. Here again, if production is not balanced by excretion (and in this case by anabolism), plasma levels will rise.

3. Isosmotic sodium reabsorption is almost total in the proximal

tubule, and less than usual is available for exchange with hydrogen ions throughout the tubule and with potassium ions in the distal tubule. This has two important results:

reduction of hydrogen ion secretion throughout the nephron. There is resultant systemic *metabolic acidosis* with *a low plasma bicarbonate concentration*. (p. 93);

reduction of potassium secretion in the distal tubule with *potassium retention*. The tendency to hyperkalaemia is aggravated by movement of potassium out of cells due to the acidosis (p. 65).

If the low GFR is due to low renal blood flow aldosterone secretion will be maximal (p. 38), and any sodium reaching the distal tubule will be almost completely reabsorbed in exchange for H^+ and K^+. *Urinary sodium* concentration will then be *low*. Thus the laboratory findings in advanced cases will be as follows:

Effects on plasma

Uraemia, high plasma creatinine concentration
Metabolic acidosis with a low plasma bicarbonate
A tendency to hyperkalaemia
Hyperuricaemia, hyperphosphataemia, hypocalcaemia (p. 242).

Effects on urine

Oliguria
Urinary osmolality (and specific gravity), urea and sodium concentrations appropriate to the clinical state. If the patient is *hypotensive or dehydrated the osmolality, specific gravity and urea concentration should be high and the sodium concentration much lower than that of plasma* (it may be less than 5 mmol/l).
N.B. These urinary findings may be reversed if the dehydration is due to urinary water loss caused by, for example, diabetes insipidus or to an osmotic diuresis (p. 49). Additional findings will depend on the cause of the low GFR.

Causes of reduced glomerular filtration rate.—The relatively uncomplicated picture is usually seen only in the early stages of the disease. Later the nephron as a whole is affected.

1. Reduction in differential hydrostatic pressure in the glomerulus.
   Reduced glomerular blood flow ("pre-renal uraemia"). Renal circulatory insufficiency due to:
   low systemic blood pressure (haemorrhage, dehydration, "shock");
   congestive cardiac failure;
   bilateral renal artery stenosis;
   acute oliguric renal failure.

Increased intraluminal pressure;
? Acute oliguric renal failure (p. 18);
Obstruction of the ureters or urethra ("post-renal uraemia").
2. Disease of the glomerulus.
   Acute glomerulonephritis
   Chronic glomerulonephritis.

1. *Renal circulatory insufficiency* is probably the commonest cause of a low GFR with relatively normal tubular function, and is most commonly due to reduction of the circulating volume by haemorrhage or dehydration. The many other causes of the "shock" syndrome include acute intra-abdominal lesions (for example, rupture of an ectopic pregnancy, acute pancreatitis and perforation of a peptic ulcer) and intravascular haemolysis (including that due to mismatched blood transfusion). If blood pressure (or, more accurately, renal blood flow) is restored within a few hours the condition is reversible: if the condition persists for longer periods of time the danger of ischaemic damage to the tubules increases, and when glomerular function is restored the picture of acute oliguric renal failure manifests itself within a few days. If the low GFR is due to hypovolaemia the patient will usually be hypotensive and possibly clinically dehydrated and, in addition to the laboratory findings listed above, there may be haemoconcentration (p. 42). Uraemia due to renal dysfunction is aggravated if there is increased protein breakdown, due either to tissue damage or to the presence of blood in the gastro-intestinal tract or large haematomas: the liberated amino acids are converted to urea in the liver, and intravenous amino acid infusion may have the same effect. Increased tissue breakdown also aggravates hyperkalaemia and acidosis.

In congestive cardiac failure circulation to the kidney may be sufficiently impaired to cause mild uraemia: the clinical findings are those of the primary condition, and haemodilution is often present. In renal artery stenosis hypertension of renal origin is usually present.

In acute oliguric renal failure (p. 18) it has been shown that renal blood flow, especially to the cortex, is reduced, and may account for some of the oliguria in this condition.

Increased intraluminal pressure reduces the hydrostatic pressure gradient essential for normal filtration. "Post-renal uraemia" may be due to obstruction by calculi, polyps or other neoplasms, strictures or prostatic hypertrophy. Long-continued back pressure may lead to renal damage, with uraemia persisting after the relief of the obstruction.

2. *Glomerulonephritis* reduces the permeability of the glomerular membrane; in chronic glomerulonephritis the tubules also are usually involved in scarring and the biochemical findings are therefore those of generalised renal dysfunction (p. 16). In acute glomerulonephritis fluid

and sodium retention occur, usually in equivalent amounts so that plasma sodium concentrations remain normal. The overloading may be aggravated by injudicious fluid administration: in such cases there is haemodilution, and if the administered fluid is hypotonic, a fall in plasma sodium concentration. There is often a history of sore throat, and protein, casts and red cells in small amounts are found in the urine. Proteinuria and occasional erythrocytes may also be found in the chronic condition.

Increased glomerular permeability occurs in the *nephrotic syndrome*. All but the highest molecular weight plasma proteins can then pass the glomerulus and proteinuria of several grams a day occurs. The main effects are on *plasma proteins* and the subject is discussed more fully in Chapter XIV. Uraemia occurs only in the late stages of the disease when many glomeruli cease to function.

Tubular Dysfunction

Consequences of generalised tubular dysfunction.—Even if the GFR is normal, damage to the tubular cells impairs the adjustment of the composition and volume of the urine.

1. The countercurrent mechanism may be impaired. Water reabsorption is reduced and large volumes of dilute urine are passed.

2. The tubules cannot secrete hydrogen ion and therefore cannot reabsorb bicarbonate normally and cannot acidify the urine.

3. Reabsorption of sodium and exchange mechanisms involving sodium are impaired and the urine contains an inappropriately high concentration of sodium relative to the state of hydration (p. 15). Failure of sodium reabsorption in the proximal tubule contributes to impairment of water reabsorption at this site.

4. Potassium reabsorption in the proximal tubule is impaired and potassium depletion may develop.

5. Reabsorption of glucose, phosphate, urate and amino acids is impaired and there is often glycosuria, phosphaturia and generalised aminoaciduria (acquired Fanconi syndrome). The plasma phosphate and urate levels may be low.

Uraemia occurs only if fluid and electrolyte depletion causes renal circulatory insufficiency.

Thus the laboratory findings in advanced cases may be:

Effects on plasma
    Metabolic acidosis with low plasma bicarbonate
    Hypokalaemia*
    Hypophosphataemia and hypouricaemia*
    Haemoconcentration (if the patient is volume depleted)
    A normal plasma urea* (unless the patient is dehydrated)

Effects on urine

Polyuria*

Inappropriately low osmolality (and specific gravity) and inappropriately low urea concentration*

An inappropriately high concentration of sodium* (more than about 30 mmol/l even if the patient is dehydrated or hypotensive).

Contrast those findings marked * with those of glomerular damage.

It must be stressed that the finding of inappropriate urinary concentrations depends on the patient being hypotensive or dehydrated at the time of the investigation. If the patient is normotensive and well hydrated measurement of urinary concentrations adds nothing to diagnostic precision (Chapter II).

In most cases there is mild proteinuria, and casts are present in the urine.

Causes of predominantly tubular dysfunction.—In most of these cases the glomeruli are involved in the later stages of the disease. A few examples only are given here.

Recovery phase of acute oliguric renal failure (p.18) due to:

prolonged renal circulatory insufficiency

proximal tubular necrosis due to various poisons (e.g. carbon tetrachloride)

Progressive damage to the tubules due to:

hypercalcaemia

hypokalaemia

hyperuricaemia

Wilson's disease (copper, p. 363)

Bence Jones protein in myeloma (p. 323)

galactosaemia (p. 202)

various poisons, especially heavy metals

early pyelonephritis

Inborn errors of specific tubular functions are discussed in the relevant chapters and summarised in Chapter XVI.

Progressive damage to the tubular cells is frequently due to precipitation of substances such as calcium around the renal tubules; prolonged potassium depletion causes vacuolation of the tubular cells. Many of the other substances listed under this heading can cause acute oliguric renal failure if present in high enough concentrations. In the early stages of these chronic conditions the picture may be that described under predominantly tubular lesions: for instance, cases of chronic hypercalcaemia can present complaining of polyuria, and are often found to have hypokalaemia. In the later stages the whole nephron is involved in scarring with impairment of glomerular function.

Pyelonephritis, while affecting the kidneys as a whole, has a predilection for the medulla. The ability to form a concentrated urine, and other tubular functions, are therefore lost early in the disease.

Generalised Renal Failure

Probably neither isolated glomerular nor isolated tubular failure occurs. However, in the conditions described in the previous pages, relatively crude methods of investigation do not detect the minor degree of dysfunction of the other part of the nephron. In most cases of "pure" tubular dysfunction, however, reduced urea or creatinine clearances may indicate minimal glomerular dysfunction in the presence of "normal" plasma urea and creatinine levels.

Between these two extremes there is a spectrum of conditions in which the proportions of tubular and glomerular dysfunction vary. The findings will depend on the relative contributions from these two factors, and an attempt has been made to indicate this in Table I. The dotted line indicates that the proportions are variable.

1. When the GFR falls below about 30 per cent of normal, substances almost unaffected by tubular action such as *urea and creatinine* will be retained with a consequent rise in their plasma concentration.

2. The degree of retention of *potassium, urate and phosphate* will depend on the balance between the degree of glomerular retention and the degree of loss due to failure of proximal tubular reabsorption: at the glomerular end of the spectrum so little is filtered that, despite failure of reabsorption, blood levels rise; at the tubular end of the spectrum glomerular retention is more than balanced by failure to reabsorb any filtered potassium, urate and phosphate. Similarly, the urine volume depends on the balance between the volume filtered, and the proportion reabsorbed by the tubules (remember that normally 99 per cent of filtered water is reabsorbed). At the glomerular end of the spectrum nothing is filtered and the patient is anuric: at the tubular end, although filtration is reduced, tubular reabsorption is so impaired that the patient suffers from polyuria.

3. While *plasma levels* of urea and creatinine depend largely on glomerular function, *urinary concentrations* depend almost entirely on tubular function. However little is filtered at the glomerulus, the composition of the filtrate is that of a plasma ultrafiltrate. Any variation from this in urine passed is due to tubular activity. The fewer tubules working, the nearer urine concentrations will be to those of plasma. Urine concentrations *inappropriate to the state of hydration* suggest tubular damage, whatever the degree of glomerular damage.

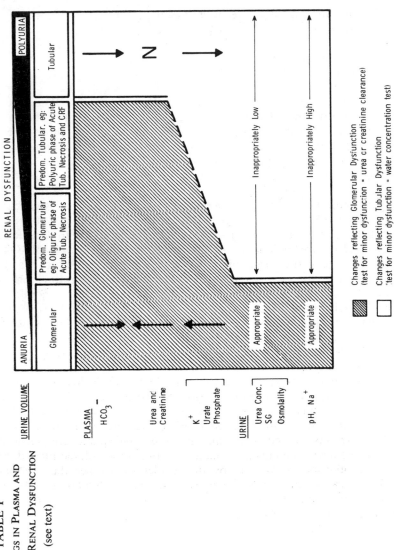

TABLE I
FINDINGS IN PLASMA AND
URINE IN RENAL DYSFUNCTION
(see text)

Causes of generalised renal dysfunction.—These can be acute or chronic.

Acute

Acute oliguric renal failure ("acute tubular necrosis"; acute renal failure).

Chronic

All the conditions described under the headings of "Glomerular Dysfunction" and "Tubular Dysfunction" can progress to generalised disease of the nephron, giving the common picture of chronic renal failure. Many other diseases (for instance chronic glomerulonephritis and polycystic disease) affect the kidney as a whole.

Acute oliguric renal failure often follows a period of reduced GFR due to renal circulatory insufficiency, and from the therapeutic point of view it is important to distinguish these two phases. The oliguria is probably due not to glomerular damage, but to reduced cortical blood flow: this may be aggravated by back-pressure on the glomeruli due to obstruction of tubular flow by oedema. At this stage, as indicated in Table I, the findings are at the glomerular end of the spectrum; the fact that the tubules are damaged is evident if the urinary concentrations are inappropriate. Water retention may cause oedema with haemodilution unless fluid intake is restricted.

As cortical blood flow increases and as the tubular oedema resolves, glomerular function recovers before that of the tubules. The findings gradually progress to the tubular end of the spectrum until they approximate to those of "pure" tubular lesions and as urinary output increases, and is further increased by the high osmotic load of urea, the polyuria may cause water depletion. Electrolyte depletion may occur, and the initial hyperkalaemia may be replaced by hypokalaemia. Mild acidosis (common to both glomerular and tubular lesions) persists until late. Finally, recovery of the tubules restores renal function to normal.

In chronic renal failure individual nephrons will be affected to different degrees. As more glomeruli are involved the rate of urea excretion falls and cannot balance the rate of production by protein catabolism: as a consequence the plasma urea level rises and the urea concentration in the filtrate of the functioning nephrons rises. This may cause an osmotic diuresis in these nephrons: in other nephrons, tubules may be damaged out of proportion to the glomeruli. Both the tubular dysfunction in nephrons with functioning glomeruli and the osmotic diuresis through intact nephrons contribute to the polyuria which may occur in chronic renal failure. If water loss is replaced by a high fluid intake, urea excretion can continue through functioning glomeruli at a high rate: the increased excretion through normal glomeruli may eventually balance the reduced permeability of others, and a new steady state is reached at a higher level of plasma urea. At this stage the findings are near the tubular end of the spectrum in Table I, the glomerular dysfunction being indicated by the high plasma urea level. If these subjects are kept well hydrated they may remain in a stable condition with a moderately raised plasma urea for years. Potassium levels are variable, but tend to be raised. If nephron destruction continues the condition approximates more and more to the glomerular end of the spectrum, until oliguria precipitates a steep rise in plasma urea and potassium concentrations. This stage is often terminal. Before assuming that the latter is the case, care should be taken to ensure that the sudden rise is not due to electrolyte and water depletion.

In prolonged chronic renal failure osteomalacia may occur and may occasionally require cautious treatment (p. 253).

Differential Diagnosis of Oliguria with Uraemia

Oliguria with uraemia may be due to renal circulatory insufficiency, to acute glomerulonephritis, or to generalised renal damage. The diagnosis of acute glomerulonephritis can usually be made on clinical grounds, and on the finding of microscopic haematuria. However, the differentiation between renal circulatory insufficiency and generalised tubular damage may be difficult, and the treatment is radically different in the two cases. In most situations the clinical history and examination will differentiate between the two. If doubt remains and *if the patient is dehydrated or hypotensive* (i.e. has a low renal blood flow), measurement of urinary sodium or urea concentration, or urinary osmolality or specific gravity may help. Under these conditions a concentrated urine, containing very little sodium, should be passed if the tubules are functioning. Failure to concentrate the urine implies true renal failure.

INVESTIGATION OF RENAL FUNCTION

Plasma urea and creatinine levels depend on the balance between their production and excretion.

Urea is the product of amino acid breakdown in the liver and is therefore derived from protein, either in the diet or in body tissues: the rate of production is accelerated by a high protein diet, or by increased endogenous catabolism due to starvation or tissue damage. The capacity of the normal kidney to excrete urea is high, and in the presence of normal renal function only extremely high protein diets can cause rises of plasma urea to levels above the "normal" range. In patients with a gross increase in tissue catabolism due to severe tissue damage or to acute starvation urea levels may rise above "normal", especially as renal function is often mildly impaired in such cases due to renal circulatory factors. (In chronic starvation plasma levels tend to be low due to protein depletion.) In spite of these reservations, a significantly high plasma urea concentration almost certainly indicates impaired renal function (this is almost certainly so at levels above 13 mmol/l [urea above 80 mg/dl: BUN above 37 mg/dl]). The probability is increased if a cause for renal dysfunction is present, or if there are protein, casts or cells in the urine. In the few cases when doubt remains, clearance studies (see below) or measurement of plasma creatinine may resolve it.

Plasma *creatinine* is not affected by diet, because it is produced endogenously by creatine breakdown. The circulating creatinine is derived from metabolism of tissue creatine: although plasma levels would be expected to increase with increased tissue breakdown, they do so less than those of urea. In theory, therefore, creatinine would seem preferable to urea estimation as an index of renal function, and for this reason is used by many laboratories for this purpose. However, in spite of an improvement in the precision and speed of creatinine estimation with the advent of automated methods, the precision of the urea method is still significantly higher, especially at near normal levels. Which of these parameters is used is largely a matter of local choice.

If urea or creatinine levels are significantly raised, and especially if they are rising, if there is oliguria or a history suggestive of renal disease, or if there is proteinuria, and the urine contains protein, casts, cells or bacteria in significant amounts, they can usually be safely assumed to be due to renal impairment. For routine clinical purposes progress can be followed merely by measuring these concentrations in the plasma, because they reflect changes in renal clearance.

Detection of Minor Degrees of Renal Damage

More than 60 per cent of the kidney must be destroyed before either plasma urea or creatinine levels are significantly raised (this is only valid for urea if the patient is taking a normal or low protein diet, and if there is no excessive protein catabolism, p. 19). Damage may be suspected for other reasons, such as evidence of renal infection, a previous episode of uraemia, or the finding of protein in the urine. If plasma urea or creatinine concentrations are normal more refined tests may be useful: by the time they are abnormal presence of disease is not in doubt. This is a general principle which also applies, for example, to the detection of diabetes mellitus (p. 191).

Renal concentrating ability.—This is a simple test and can be carried out on the ward. The ability to form a concentrated urine in response to water deprivation depends on normal tubular function (countercurrent multiplication), and on the presence of ADH. Failure of this ability is usually due to renal disease, but if there is any doubt the test can be repeated after administration of ADH (pitressin) (Appendix, p. 28).

Clearance tests of glomerular function.—Clearance tests measure the amount of blood which could theoretically be completely cleared of a substance per minute. This figure can be calculated if the amount excreted per minute is divided by its concentration in the blood: the amount excreted per minute can be calculated on a timed urine collection by multiplying the concentration of the substance in the specimen by the volume passed in the time, and dividing by the time in minutes. In the case of urea therefore:

$$\text{Urea Clearance (ml/minute)} = \frac{\text{Urinary [urea]} \times \text{Urine volume (ml)}}{\text{Plasma [urea]} \times \text{Time of collection (minutes)}}$$

Care should be taken that plasma and urinary urea concentrations are expressed in the same units (e.g. mmol/l).

If a true estimate of GFR is to be made a substance should be chosen which is excreted solely by glomerular filtration and is not reabsorbed or secreted by the tubules. *Inulin* is thought to be such a substance: inulin is not produced in the mammalian body and measurement of its clearance requires administration either by constant infusion to maintain its level in the blood steady during the period of the test, or by a single injection followed by serial blood sampling so that its concentration at the mid-point of urine collection can be calculated. Such exogenous clearances are not very practicable for routine use.

Substances produced in the body are usually present in a fairly steady circulating concentration for the period of the test, and blood need only be taken at the mid-point of the urine collection. Urea and creatinine clearances are commonly used for clinical purposes. Neither of these substances fulfils the criterion that they are not reabsorbed from or added to the glomerular filtrate by the tubular cells. Small amounts of *urea* diffuse back into the blood stream from the proximal tubule, and urea clearance values are lower than those of inulin: small amounts of *creatinine* are secreted by the tubule and give clearance values higher than those of inulin. Either clearance gives an approximation to the GFR adequate for clinical purposes, and although the estimation of urea is more precise than that of creatinine, there is little to choose between them. It should be noted that the rate of protein breakdown, while it may affect levels of plasma urea, does *not* affect its rate of clearance by the kidney.

If urea clearance is chosen, care should be taken to inhibit the urea-splitting activity of organisms such as *Proteus vulgaris*, which may be present in the urine, by the use of mercury salts as a preservative. If such organisms are present, short periods of collection (for example, three collections of 1 hour) are preferable to a long one (for example, 24 hours). Creatinine can also be destroyed by bacterial action, and estimations should always be done soon after the collection is complete.

The biggest error of any clearance method is in the timed urine collection.

It should be noted that clearance tests will give equally low values with renal circulatory insufficiency, or with "post-renal" causes, as with true renal damage, and cannot distinguish between them.

Incidental Abnormal Findings in Renal Failure

In assessing the severity and progress of renal failure it is usual to estimate plasma urea or creatinine, electrolytes (especially potassium) and bicarbonate concentrations. Other abnormalities occur which, although not useful in diagnosis or assessment of renal dysfunction, may be misinterpreted if the cause is not recognised. Plasma *urate* levels rise in parallel with plasma urea, and a high level does not necessarily indicate primary hyperuricaemia (p. 376). Plasma *phosphate* levels also rise, and those of *calcium* fall (p. 242). In chronic renal failure secondary hyperparathyroidism may lead to bone disease with a high level of *alkaline phosphatase.* Hypocalcaemia should only be treated if there is evidence of bone disease (p. 253). *Anaemia* is commonly present, and is normocytic and normochromic in type: it does not respond to iron therapy.

The investigation of renal function is summarised on p. 30.

BIOCHEMICAL PRINCIPLES OF TREATMENT OF
RENAL DYSFUNCTION

Oliguric renal failure.—The oliguria of dehydration or haemorrhage, which is due to a reduction of glomerular filtration rate only, should be treated with the appropriate fluid (p. 61).

In oliguric renal failure due to parenchymal damage the aims are:

to restrict fluid and sodium, giving only enough fluid to replace that lost in the previous 24 hours (p. 43);

to provide an adequate non-protein energy source to prevent aggravation of uraemia and hyperkalaemia by increased endogenous catabolism; to prevent dangerous hyperkalaemia (p. 73).

Diuretics tend to increase renal blood flow. In some cases of acute oliguric renal failure recovery may be hastened by administration of an osmotic diuretic, such as mannitol, or other diuretics such as frusemide (furosemide) or ethacrynic acid in large doses.

In chronic renal failure with polyuria the aim is to replace fluid and electrolytes lost. Sodium and water depletion may aggravate the uraemia.

Haemodialysis or peritoneal dialysis removes urea and toxic substances from the blood stream, and corrects electrolyte balance, by dialysing the patient's blood against fluid containing no urea, and normal plasma concentrations of electrolytes, *ionised* calcium and other plasma constituents. The blood is either passed over a dialysing membrane before being returned to the body, or the folds of the peritoneum are used as a dialysing membrane with their capillaries on one side, and suitable fluid injected into the peritoneal cavity on the other. A relatively slow reduction of urea concentration is preferable to a rapid one because of the danger of cellular overhydration when extracellular osmolality falls abruptly (p. 35). Dialysis is used in cases of potentially recoverable acute oliguric renal failure, to tide the patient over a crisis, or as a regularly repeated procedure in suitable cases of chronic renal failure. It may also be used to prepare patients for renal transplantation, and to maintain them until the transplant functions adequately.

RENAL CALCULI

Renal stones are usually composed of normal products of metabolism which are present in the normal glomerular filtrate, often at concentrations near their maximum solubility: quite minor changes in urinary composition may cause precipitation of such constituents, whether in the substance of the kidney (see section on tubular damage, p. 15), as crystals, or as calculi. Although this discussion concerns stone formation it should be remembered that crystalluria and parenchymal damage can occur under the same circumstances, and that the treatment of all such conditions is the same.

Conditions Favouring Calculus Formation

A high urinary concentration of one or more constituents of the glomerular filtrate.

(a) *A low urinary volume*, with normal renal function, due to restricted fluid intake or excessive fluid loss over long periods of time (this is particularly common in the tropics). This condition favours formation of most types of stone, especially if one of the other conditions listed below is also present.

(b) An abnormally *high rate of excretion* of the metabolic product forming the stone, due either to an increased level in the glomerular filtrate (secondary to high plasma concentrations) or to a failure of normal tubular reabsorption from the filtrate.

Changes in pH of the urine, sometimes due to bacterial infection, which favour precipitation of different salts at different hydrogen ion concentrations.

Urinary stagnation due to obstruction to urine outflow.

Lack of normal inhibitors.—It has been suggested that normal urine contains an inhibitor, or inhibitors, of calcium oxalate crystal growth that are absent in the urine of some patients with a liability to recurrent calcium stone formation.

COMPOSITION OF URINARY CALCULI

Calcium-containing stones
    Calcium oxalate   ⎫ with or without magnesium
    Calcium phosphate ⎭ ammonium phosphate.
Uric acid-containing stones.
Cystine-containing stones.
Xanthine-containing stones.

Calculi Composed of Calcium Salts

These account for between 70 and 90 per cent of all renal stones. Precipitation of calcium is favoured by hypercalcuria, and the type of salt depends on urinary pH and on the availability of oxalate or phosphate. All patients presenting with renal calculi should have a plasma calcium estimation performed, and if this is normal it should be repeated at regular intervals.

Hypercalcaemia causes hypercalcuria if renal function is normal, and estimation of urinary calcium in such cases does not help in the diagnosis. The causes and differential diagnosis of hypercalcaemia are discussed on p. 246.

In many subjects with calcium-containing renal calculi the plasma calcium level is normal. It is in such cases that the estimation of the daily excretion of urinary calcium may be useful. The commonest cause of *hypercalcuria with normocalcaemia* is the so-called *idiopathic hypercalcuria*, a name which reflects our ignorance of the aetiology of the condition: because some of these cases may represent an early stage of primary hyperparathyroidism, plasma calcium estimations should be carried out at regular intervals, especially if the plasma phosphate level is low. Any *increased release of calcium from bone*, as in actively progressing osteoporosis (in which loss of matrix causes secondary decalcification) or in prolonged acidosis (in which ionisation of calcium salts is increased) causes hypercalcuria, but rarely hypercalcaemia. One type of renal tubular acidosis (p. 98) not only increases the renal load of calcium but, because of the relative alkalinity of the urine, favours its precipitation in the kidney and renal tract: this is a *rare* cause.

An increased excretion of *oxalate* favours the formation of the very insoluble calcium oxalate, even if calcium excretion is normal. The source of the increased oxalate may be the diet: the very rare inborn error of oxalate metabolism, primary hyperoxaluria, should be considered if renal calculi occur in childhood.

It has already been mentioned that *alkaline conditions* favour calcium precipitation, and whereas calcium oxalate stones form at any urinary pH, a high pH favours formation of calcium phosphate: this type of stone is particularly common in chronic renal infection with urease-containing (urea-splitting) organisms (for example, *Proteus vulgaris*), which convert urea to ammonia.

A significant proportion of cases remains in which there is no apparent cause for the calcium precipitation.

Calcium-containing calculi are usually *hard and white*. They are *radio-opaque*. Calcium phosphate stones are particularly prone to form "staghorn" calculi in the renal pelvis, easily visualised by straight x-ray.

Treatment of calcium-containing calculi.—This depends on the cause. Excretion of calcium should be reduced

by treating the primary condition (especially hypercalcaemia);

if this is not possible, by reducing intake of calcium in the diet, and possibly by decreasing calcium absorption by administration of oral phosphate (p. 252);

by reducing the concentration of urinary calcium by maintaining a high fluid intake (unless renal failure is present). It is the concentration rather than the total 24-hour output which determines the tendency to precipitation.

Uric Acid Stones

These account for about 10 per cent of all renal calculi, and are sometimes associated with *hyperuricaemia* (with or without clinical gout). Precipitation is favoured in an *acid urine*. In a large proportion of cases no predisposing cause can be found.

Uric acid stones are usually *small, friable* and *yellowish-brown* in colour, but can be large enough to form "staghorn" calculi. They are *radiotranslucent*, but may be visualised by an intravenous pyelogram.

Treatment of hyperuricaemia is discussed on p. 374. If the plasma urate concentration is normal, fluid intake should be kept high and the urine alkalinised. A low purine diet (p. 374) may help to reduce urate production and therefore excretion.

Cystine Stones

These are rare. In normal subjects the concentration of urinary cystine is well within its solubility. In severe cases of the inborn error cystinuria (p. 357) the solubility may be exceeded and the patient may present with *radio-opaque* renal calculi. Like urate, cystine is more soluble in alkaline than acid urine and the principles of treatment are the same as for uric acid stones. Penicillamine can also be used in therapy (p. 357).

Xanthine Stones

These are very uncommon and may be the result of the rare inborn error, xanthinuria (p. 376). Xanthine stones following the use of xanthine oxidase inhibitors, such as allopurinol, have not been reported.

SUMMARY

THE KIDNEYS

1. Normal renal function depends on a normal glomerular filtration rate (GFR) and normal tubular function.

2. In most cases of renal disease glomerular and tubular dysfunction coexist.

3. A low GFR leads to:
   oliguria;
   uraemia and retention of other nitrogenous end-products, includ-
      ing urate, and of phosphate;
   metabolic acidosis;
   hyperkalaemia.

4. A low GFR with normal tubular function is most commonly due to renal circulatory insufficiency. It can also be the result of obstruction to urinary outflow, or of glomerular damage.

5. In the nephrotic syndrome the glomerulus is more permeable than normal to proteins.

6. Tubular damage leads to:
   polyuria. The urine is inappropriately dilute and contains an in-
      appropriately high sodium concentration in relation to the
      patient's state of hydration;
   metabolic acidosis;
   hypokalaemia;
   hypophosphataemia and hypo-uricaemia.

7. In acute oliguric renal failure there is a reduced GFR and the findings are usually those associated with this.

8. The differentiation between the oliguria of a primary glomerular lesion with normal tubular function and of acute tubular necrosis is best made on clinical grounds and, if necessary, by examining the urine.

9. In most cases plasma urea or creatinine levels reflect changes in renal clearance and are adequate for diagnosing and following up cases of renal disease. Tubular function may be tested by assessing the ability of the kidney to concentrate urine.

10. Clearance tests are valuable if a minor degree of renal damage is suspected in spite of a plasma urea or creatinine level within the "normal" range.

RENAL CALCULI

1. The formation of renal calculi is favoured by:
   A high urinary concentration of the constituents of the calculi. This may be due to oliguria, or a high rate of excretion of the relevant substances.
   A pH of the urine which favours precipitation of the constituents of the calculi.
   Urinary stagnation.
2. Calcium containing calculi account for 70 to 90 per cent of all renal stones. They are most commonly idiopathic in origin but hypercalcaemia, especially that of primary hyperparathyroidism, should be excluded as a cause.
3. Uric acid stones account for about a further 10 per cent of renal calculi.
4. Rare causes are cystinuria, xanthinuria and hyperoxaluria.

FURTHER READING

De Wardener, H. E. (1973). *The Kidney*, 4th edit. London: J. & A. Churchill.
Kassirer, J. P. (1971). Clinical evaluation of kidney function—Tubular function. *New Engl. J. Med.*, **285**, 499.
Wills, M. R. (1978). *Metabolic Consequences of Chronic Renal Failure.* 2nd edit. Aylesbury: Harvey, Miller and Medcalf.
Smith, L. H. (1972). Nephrolithiasis. Current concepts in pathogenesis and management. *Postgrad. Med.*, **52**, 165.

APPENDIX

URINE CONCENTRATION TEST

In the normal subject restriction of water intake for a period of hours results in maximal stimulation of ADH secretion (p. 40). ADH acts on the collecting ducts, water is reabsorbed and a concentrated urine is passed. If the counter-current multiplication mechanism is impaired (p. 5) maximal water re-absorption cannot take place, and if ADH levels are low the effect is similar.

If the feedback mechanism is intact ADH levels are sufficient to produce maximal effect. Under these circumstances administration of exogenous ADH will not improve renal concentrating power. If, however, the primary disease is diabetes insipidus with normal tubular function, administration of ADH will convert this to normal.

This test should not be performed if the patient is already dehydrated. In such cases the demonstration of a low urine to plasma osmolality ratio (see below) is diagnostic without resort to fluid restriction. *The patient should be kept under observation during the test*, which should be terminated, after collecting blood and urine, if he becomes distressed.

Procedure

The patient is allowed no food or water after 6 p.m. on the night before the test.

On the day of the test:—

7 a.m.—The bladder is emptied, and the *specimen discarded.*

8 a.m.—The bladder is emptied. The osmolality (or specific gravity) of this specimen is measured. If this reading is above 850 mmol/kg (specific gravity 1·022) the test may be terminated: if it is below this figure the osmolality (or specific gravity) of urine passed at 9 a.m. should be measured.

Blood may also be taken for determination of plasma osmolality.

Interpretation

A maximum osmolality of less than 850 mmol/kg, or a specific gravity less than 1·022 indicates impaired renal concentrating power, either due to tubular disease or to diabetes insipidus: the urine to plasma osmolality ratio should be above 3·0 in normal subjects. In most cases it is clear which of these possibilities is implicated. Where this is still in doubt the pitressin test may be carried out.

PITRESSIN TEST

The above procedure is followed, but 5 units of the oily suspension of vasopressin (pitressin) tannate is injected intramuscularly at 7 p.m. on the evening before the test. If the failure to concentrate is due to tubular disease this will not be improved by the pitressin: if it is due to diabetes insipidus the urine will now be concentrated normally.

Caution.—1. Hydrometers are frequently inaccurate. The specific gravity of distilled water should read 1·000: if it does not, a suitable correction should be made.

2. The urine should be at the temperature to which the hydrometer is calibrated, usually room temperature. Significant errors can be due to measuring specific gravity on freshly passed urine.

3. The hydrometer should be floating freely when the reading is taken. If it touches the wall of the container false values will be obtained. The procedure should be carried out by someone experienced with it.

Urinary urea concentration may be a more valuable estimation. Values above 330 mmol/l (2 g/dl) usually indicate good renal function. If an osmometer is available, measurement of urinary osmolality gives more precise results.

4. Sugar, protein and contrast media used in intravenous pyelography contribute to urinary specific gravity and in their presence high readings are not necessarily indicative of normal renal function. Protein, because of its high molecular weight, contributes very little to osmolality. Glucose does, however, significantly affect it: 150 mmol/l (2·7 g/dl) of glucose in the urine adds 0·001 to the specific gravity and 150 mmol/kg to osmolality. In the presence of glycosuria, estimation of urinary urea concentration may be valuable.

INVESTIGATION OF RENAL FUNCTION

1. Measurement of urinary volume helps to distinguish between predominant tubular or predominant glomerular dysfunction (Table I).

2. *Either* plasma urea *or* creatinine levels may be used to diagnose or to follow unequivocal glomerular dysfunction, whether or not associated with tubular damage (p. 19).

3. Potassium levels should be measured regularly because of the danger of hyperkalaemia in predominant glomerular failure and hypokalaemia in predominant tubular failure.

In most cases these are the only estimations required to make a diagnosis and to follow treatment. However:

1. if plasma urea/creatinine levels are *normal* clearance tests may indicate minor degrees of *chronic glomerular dysfunction* (p. 20);

2. the urine concentration test may indicate *chronic tubular dysfunction* (p. 28);

3. *if the patient with acute oliguric renal failure is dehydrated or hypotensive* estimation of urinary osmolality *or* specific gravity, *or* urinary sodium, *or* urea concentration may help to distinguish between renal circulatory insufficiency and renal damage (p. 19).

Microscopic examination of the urine for casts, and bacteriological examination should be carried out in most cases.

INVESTIGATION OF THE PATIENT WITH RENAL CALCULI

1. If the stone is available, send it to the laboratory for analysis.

2. Exclude *hypercalcaemia* and *hyperuricaemia*. If either is present it should be treated.

3. If the *plasma calcium is normal*, collect a 24-hour specimen of urine (in a container containing acid to keep calcium in solution) for *urinary calcium estimation*. If hypercalcuria is present it should be treated.

4. *If all these tests are negative*, screen the urine for *cystine*. This is especially important if there is a family history. If the qualitative test is positive, cystine should be estimated quantitatively on a 24-hour specimen of urine (p. 357).

5. If the patient is acidotic and the urine is alkaline, perform an ammonium chloride load test (p. 109).

6. If renal calculi occur in *childhood*, a low *plasma urate* and high *urinary xanthine* suggest xanthinuria. If these are normal the 24-hour excretion of *oxalate* should be estimated to exclude primary hyperoxaluria as a cause.

Chapter II

SODIUM AND WATER METABOLISM

THE control of sodium and of water balance are so closely linked that an understanding of one is impossible without an understanding of the other.

Sodium is normally the most abundant extracellular cation and it accounts, together with its associated anions, for most of the osmotic activity of the extracellular fluid (ECF). Osmotic activity depends on *concentration*, and is determined by the relative amounts of water and sodium, rather than the absolute amounts of either. Hypo- or hypernatraemia is caused by imbalance between the two, and the clinical picture is due to the consequent osmotic changes.

If sodium and water are lost in equivalent amounts the osmolal concentration is unaffected: symptoms are then those of inadequate circulatory volume. Of course, osmotic and volume changes usually occur together.

Transport of sodium, of potassium and of hydrogen ions across cell membranes is interdependent, and disturbances of potassium and hydrogen ion homeostasis often accompany those of sodium: changes in associated anions such as chloride and bicarbonate often occur at the same time. The separation of the contents of this chapter from those on potassium and hydrogen ion homeostasis is arbitrary, and for ease of discussion only. A clinical situation should be assessed with all these factors in mind.

WATER AND SODIUM BALANCE

A 70 kg man contains approximately 45 litres of water and 3000 mmol of osmotically active sodium. Maintenance of the total amount depends on the balance between intake and loss. Water and electrolytes are taken in food and drink, and are lost in urine, faeces and sweat: in addition, about 500 ml of water is lost daily in expired air.

Loss Through the Kidneys and Intestinal Tract

Renal loss of sodium and water depends on the balance between that filtered at the glomerulus and that reabsorbed during passage through the tubules, and therefore on normal glomerular and tubular function. *Faecal loss* is mainly dependent on the balance between the sodium and water secreted in bile, saliva and gastric, pancreatic and intestinal secretions, and reabsorbed during passage through the intestine: the contribution from oral intake is small. This loss therefore depends on the integrity of the intestinal epithelial cells, and on the time for which intestinal contents are in contact with them. Intestinal hurry or loss of secretions from the upper gastro-intestinal tract by vomiting or through fistulae are important causes of sodium and water depletion.

It is a sobering thought that approximately 200 litres of water and 30 000 mmol of sodium are filtered through the glomerulus, and a further 10 litres of water and 1500 mmol of sodium are secreted into the intestinal tract each day. In the absence of homeostatic mechanisms for reabsorption, the whole of the extracellular water and sodium could be lost in about 2 hours. It is, therefore, not surprising that failure of these mechanisms causes such extreme disturbances of water and sodium balance. Normally 99 per cent of this initial loss is reabsorbed, and net daily losses amount to about 1·5 to 2 litres of water and 100 mmol of sodium in the urine and 100 ml and 15 mmol in faeces.

Loss in Sweat and Expired Air

Normal daily water loss in sweat and expired air amounts to about 900 ml: less than 30 mmol of sodium is lost a day in sweat. Although anti-diuretic hormone (ADH) and aldosterone have some effect on the composition of sweat, this is relatively unimportant, and sweat loss is primarily controlled by body temperature. Respiratory water loss depends on respiratory rate, and control of this bears no relation to the body requirements for water. Normally loss by sweat and respiration is rapidly corrected by changes in renal and intestinal loss. However, as neither can be controlled to meet sodium and water requirements, they may contribute considerably to abnormal balance when homeostatic mechanisms fail, or in the presence of gross depletion, whether due to poor intake or to excessive loss by other routes.

DISTRIBUTION OF WATER AND SODIUM IN THE BODY

In mild disturbances of water and electrolyte metabolism the total amount of these in the body is of less importance than their distribution within it.

Distribution of Electrolytes

The body has two main fluid compartments of very different electrolyte composition. The compartments are:

the **intracellular compartment**, in which *potassium* is the predominant cation;

the **extracellular compartment**, in which *sodium* is the predominant cation.

The extracellular fluid can be subdivided into:

interstitial fluid which is of very low protein concentration;

intravascular fluid (plasma) which contains protein in high concentration.

Distribution of Electrolytes Between Cells and Extracellular Fluid

The intracellular concentration of sodium is less than a tenth of that in the extracellular fluid (ECF), while that of potassium is about thirty times as much. In absolute amounts about 95 per cent of the metabolically active sodium in the body is outside cells, and about the same proportion of potassium is intracellular.

Other ions tend to move across cell walls with sodium and potassium. Hydrogen ion has already been mentioned. For example, magnesium and phosphate are predominantly intracellular, and chloride extracellular, ions. The distribution of all these, and of bicarbonate, will be affected by the same conditions as those mentioned.

Distribution of Electrolytes Between Plasma and Interstitial Fluid

The vascular wall, unlike that of the cell, is freely permeable to small ions. Plasma contains protein in significant concentration and interstitial fluid does not. Electrolyte concentrations are therefore very slightly higher in the latter to balance the osmotic effect of the protein concentration inside vessels. The difference is small and clinically insignificant, and for practical purposes one can assume that plasma electrolytes are representative of those in the extracellular fluid as a whole.

DISTRIBUTION OF WATER IN THE BODY

A little over half the body water is inside cells. Of the extracellular water about 15–20 per cent is in the plasma. The remainder makes up the extravascular, extracellular interstitial fluid.

The distribution of water across cell walls depends on the *in vivo* osmotic difference between intra- and extracellular fluid: that across blood vessel walls is determined by the balance between the *in vivo* effective osmotic (or oncotic, p. 36) pressure of plasma and the net outward hydrostatic pressure. Correct interpretation of plasma electrolyte results depends on a clear understanding of these factors.

Osmotic Pressure

Net movement of water across a membrane permeable only to water depends on the concentration difference of dissolved particles on the two sides. For any given weight/volume concentration, the larger the particle (the higher the molecular weight) the fewer there are in unit volume, and the less osmotic effect they will exert. However, if the membrane is freely permeable to smaller particles, as well as to water, these exert no osmotic effect, and the larger molecules become more impoı ,ant in affecting movement of water. To explain water distribution in the body it is important to appreciate these two factors:

  number of particles per unit volume;

  particle size related to membrane permeability.

Plasma Osmotic Pressure: Distribution of Water across Cell Walls

For the definitions of osmolarity and osmolality see p. 37.

Measured plasma osmolality.—Plasma osmotic pressure is usually determined by measuring the depression of its freezing point below that of pure water. The result is a measure of its *total* osmotic pressure—the osmotic effect which would be exerted by the sum of all the dissolved molecules and ions across a membrane permeable only to water. Table II shows that by far the largest contribution to this (90 per cent or more) comes from *sodium and its associated anions*, the effect of protein being negligible. The only major difference between extravascular fluid and plasma is in their protein content: thus *total plasma osmolality is almost identical with the osmolality of the fluid bathing cells.*

TABLE II

APPROXIMATE CONTRIBUTIONS OF PLASMA CONSTITUENTS TO PLASMA OSMOLARITY

	Concentration (mmol/l unless otherwise stated)	Osmolarity (mmol/l)
Sodium	135 ⎱	270
Associated anions	135 ⎰	
Potassium	3·5 ⎱	7
Associated anions	3·5 ⎰	
Calcium	2·5 (10 mg/dl) Assuming 40% protein bound (p. 232) ionised calcium = 1·5	1·5
Associated anions		1·5 or 3
Magnesium	1	1
Associated anions		1 or 2
Urea	5 (30 mg/dl)	5
Glucose	5 (90 mg/dl)	5
Protein	70 g/l	Approximately 1
Total		Approximately 295

Hydrostatic pressure differences across cell walls are negligible, and cell hydration depends on the osmotic difference between intra- and extracellular fluid. The cell membrane is freely permeable to water, but although different substances diffuse, or are actively transported, across it at different rates, solute always enters cells more slowly than does water. Normally the total intracellular osmolality, due predominantly to potassium and associated anions, equals that of the ECF, mostly due to sodium salts, and there is no *net* movement of water in or out of cells. In pathological circumstances rapid changes of extracellular solute concentration affect cell hydration: slower changes, by allowing time for redistribution of solute, have less effect. The speed of change is even more important than the magnitude, and over-zealous treatment of abnormal extracellular osmolality may produce more dangerous alteration in cell hydration than the initial abnormality.

Because *sodium* and its associated anions account for about 90 per cent of plasma osmolality in the normal subject (Table II), rapid changes of sodium concentration affect cell hydration, a rise causing cellular dehydration and a fall cellular overhydration. Normal levels of *urea* and *glucose* contribute very little to measured plasma osmolality. However, concentrations 15-fold or more above normal can occur in severe uraemia and hyperglycaemia, and then these solutes contribute to it very significantly. Urea diffuses into cells slowly (about 100 000 times more slowly than water), and during prolonged uraemia its osmotic *effect* is reduced by this fact: acute changes can alter cell hydration. Glucose is actively transported into cells, but once there is rapidly metabolised: intracellular levels remain low, and severe hyperglycaemia has a marked influence on cell hydration. Although uraemia and hyperglycaemia can cause cellular dehydration, the contribution of normal concentrations of urea and glucose to plasma osmolality is so small that reduced levels of these solutes, unlike those of sodium, do not cause cellular overhydration.

Because the *speed* of change is even more important than the absolute level, over-zealous treatment of established hypernatraemia (with hypotonic fluids), uraemia (by haemodialysis), or hyperglycaemia (with large doses of insulin) may produce dangerous cerebral cellular overhydration.

Concentrations of other solutes of low osmolal concentration, such as those of calcium, potassium or magnesium, rarely vary by a factor of more than 3, even in pathological states; they do not cause significant osmolality changes.

Substances not transported into cells, such as mannitol, can be infused to reduce cerebral oedema. They can also be used as osmotic diuretics. (Hypertonic glucose or urea can also be used in this way.) They are included in total plasma osmolality measurements.

A plasma alcohol level of 80 mg/dl contributes about 17 mmol/kg to osmolality.

Calculated plasma osmolarity.—Under most circumstances plasma osmolarity can be calculated sufficiently accurately for clinical purposes if plasma sodium, potassium, urea and glucose concentrations are known. If all measurements are in mmol/l, the *approximate* total osmolarity will be:

$$2[Na^+] + 2[K^+] + [urea] + [glucose]$$

The factor of 2 applied to sodium and potassium concentrations allows for associated anions, and assumes complete ionisation.

This calculation is *not* valid:

if an unmeasured osmotically active substance, such as mannitol, or alcohol is circulating;

if there is gross hyperproteinaemia or lipaemia (see section on "Osmolarity and Osmolality").

Plasma Colloid Osmotic Pressure: Distribution of Water across Capillary Walls

The distribution of water across capillary walls is unaffected by electrolyte concentration, but affected by the osmotic effect of *plasma proteins.*

The maintenance of blood pressure depends on the retention of intravascular water at a pressure higher than that of interstitial fluid. Thus hydrostatic pressure is tending to force fluid into the extravascular space. In the absence of any opposing effective osmotic pressure across vascular walls water would be lost rapidly from the intravascular compartment.

Unlike the cell membrane, the capillary wall is freely and rapidly permeable to small molecules; sodium exerts almost no osmotic effect at this site. The smallest molecule present intravascularly at significant concentration, and which is virtually absent extravascularly, is albumin (MW about 70 000). The capillary wall is almost impermeable to it. Albumin concentration is therefore the most important factor opposing the net outward hydrostatic pressure: the higher molecular weight proteins although together present in much the same weight/volume concentration as albumin, contribute much less to this effect because of their larger size. This effective osmotic pressure across blood vessel walls is the *colloid osmotic* or *oncotic pressure.*

Because proteins contribute negligibly to measured plasma osmolality (Table II), *measurement of osmolality cannot be used to assess osmotic effects across capillary walls.*

Units of Measurement of Osmotic Pressure: Osmolarity and Osmolality

Concentrations of molecules can be expressed in two ways:

in molarity, the number of moles (or mmol) *per litre of solution*;

in molality, the number of moles (or mmol) *per kg of solvent*.

If the molecules are dissolved in pure water at concentrations such as are found in biological fluids, these two figures will differ very little. Plasma, however, is a complex solution. It also contains large molecules such as proteins, and the total volume of solution (water+protein) is greater than that of solvent (water only). The small molecules are dissolved only in the water, and at a protein concentration of 70 g/l the volume of water is about 6 per cent less than that of total solution (that is, the molality will be about 6 per cent greater than the molarity). Most methods of measuring individual ions such as sodium, measure them in molarity (mmol/l).

Osmotic pressure can also be expressed in two ways:

in *osmolarity* expressed as mmol *per litre of solution*;

in *osmolality* expressed as mmol *per kg of solvent*.

(The term milliosmol (mosmol) has been used to express osmolarity and osmolality: it has been recommended that this terminology be abandoned.)

The usual method of measuring plasma osmotic pressure by freezing point depression depends on osmo*la*lity, whereas calculated osmotic pressure from dissolved ions and molecules is expressed in osmo*la*rity. The osmometer therefore measures osmotic concentration in *extracellular water*. It is this osmo*la*l, rather than osmo*lar*, concentration which exerts its effect across cell walls and which is concerned in homeostatic mechanisms.

Under normal circumstances, although measured osmolality should be higher than calculated osmolarity (because of the protein content of plasma), little significant difference is found between the two figures. This is because incomplete dissociation of molecules to ions reduces the osmotic effect by about the same amount as the volume occupied by protein raises it. Calculated osmolarity is then adequate for clinical purposes. *This is not true in gross hyperlipaemia or hyperproteinaemia* when protein or lipid contribute much more than 6 per cent to measured plasma volume. Measured sodium levels and calculated osmolarity may then give results significantly lower than the true concentrations in plasma water and measurement of osmolality is advantageous. Moreover, as urine does not normally contain either protein or lipids, *direct comparison between urine and blood can only be made by measurement of osmolality of both.* Calculation of urinary osmolarity is not feasible because of the considerable variation in concentration of different solutes, and osmolality should always be measured.

CONTROL OF SODIUM AND WATER METABOLISM

The following is a simplified account of a complex situation and the student should bear this in mind.

CONTROL OF SODIUM

There seems to be no mechanism controlling sodium intake comparable to the thirst mechanism for water. The most important factor controlling loss is the mineralocorticoid hormone, aldosterone.

Aldosterone

Aldosterone is secreted by the zona glomerulosa of the adrenal cortex (p. 133). It affects sodium-potassium (and possibly sodium-hydrogen ion) exchange across *all* cell membranes. We shall concentrate on its effect on renal tubular cells, but we should bear in mind that it also affects faecal sodium loss, and the distribution of electrolytes in the body.

In the distal renal tubule aldosterone increases sodium reabsorption from the glomerular filtrate in exchange for potassium or hydrogen ion. The net result is retention of sodium, and loss of potassium and hydrogen ions. In the presence of high levels of circulating aldosterone *urinary sodium concentrations are low*.

Many factors have been implicated in the feed-back control of aldosterone secretion. Such factors as local electrolyte concentration in the adrenal and the kidney are probably of little clinical or physiological importance compared with the effect on the adrenal of the renin-angiotensin system, although this has been questioned by some authorities.

The Renin-Angiotensin System

Renin is a proteolytic enzyme secreted by a cellular complex situated near the renal glomeruli (and therefore called the juxtaglomerular apparatus). In the blood stream it acts on a renin substrate (an α_2-globulin) to form *angiotensin I*. This decapeptide is further split by a peptidase, located predominantly in the lungs, to *angiotensin II*. This peptide hormone has two main systemic actions:

it acts directly on blood vessel walls, causing vasoconstriction. It therefore helps to maintain blood pressure;

it stimulates the cells of the zona glomerulosa to secrete aldosterone.

The most important stimulus to renin production seems to be reduced renal blood flow (possibly changes in the mean blood pressure in the renal afferent arterioles are the actual stimuli). Poor renal blood flow is often associated with an inadequate systemic blood pressure and the two effects of angiotensin II ensure that this is corrected:

vasoconstriction raises the blood pressure before the circulating volume can be restored;

sodium retention occurs due to the action of aldosterone: as we shall see later, this will usually be accompanied by water retention with restoration of the circulating volume.

We may make an oversimplified statement based on the effect and control of aldosterone secretion. Aldosterone causes sodium retention. The stimulus to its secretion is renin, which is controlled, effectively, by circulating blood volume. Therefore, *blood volume controls net sodium retention.*

Non-Aldosterone Factors Affecting Renal Sodium Excretion

Aldosterone is probably the most important factor affecting sodium excretion. However, proximal tubular factors also affect it. These are:

changes in *glomerular filtration rate*;

a very *low GFR decreases sodium excretion*;

a very *high GFR causes natriuresis*.

Changes in GFR, by altering the amount of sodium filtered in unit time, vary the amount presented to the proximal tubule. If this is much reduced sodium will be almost completely reabsorbed in the proximal tubule; if much increased, proximal tubular mechanisms are swamped.

changes in *renal blood flow*.

A *low tubular blood flow* has been shown to increase proximal tubular reabsorption and *decrease sodium excretion*.

An *increased tubular blood flow* has been shown to decrease proximal tubular reabsorption and *increase sodium excretion*.

changes in *oncotic pressure of plasma in tubular blood vessels*.

Haemoconcentration (p. 42) increases proximal tubular reabsorption and *decreases sodium excretion*.

Haemodilution (p. 42) decreases proximal tubular reabsorption and *increases sodium excretion*.

Note that, in general, these effects potentiate each other and that of aldosterone. Thus, expansion of the plasma volume increases renal blood flow, cutting off aldosterone secretion, increasing the GFR, and increasing the tubular flow: if the expansion is with protein-free fluid there will be haemodilution. All these factors cause natriuresis.

There is some evidence that a third factor, probably hormonal, may be concerned in the control of sodium excretion in certain circumstances —*natriuretic hormone* or "third factor" (the GFR is the "first factor" and aldosterone the "second factor"). This hormone, which is said to be secreted in response to an expansion of plasma volume, may inhibit sodium reabsorption in the proximal tubule. It has been suggested that it fails to be secreted in oedematous states.

In certain inborn errors of renal tubular function the kidneys cannot respond to aldosterone; this may also be the case in the recovery phase of acute tubular necrosis, and with tubular damage of any kind: for instance, prolonged hypokalaemia in primary aldosteronism may prevent maximal tubular response to aldosterone.

In the later discussion we will assume the renin-aldosterone mechanism to be of overriding importance in the control of sodium excretion.

CONTROL OF WATER

Intake of water is controlled by thirst: absorption from the intestine is usually almost complete. The hypothalamic thirst centre responds to increased osmolality in the blood circulating through it (or, more accurately, to the osmotic difference between this and the cells of the centre), and therefore, usually to changes in sodium concentration. This occurs whether or not the fluid volume is normal.

Loss of water through the kidney is also controlled via the hypothalamus in response to osmolality changes. Increased plasma osmolality (usually due to increased sodium concentration) stimulates secretion of antidiuretic hormone (ADH; Vasopressin). The mechanism is very sensitive and a rise of osmotic pressure of only 2 per cent quadruples ADH output, while a similar fall cuts it off completely. Such changes (of about 3 mmol/l of sodium) may *not* be obvious using relatively crude laboratory methods.

There is some evidence that an extreme reduction in circulating volume may stimulate ADH secretion, and that this mechanism may even override the negative feedback due to hypo-osmolality.

Antidiuretic hormone is a peptide produced by the hypothalamus and secreted by the posterior pituitary gland. By increasing the permeability of the cells of the collecting ducts it increases passive water reabsorption along the osmotic gradient produced by the countercurrent mechanism (p. 7). Under its influence a *concentrated urine of high osmolality and specific gravity is passed.*

Water intake and loss are both controlled by plasma osmolality. Plasma osmolality normally depends largely on its sodium concentration. We have already pointed out that fluid volume controls sodium retention. We can now make the further *simplified* statement that *sodium concentration controls the amount of water in the body.* (It must be stressed that in *uraemia or hyperglycaemia,* or *after infusion of substances such as mannitol,* this is not true.)

Non-ADH Factors Affecting Water Excretion

The distal nephron has a limited capacity for reabsorption of water, even in the presence of a maximal amount of ADH: thus if a large volume reaches this site, water diuresis will occur.

The glomerular filtrate may contain high concentrations of osmotically active substances which are not completely reabsorbed during passage through the proximal tubules and loop of Henle: these inhibit water reabsorption at these sites (p. 10) and a larger volume than normal reaches the distal nephron. This is the basis of action of osmotic diuretics such as mannitol and, to a lesser extent, urea and hypertonic glucose. Patients being fed intravenously, or who, because of tissue damage, are breaking down larger than usual quantities of protein and producing excessive amounts of urea from the released amino acids, may become water-depleted even in the presence of adequate amounts of ADH (p. 48): urinary osmolality will be high.

As in the case of sodium and aldosterone, such non-ADH effects are usually relatively unimportant.

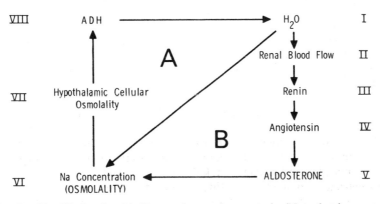

FIG. 2.—Simplified cycle of sodium and water homeostasis. (Note that in uraemia, hyperglycaemia and after infusion of hyperosmolal solutions such as mannitol or amino acids, factors other than sodium cause significant changes in osmolality.)

INTERRELATIONSHIP BETWEEN SODIUM AND WATER HOMEOSTASIS

Let us now look more closely at the simplified statements made above concerning the control of ADH and aldosterone secretion. We said that, effectively, sodium controls ADH secretion and water controls aldosterone secretion. But ADH controls water loss and aldosterone controls sodium loss. The homeostasis of sodium and water is interdependent and this simplified scheme is shown in Fig. 2. The thirst mechanism is not shown here but it should be remembered that an increase of osmolality not only reduces water loss but increases thirst and therefore intake. The diagonal line which divides the rectangle into two triangles, A and B, indicates that changes in hydration can also directly alter sodium concentration.

We shall refer to this scheme repeatedly while explaining changes occurring in pathological states. The reader should bear in mind that it is a simplified scheme and does not allow for the effect of fluid shifts across cell walls on extracellular concentrations and volume: nor does it allow for changes in osmotically active solute other than sodium.

ASSESSMENT OF SODIUM AND WATER BALANCE

Although it is easy to measure the intake of water and sodium of patients receiving liquid feeds (either oral or intravenous), it is less easy to do so when a solid diet is being taken. Fortunately, accurate measurement is rarely necessary in such cases.

Renal loss is also easy to measure but that in formed faeces, sweat and expired air ("insensible loss") is more difficult to assess and may be important when homeostatic mechanisms have failed, in the presence of abnormal losses by extrarenal routes, in unconscious patients and in infants (see p. 48). The aim should be to ensure that such subjects are normally hydrated, and then to keep them "in balance". The medical attendant has the difficult task of imitating normal homeostasis without the aid of the sensitive and interlinked mechanisms of the body.

ASSESSMENT OF THE STATE OF HYDRATION

Assessment of the state of hydration of a patient depends on observation of his clinical state, and on laboratory evidence of haemoconcentration or haemodilution. It must be stressed that both these methods are crude, and that quite severe disturbances of water balance can occur before they are obvious clinically or by laboratory methods.

In extracellular fluid loss (other than that due to haemorrhage), water and often electrolytes are lost from the vascular compartment, and the concentrations of large molecules and of cells rise: there is therefore a rise in all plasma protein fractions, in haemoglobin level, and in the haematocrit reading (*haemoconcentration*). Conversely, in overhydration these concentrations fall (*haemodilution*). These findings, of course, may also be affected by pre-existing abnormalities of protein or red cell concentrations, and changing concentrations may be more informative than single readings.

The problem of assessing hydration can be a difficult one, but can usually be resolved if the history and clinical and laboratory findings are all taken into account.

Assessment of Fluid Balance

By far the most important measurement in assessing changes in day-to-day fluid balance is that of intake and output. "Insensible loss" is usually assumed to be about 900 ml per day, but this must be balanced against "insensible" production of about 500 ml water daily by metabolism. The *net* "insensible loss" is therefore the difference between these two—about 400 ml a day. A normally hydrated patient, unable to control his own balance, should be given this basic volume of fluid daily with, in addition, the volume of measured losses (urine, vomitus, etc.) during the preceding 24 hours. If he is thought to be abnormally hydrated, fluid intake should be adjusted until hydration is restored. Inevitably the intake for any day must be calculated from the output in the preceding 24 hours. This is adequate when the patient starts in a normal state of hydration.

It should be remembered that a pyrexial patient may lose a litre or more of fluid in sweat, and that if he is also overbreathing, or on a respirator, respiratory water loss can be considerable. In such cases the allowance of 400 ml daily for insensible loss may not be enough.

Many very ill patients are incontinent of urine, and measurement of even this volume may be impossible. Changes in body fluid may be assessed by daily weighing (1 litre of water weighs 1 kg). Over short periods of time changes in solid body weight will be small, and alteration in weight can be assumed to be due to changes in fluid balance. Unfortunately, very ill patients may be unable to sit in weighing chairs. Some hospitals have weighing beds, in which the patient is weighed with the bed and bedclothes. In this situation, care must be taken to keep the weight of the bedclothes constant.

Assessment of fluid balance in severely ill patients may present grave problems. Every attempt must be made to make *accurate* measurements of fluid intake and loss. In most circumstances carefully kept fluid charts and daily weighing are of more importance than frequent plasma electrolyte estimations. *Inaccurate charting* is useless, and *may be dangerous.*

Assessment of Sodium Balance

If fluid balance is accurately controlled *in a normally hydrated patient* plasma sodium concentrations are the best means of assessing sodium balance. As we shall see later, the proviso of normal hydration is important, and plasma electrolyte values *alone* should never be used to assess the need for therapy. They must be correlated with the history and the present state of hydration.

CLINICAL FEATURES OF WATER AND SODIUM DISTURBANCES

We are now in a position to explain the immediate clinical consequences of water and sodium disturbances. These depend on changes in extracellular osmolality and hence in cellular hydration (sodium) and changes in circulating volume (water).

CLINICAL FEATURES OF DISTURBANCES OF SODIUM CONCENTRATION

Measuring sodium concentration is a poor man's substitute for measuring osmolality. Sodium levels *per se* are not important to the body, but osmolality is. It is important to understand that the one does not always reflect the other.

If sodium concentration changes without change of other extracellular solute most of the immediate clinical features are due to an osmotic difference across cell walls. Gradual changes, by allowing time for redistribution of diffusible solute such as urea, and therefore for equalisation of osmolality without major shifts of water, may produce little clinical effect.

Hyponatraemia *sometimes* reflects extracellular *hypo-osmolality*, and may therefore cause cellular overhydration. However, it may be appropriate to the osmolal state. For example, acute uraemia, hyperglycaemia, infusion of amino acids or mannitol or high plasma alcohol levels increase extracellular osmolality: the consequent homeostatic dilution of *total* extracellular solute towards a normal osmolality results in hyponatraemia. This therefore reflects partial or complete compensation of hyperosmolality, not hypo-osmolality. In gross hyperlipaemia (especially that due to intravenous lipid feeding) or hyperproteinaemia, "hyponatraemia" found in whole plasma may coexist with a normal sodium concentration in plasma water, and therefore with normal osmolality at cell walls (p. 35). *Infusion of sodium salts into patients with appropriate hyponatraemia is dangerous.* It is also important to exclude artefactual "hyponatraemia" caused by taking blood from the limb into which fluid of low sodium concentration is being infused (p. 456).

In the cases in which hyponatraemia reflects true hypo-osmolality overhydration of cerebral cells may cause *headache, confusion* and, later, *fits*.

Hypernatraemia *always* reflects extracellular *hyperosmolality* with the danger of cellular dehydration. Because solute other than sodium is present in very low concentration in normal plasma (Table II, p. 34) severe hypoglycaemia for example, could only reduce plasma osmolality by about 5 in 280 mmol/l. (Hyperglycaemia, by contrast, can increase it by as much as 50 mmol/l). Significant hypernatraemia cannot, therefore, be appropriate to the osmolal state.

The clinical effects of dehydration of cerebral cells are *thirst, mental confusion* and, later, *coma*.

Hyponatraemia is a much more common finding than hypernatraemia and is often appropriate. The number of cases in which it indicates true hypo-osmolality is probably similar to the incidence of hypernatraemia, which always reflects hyperosmolality.

CLINICAL FEATURES OF DISTURBANCES OF FLUID VOLUME

The clinical features of isosmolal disturbances of fluid balance are due to changes in circulating volume.

Fluid depletion causes hypovolaemia. If isosmolal fluid is lost redistribution of water does not occur across cell membranes and clinical signs of "dehydration" and thirst may be absent. *Hypotension and collapse* follow, and death may result from circulatory insufficiency. Laboratory findings are those of *haemoconcentration* with a raised plasma protein concentration and haematocrit. Because renal perfusion is poor, glomerular filtration rate may fall and there may be *uraemia* and other signs of nitrogen retention.

Fluid excess with no change in osmotic difference across cell walls, causes *hypertension* and overloading of the heart, with *cardiac failure.* Oedema occurs when albumin levels are low, because the reduced oncotic pressure together with increased intravascular hydrostatic pressure causes passage of water into interstitial fluid. Laboratory findings are characteristic of *haemodilution* and, unless there is renal glomerular failure, the plasma urea level tends to be low.

DISTURBANCES OF SODIUM AND WATER METABOLISM

Disturbances of sodium and water metabolism are most commonly due to excessive losses from the body. Sometimes inadequate intake contributes to the deficiency. In the presence of normal homeostatic mechanisms excessive intake is rarely of clinical importance.

Disturbances can also be due to abnormal amounts of the homeostatic hormones, aldosterone and ADH, or to failure of the end organs, particularly the kidney and intestine, to respond to them.

WATER AND SODIUM DEFICIENCY

Pure water and pure sodium deficiency are rare. However, conditions in which there is imbalance between loss of water and sodium are relatively common. Predominant sodium depletion is almost always accompanied by some water depletion and *vice versa*.

TABLE III

APPROXIMATE SODIUM CONCENTRATIONS IN BODY FLUIDS (mmol/l)

Plasma	Gastric	Biliary and pan-creatic	Small intes-tinal	Ileal	Ileostomy (new)	Diarrhoea	Sweat
140	60	140	110	120	130	60	60

Predominant Water Depletion

This syndrome is due to loss of water in excess of loss of sodium. It is usually the result of the loss of fluid with a lower concentration of sodium than that of plasma, or of deficient water intake. Sweat and gastric juice contain concentrations of sodium much lower than those in plasma (see Table III); in diabetes insipidus, during an osmotic diuresis, or in the rare inborn error associated with failure of the renal tubules to respond to ADH, urine of low sodium concentration is passed. *As hyperosmolality due to predominant water loss causes thirst, effects are only seen if water is not available or cannot be taken in adequate quantities.*

The clinical situations associated with predominant water loss are: *water deficiency in the presence of normal homeostatic mechanisms.*

Excessive fluid loss—

Loss of excessive amounts of sweat.

Loss of gastric juice.

Loss of fluid stools of low sodium content (usually in infantile gastroenteritis).

Excessive respiratory loss, especially on a respirator.

Loss of fluid from extensive burns.

Deficiency of fluid intake—

Inadequate water supply, or mechanical obstruction to its intake.

failure of homeostatic mechanisms for water retention.

Inadequate response to thirst mechanism (for example in comatose patients and infants). It is doubtful if significant water depletion can develop if thirst mechanisms are normal and water is available.

Deficiency of ADH (diabetes insipidus).

Overriding of ADH action by osmotic diuresis.

Failure of renal tubular cells to respond to high levels of ADH (nephrogenic diabetes insipidus).

In most clinical states associated with predominant water depletion more than one of these factors is responsible.

Predominant water depletion with normal homeostatic mechanisms.— The reader should refer to Fig. 2 during the following explanation.

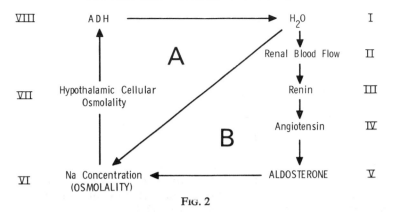

FIG. 2

Triangle B—Immediate Effects:

loss of water in excess of sodium increases the plasma sodium concentration and osmolality (diagonal line I–VI);

reduction of circulating volume reduces renal blood flow and stimulates aldosterone production (I–V). Sodium is retained and the rise in plasma sodium is aggravated (V–VI).

Triangle A—Compensatory Effects:

Increased plasma osmolality stimulates:

thirst, increasing water intake if water is available and if the patient can respond to it (not shown on diagram).

ADH secretion (VI–VIII). Urinary volume falls (VIII–I) and water loss through the kidney is minimised.

If adequate amounts of water are available depletion is rapidly corrected. In the absence of intake, or in the presence of continued loss, the mechanisms in Triangle A cannot function effectively and homeostasis breaks down. Hypernatraemia is therefore an early finding in water depletion: it is important to realise that it may be present before clinical signs of fluid depletion are obvious.

The clinical signs are those of:

hypernatraemia (p. 44);
later, fluid depletion (p. 45);
oliguria due to ADH secretion.

The laboratory findings are:

hypernatraemia;
haemoconcentration (due to fluid depletion);
mild uraemia (due to fluid depletion and hence low GFR);
a urine of low volume, high osmolality and specific gravity and high urea concentration (due to the action of ADH);

low urinary sodium concentration (in response to high aldosterone levels).

Failure of homeostatic mechanisms for water retention.—*Diabetes insipidus* may be due to pituitary or hypothalamic damage caused by head injury or during hypophysectomy, or to invasion of the region by tumour. It may be idiopathic in origin.

Hereditary nephrogenic diabetes insipidus is a rare inborn error of renal tubular function in which ADH levels are high but the tubules cannot respond to it (compare pseudohypoparathyroidism, p. 242): the newborn infant passes large volumes of urine and rapidly becomes dehydrated. Urinary loss is difficult to assess at this age, and the cries of thirst may be misinterpreted. During the *recovery phase of acute oliguric renal failure*, or in other conditions with predominant tubular damage, there may also be failure to respond to ADH.

The reader should refer again to Fig. 2 and to the immediately preceding section.

The first stage is identical with that described in the previous section. Because the mechanisms shown in Triangle A are not functioning, compensation cannot occur, and the point is soon reached at which intake cannot adequately replace loss.

The clinical signs and findings are those described in the previous section, with the exception that there is *polyuria*, not oliguria, and that the urine is of *low osmolality, specific gravity and urea concentration*. These findings are the result of ADH deficiency. If any doubt remains as to the diagnosis, water may be deliberately withheld (see Appendix to Chapter I). In the absence of ADH, or when the tubules fail to respond to it, concentration of urine fails to occur. In the case of true ADH deficiency a concentrated urine will be passed if exogenous ADH (pitressin) is given. In nephrogenic diabetes insipidus maximal amounts of ADH are already circulating, and administration of it will not influence water reabsorption.

The unconscious patient.—The syndrome of predominant water depletion with hypernatraemia is seen most commonly in the unconscious or confused patient or in the infant with gastro-enteritis or pneumonia. In such subjects there is usually more than one cause of water depletion.

1. They are often pyrexial. Loss of hypotonic *sweat* is increased.

2. They may be *overbreathing* because of pneumonia, acidosis or brain stem damage. Water loss is increased.

3. Humidifiers on *respirators* may be inadequate and artificially respired patients, who tend to be hyperventilated, may become water depleted.

4. They may be given hypertonic intravenous infusions, either to provide nutrient (dextrose or amino acids) or to produce osmotic diuresis in cases of poisoning (thus increasing the urinary loss of poison).

There may be tissue damage and hence breakdown of protein to urea. The coma may be due to diabetes mellitus with glycosuria. All these factors will contribute further to plasma hyperosmolality and will cause an *osmotic diuresis* overriding the effect of ADH.

5. There may be true diabetes insipidus due to head injury.

6. *The subject cannot respond to hyperosmolality by drinking.*

In the presence of factors 4 and 5 a high urine volume contributes to dehydration (and is *not* an indication of "good" hydration) and the extent of loss may not be realised if the subject is incontinent. Homeostatic mechanisms may not be capable of responding to hyperosmolality (5 and 6) and hypernatraemia is an early finding. It may occur before clinical signs of volume depletion are evident and is dangerous because of the resulting cellular dehydration.

Such a syndrome is also seen in unconscious or confused patients with extensive *burns*. The water in the exuded ECF evaporates, while some of the electrolyte is reabsorbed into the circulation. Tissue breakdown and intravenous feeding contribute to an osmotic diuresis.

Hyperosmolar saline should never be used as an emetic in cases of poisoning. Movement of water into the gut along the osmotic gradient, and absorption of some of the sodium, can cause marked hypernatraemia: the vomiting or unconscious patient cannot respond to the hyperosmolality by drinking. Death is a common consequence.

Predominant Sodium Depletion

Some body fluids are hypo-osmolal, and we have seen that their loss can cause predominant water depletion with hypernatraemia. By contrast, no body fluid has a sodium concentration much higher than that of plasma (Table III, p. 46). Untreated loss of isosmolar fluid, as in diarrhoea or through fistulae, cannot cause predominant sodium depletion with hyponatraemia: the effects of such loss are due to hypovolaemia (p. 45). However, it is very common for a patient or his doctor to replace isotonic loss with hypotonic fluid, or with water alone. The consequent hyponatraemia reflects a state of predominant sodium depletion directly comparable to the hypernatraemia of predominant water depletion.

The clinical conditions accompanied by predominant sodium deficiency are:

sodium deficiency in the presence of normal homeostatic mechanisms.

Vomiting ⎫
Diarrhoea ⎪ followed by replacement with fluid
Loss through fistulae ⎬ low in sodium
Excessive sweating ⎭

failure of homeostatic mechanisms for sodium retention.

Addison's disease (absence of aldosterone);

"pseudo" Addison's disease (failure of the otherwise normal renal tubular cells to respond to aldosterone). This is very rare.

Renal tubular failure.

Predominant sodium depletion with normal homeostatic mechanisms.— The reader should refer to Fig. 2.

Triangle A. These mechanisms tend to correct hypo-osmolality (and therefore to safeguard normal cell hydration).

ADH secretion is cut off in response to hypo-osmolality (VI–VIII).

Water is lost in the urine (VIII–I).

This would tend to restore osmolality at the expense of volume, as shown in the diagonal line I–VI. However, if the volume is artificially maintained by replacing this urinary loss with hypotonic fluid, hypo-osmolality (as shown by hyponatraemia) persists.

Sodium depletion may even be aggravated by this regime. If the urinary loss of fluid were uncorrected loss of fluid would stimulate aldosterone secretion (I–V): sodium would be retained (V–VI), and the sequence of events depicted in Triangle A would be reversed, so that the balance of both sodium and water would be corrected. Expanding plasma volume by infusion cuts off aldosterone secretion, and sodium is lost in the urine despite sodium depletion and hypo-osmolality. The net effect is restoration of circulating volume by infusion, but cellular overhydration due to hypo-osmolality.

The clinical signs are those of hypo-osmolality (p. 44).

The laboratory findings are:

hyponatraemia;

passage of a large volume of dilute urine (due to inhibition of ADH secretion);

and if fluid intake is excessive

haemodilution;

a low plasma urea concentration due to the high GFR (intravenous infusion is one of the commonest causes of a low plasma urea);

a high urinary sodium concentration (due to inhibition of aldosterone secretion).

It is a common practice to infuse patients post-operatively with fluid such as "dextrose saline", which contains about 30 mmol/l of sodium. It is not, therefore, surprising that post-operative hyponatraemia is so often found. No serious harm is done in subjects with normal renal function, because the mechanisms described above rapidly correct both osmolality and sodium and water balance when the infusion is stopped. However, by alternating "dextrose saline" and isotonic saline, it is our experience that most patients can maintain relatively normal plasma sodium levels, whilst undergoing a reassuring diuresis: the danger of overloading the circulation is minimal if renal function is normal.

In seriously ill patients, with impairment of homeostatic mechanisms, infusion of "dextrose saline" is even more likely to cause hyponatraemia, together with fluid overload.

Failure of homeostatic mechanisms for sodium retention.—Addison's disease with hypoaldosteronism is by far the most important cause of this type of disturbance. Primary renin deficiency has been reported. During the recovery phase of acute oliguric renal failure large amounts of sodium can be lost in urine due to residual tubular dysfunction.

The reader should again refer to Fig. 2 and to the immediately preceding section.

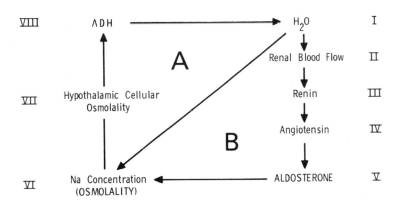

FIG. 2

The first stage is identical with that described in the preceding section. The homeostatic mechanisms of Triangle B are not functioning. Osmolality is maintained until late by water loss; in advanced cases the stimulatory effect of hypovolaemia on ADH secretion (p. 40) may override the normal physiological cycle, and hypo-osmolality supervenes.

The clinical signs are:

    those of fluid depletion (p. 45);

    later those of hypo-osmolality (p. 44).

The laboratory findings are:

    haemoconcentration (due to fluid depletion);

    renal circulatory insufficiency with mild uraemia (due to fluid depletion);

    later hyponatraemia;

    inappropriately high urinary sodium concentration in the face of volume depletion (due to aldosterone deficiency).

WATER AND SODIUM EXCESS

In the presence of normal homeostatic mechanisms an excess of water and sodium is of relatively little importance, because it is rapidly corrected. These syndromes are commonly associated with failure of homeostatic mechanisms.

Predominant Excess of Water

Water overloading may occur in circumstances in which normal homeostasis has failed, or is over-ridden:

in *renal failure*, when fluid of low sodium concentration has been replaced in excess of that lost. Fluid balance should be carefully controlled in such patients;

in the presence of *"inappropriate" ADH secretion* (the term "inappropriate" is used in this book to describe continued secretion of a hormone under conditions in which it should normally be cut off). ADH is one of the peptide hormones which can be manufactured by malignant tissue of non-endocrine origin (see Chapter XXII for a further discussion). "Inappropriate" secretion (possibly from the pituitary or hypothalamus itself) occurs in a variety of other conditions, including infections. Such ADH production is not under normal feedback control and therefore continues in circumstances in which its secretion should be completely cut off (in the presence of low extracellular osmolality). This fact is evidenced by the production of a relatively concentrated urine in the presence of dilute plasma;

during intravenous administration of the posterior pituitary hormone *oxytocin* ("Syntocinon", "Pitocin") to induce labour. Oxytocin has an antidiuretic effect similar to that due to ADH. Glucose (5 per cent) is often used as a carrier, and, as the glucose is metabolised, the net effect is water retention. If infusion is prolonged dangerous hyponatraemia and hypo-osmolality may develop: a careful watch should be kept on the patient's fluid balance and plasma sodium concentration, and some of the oxytocin should be given in isotonic saline. If we refer again to Fig. 2 we can see how excessive intake is normally corrected.

Triangle B. Immediate effects:

excess water tends to *lower plasma sodium* (diagonal arrow), (I–VI); *increased renal blood flow* cuts off aldosterone production, increasing urinary sodium loss and therefore further decreasing plasma sodium (V–VI).

Triangle A. Compensatory effects:

ADH is cut off (VI–VIII) and large volumes of *dilute urine* are passed (VIII–I).

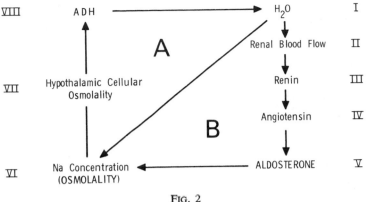

FIG. 2

When renal glomerular function is poor, in the presence of "inappropriate" ADH secretion (which is not subject to feedback control), or if the fluid contains oxytocin, Triangle A is out of action. The second stage cannot take place.

The clinical consequences are:
    those of water excess;
    if overhydration is rapid, those of hypo-osmolality.

In patients with "inappropriate" ADH secretion the hypo-osmolality is usually of gradual onset, allowing time for redistribution of solute across cell walls, and clinical symptoms may be absent despite a very low plasma sodium concentration. Administration of oxytocin in 5 per cent glucose, by contrast, reduces osmolality within a few hours, with the serious danger of cerebral overhydration (p. 35).

The findings are:
    haemodilution;
    hyponatraemia.

If the cause is glomerular failure there will be uraemia. If it is due to "inappropriate" ADH secretion the plasma urea level will tend to be low.

Predominant Excess of Sodium

Predominant sodium excess is most commonly due to an "inappropriate" secretion of excess aldosterone or other corticoids in *Conn's syndrome* (*primary aldosteronism*), or in *Cushing's syndrome*, or to excessive stimulation of a normally functioning adrenal in secondary aldosteronism. Cushing's syndrome is described more fully on p. 136.

Primary aldosteronism (Conn's syndrome).—About half the cases of primary aldosteronism (excess aldosterone secretion not subject to normal feedback control) are due to a benign aldosterone-secreting adenoma of the adrenal cortex; about 10 per cent are multiple. Most of the remaining cases are associated with bilateral nodular hyperplasia of the adrenal glands.

The reader should refer again to Fig. 2:

excess aldosterone causes *urinary sodium retention* (V–VI);

the *increased sodium concentration* stimulates ADH secretion. (VI–VIII) and water retention (VIII–I);

water retention tends to return plasma sodium concentrations to normal (diagonal arrow, I–VI);

triangle B is out of action and aldosterone secretion cannot be cut off. This tends to maintain *plasma sodium levels at or near the upper end of the normal range*;

continued action of aldosterone causes sodium retention at the expense of potassium loss, and *potassium depletion occurs*.

The clinical features are:

those of volume excess (p. 45). Patients are hypertensive but rarely oedematous (one normotensive case has been reported);

those of hypokalaemia (p. 72).

Findings are:

hypokalaemia (due to excess aldosterone);

a high plasma bicarbonate (for explanation see p. 67);

a plasma sodium in the high normal range or just above it;

a low urinary sodium concentration in the early stages. However, sodium excretion may rise later, possibly because of hypokalaemic tubular damage (p. 15), and consequent unresponsiveness to aldosterone, or because of the effect of the expanded plasma volume on non-aldosterone factors which affect sodium excretion (p. 39).

Hypokalaemic alkalosis occurs in potassium depletion from any cause. However, the association of these findings in a patient without an obvious cause (such as administration of purgatives or diuretics) for potassium loss and with hypertension should suggest the diagnosis of primary aldosteronism. Further suggestive evidence would be the finding of high aldosterone with low renin levels (in all cases of *secondary* aldosteronism *both* are high). These estimations are not available in all routine laboratories, but in the rare cases in which the diagnosis of primary aldosteronism remains a possibility after careful clinical assessment they can be performed in special centres.

Secondary aldosteronism: oedema.—Any of the conditions already described in the sections on water and sodium depletion, in which aldosterone secretion is stimulated following reduction in renal blood flow could, strictly, be called secondary aldosteronism. The term is more commonly used to indicate the conditions in which, because the initial abnormality is not corrected, long-standing increased aldosterone secretion itself produces abnormalities.

Aldosterone is secreted following stimulation of the renin-angiotensin system by a low renal blood flow. This may occur either because of local abnormalities in renal vessels or because of a reduced circulating volume.

Secondary aldosteronism, in the usually accepted sense of the word, occurs in the following conditions:

1. Redistribution of extracellular fluid, leading to a reduction of plasma volume in the presence of normal or high total extracellular fluid volume. These conditions are due to a reduced plasma oncotic pressure, and are therefore associated with low plasma albumin levels. Oedema is present. Such conditions are:

> liver disease (impaired aldosterone catabolism aggravates the hyperaldosteronism);
> nephrotic syndrome;
> protein malnutrition.

2. Damage to the renal vessels, reducing renal blood flow. These conditions are usually not associated with oedema:

> essential hypertension;
> malignant hypertension;
> renal hypertension (e.g. renal artery stenosis).

3. Cardiac failure. In this case two factors may cause low renal blood flow. Firstly, the cardiac output may be low, with poor renal perfusion pressure. Secondly, high intravascular hydrostatic pressure on the venous side of the circulation may cause redistribution of fluid and oedema. Aldosterone catabolism is also impaired.

The mechanisms in Fig. 2 are brought into play:

reduced renal blood flow stimulates aldosterone secretion (I–V);

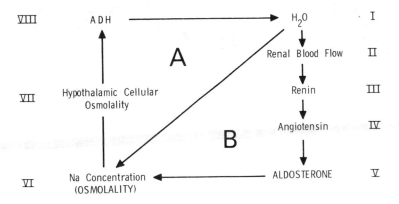

FIG. 2

TABLE IV

Hypothetical Numerical Examples to Illustrate Conditions Associated with Normal and Abnormal Plasma Sodium Concentrations

	Example	ECF (litres)	E.C. Na (mmol)	Plasma [Na] (mmol/l)	Haemoconc. Dilution	Clinical Features	Principle of Treatment
Normonatraemia (Equivalent Na and H_2O changes)							
Normal	—	20	2800	140	None	—	—
$Na+H_2O$ depletion	Early Addison's disease	15↓	2100↓	140	Conc. (Urea ↑)	Volume depletion (p. 45)	Treat cause Isosmolal saline
$Na+H_2O$ excess	Non-oedematous secondary aldosteronism	30↑	4200↑	140	Dil.	Volume excess (p. 45)	Treat cause Restrict $Na+H_2O$
Hyponatraemia (Relative H_2O excess)							
H_2O excess	Inappropriate ADH secretion	25↑	2800	112↓	Dil. (Urea N or ↓)	Volume excess (p. 45) Hypo-osmolality (p. 44)	Restrict H_2O

Na depletion	Diarrhoea with H₂O replacement	*20*	*2240↓*	112↓	None	Hypo-osmolality (p. 44)	Give isosmolal or hyper-osmolal saline
Na+H₂O excess	Oedematous secondary aldosteronism (commonest form)	30↑	3360↑	112↓	Dil.	Volume excess (p. 45) Hypo-osmolality (p. 44)	Restrict Na and H₂O Diuretics.
Na+H₂O depletion	Late Addison's disease	15↓	1680↓	112↓	Conc. (Urea ↑)	Volume deficiency (p. 45) Hypo-osmolality (p. 44)	Give steroids +normos-molal or hyperosmolal saline
Hypernatraemia (Relative Na excess) — H₂O depletion	Unconscious patient	18↓	*2800*	155↑	Mild conc.	Hyper-osmolality (p. 44)	Give hypo-osmolal fluid *slowly*
Sodium excess	Excessive intake usually in infants (*Very* rare)	*20*	3100↑	155↑	None	Hyper-osmolality (p. 44)	Remove Na by dialysis

Values in first line taken as "normals" to which other are related. "Normal" values in italics throughout.
No account has been taken of shifts of fluid across cell walls which would slightly reduce changes in plasma sodium levels.
Note especially relatively slight reduction in volume associated with hypernatraemia.

sodium retention stimulates ADH secretion (V–VIII) and therefore *water retention* (VIII–I).

Intravascular volume tends to be restored, but in conditions with hypo-albuminaemia or in cardiac failure more fluid passes into the interstitial fluid and the cycle restarts. A vicious circle is set up in which circulating volume can only be maintained by water retention and oedema results. In non-oedematous states hypertension occurs.

This cycle should only stimulate water retention in parallel with sodium retention. However, many of these patients have hyponatraemia. The reasons for this are not clear, but from the therapeutic point of view it must be remembered that in the presence of oedema the total amount of sodium in the body is high, even if there is hyponatraemia (see Table IV).

Hypokalaemia may be present, but is a less common finding than in primary aldosteronism. Again the reason is not clear, but it may be due to a redistribution of potassium between cells and extracellular fluid. Hypokalaemia is, however, more readily precipitated by diuretic therapy in secondary hyperaldosteronism than in normal subjects.

The clinical features of these cases are those of the primary condition. The findings are:

a normal or low plasma sodium concentration;

a low urinary sodium excretion;

findings due to the primary abnormality, e.g. hypo-albuminaemia, uraemia, etc.

THE DIAGNOSTIC VALUE OF URINARY SODIUM CONCENTRATION

It will be noticed that urinary sodium excretion is low when aldosterone secretion is high. In the absence of primary aldosteronism this finding indicates low renal blood flow. Similarly, except in Addison's disease, high urinary sodium excretion indicates high renal blood flow. The estimation gives no information about the state of body sodium stores. In most cases the state of hydration and sodium metabolism can be assessed by noting the clinical picture and plasma findings. Only in the rare cases in which doubt still remains or in the differential diagnosis of renal tubular and glomerular lesions (p. 19) is urinary sodium estimation indicated.

THE CLINICAL SIGNIFICANCE OF PLASMA SODIUM CONCENTRATIONS

Table IV summarises the situations in which hypernatraemia and hyponatraemia may be found, and indicates some aspects of the clinical picture which may help to differentiate the causes. The two important points to remember are:

hypernatraemia almost invariably indicates water depletion;

hyponatraemia is much more commonly due to water excess than to sodium depletion: if this is so, the body content of sodium may be normal, but is more often high. If it is due to sodium depletion the patient will be dehydrated, both on clinical and laboratory findings, with mild to moderate uraemia.

Plasma sodium levels down to about 120 mmol/l occur in many ill patients, and unless there is evidence of dehydration, require no treatment. In states in which plasma osmolality is contributed to significantly by other substances, as in severe uraemia and hyperglycaemia, mild hyponatraemia may be "appropriate".

(The student is urged to read the indications for electrolyte estimation on p. 77.)

BIOCHEMICAL BASIS OF TREATMENT OF SODIUM AND WATER DISTURBANCES

TREATMENT OF SODIUM AND WATER DISTURBANCES

It cannot be stressed too strongly that treatment should not be based on plasma sodium concentrations alone. Hyponatraemia *per se* only rarely requires treatment. Assessment of the clinical history, findings indicating haemoconcentration or haemodilution, and plasma urea level help to unravel the cause of the low sodium concentration by indicating the state of hydration and renal function; this knowledge will determine whether correction of the sodium level is indicated. Rarely it may be useful to know the urinary sodium or urea concentration, or osmolality or specific gravity. Hypernatraemia, on the other hand, should always be treated by *slow* infusion of hypotonic fluid.

Table IV uses hypothetical values to illustrate the various combinations of disturbances of sodium and water metabolism and their treatment. Solutions available for intravenous use are given in the Appendix. These rules can only be a guide to treatment in often complicated clinical situations. They may, however, help to avoid some of the more dangerous errors of electrolyte therapy.

When homeostatic mechanisms have failed, especially in renal failure, normal hydration should be maintained according to the principles outlined on p. 43.

SUMMARY

1. Homeostatic mechanisms for sodium and water are interlinked. Potassium and hydrogen ions often take part in exchange mechanisms with sodium.

2. Distribution of fluid between cells and extracellular fluid depends

on osmotic difference between the intra- and extracellular fluid. Changes in this are usually due to changes in sodium concentrations.

3. Distribution of fluid between the intravascular and interstitial compartments depends on the balance between the hydrostatic pressure and the effective plasma osmotic pressure: the latter depends largely on albumin concentration.

4. Aldosterone secretion is the most important factor affecting body sodium.

5. Aldosterone secretion is controlled by the renin-angiotensin mechanism, which responds to changes in renal blood flow.

6. ADH secretion is the most important factor affecting body water.

7. ADH secretion is controlled by plasma osmolality. Plasma osmolality depends mainly on sodium concentration.

8. Clinical effects of disturbances of water and sodium metabolism are due to:

(a) Changes in extracellular osmolality, dependent largely on sodium concentration. In pathological conditions urea and glucose concentrations can be important.

(b) Changes in circulating volume.

FURTHER READING

SWALES, J. D. (1975). *Sodium Metabolism in Disease*. London: Lloyd-Luke.
Ross, E. J. (1975). *Aldosterone and Aldosteronism*. London: Lloyd-Luke.
WILLIAMS, G. H. and DLUHY, R. G. (1972). Aldosterone biosynthesis. Inter-relationship of regulatory factors. *Amer. J. Med.*, **53**, 595.
MULROW, P. J. and FORMAN, B. H. (1972). Tissue effects of mineralocorticoids. *Amer. J. Med.*, **53**, 561.
PEART, W. S. (1977). The kidney as an endocrine organ. *Lancet*, **2**, 542.
Natriuretic Hormone. Leading article in *Lancet* (1977). **2**, 537.

APPENDIX

TABLE V

Some Sodium-Containing Fluids for Intravenous Administration

	Electrolytes (mmol/l)				Glucose (mmol/l with g/dl in brackets)	Ca (mmol/l)	Approximate osmolarity relative to plasma
	Na	K	Cl	HCO$_3$			
Saline							
"Normal" (physiological) saline	154	—	154	—	—	—	×1
Twice "Normal"	308	—	308	—	—	—	×2
Half "Normal"	77	—	77	—	—	—	×$\frac{1}{2}$
Fifth "Normal"	31	—	31	—	—	—	×$\frac{1}{5}$
"Dextrose" Saline							
	77	—	77	—	149 (2·69)	—	×1
	31	—	31	—	240 (4·3)	—	×1
	77	—	77	—	278 (5·0)	—	×1$\frac{1}{2}$
	154	—	154	—	278 (5·0)	—	×2
	154	—	154	—	556 (10·0)	—	×3
Sodium Bicarbonate							
1·4%	167	—	—	167	—	—	×1
2·8%	334	—	—	334	—	—	×2
*8·4%	1000	—	—	1000	—	—	×6

* Most commonly used bicarbonate solution. Note marked hyperosmolarity. It should only be used if very strongly indicated.

TABLE V (continued)

	Electrolytes (mmol/l)				Glucose (mmol/l with g/dl in brackets)	Ca (mmol/l)	Approximate osmolarity relative to plasma
	Na	K	Cl	HCO₃			
Complex Solutions							
Ringer's Solution	147	4·2	156	—	—	2·2	×1
Darrow's Solution	122	35·8	104	53 (as lactate)	—	—	×1
Hartmann's Solution	131	5·4	112	29 (as lactate)	—	1·8	×1
Sodium lactate 1/6 molar	167	—	—	167 (as lactate)	—	—	×1
Amino acid solutions							
"Amigen" 5%	35	18	22				Hyperosmolar
"Aminosol 3·3%"	53	0·15			278 (5·0)	2·5	Hyperosmolar
"Aminosol 10%"	160	0·5					Hyperosmolar
"Trophysan 5"	6	8					Hyperosmolar
"Vamin Glucose"	50	20	55		557 (10)	2·5	Hyperosmolar

TABLE VI

Sodium-Free Dextrose Solutions for Intravenous Administration (Usually Used as an Energy Source)

Dextrose	Glucose (mmol/l)	Osmolarity	kJ/l (Calories/l in brackets)
5%	278	Isosmolar	860 (205)
10%	556	Hyperosmolar	1720 (410)
20%	1112	Hyperosmolar	3440 (820)
40%	2224	Hyperosmolar	6880 (1640)

TABLE VII

Other Hyperosmolar Solutions (Usually Used as Osmotic Diuretics)

	Concentration (mmol/l)	kJ/l (Calories/l in brackets)
Mannitol 10%	550	0
20%	1100	0
Urea 4%	667	0 (But may also contain fructose
30%	5000	0 10% giving 1720 kJ/l)
Sorbitol 20%	1100	3360 (800)
30%	1650	5040 (1200)
Fructose 10%	556	1720 (410)
20%	1112	3440 (820)
40%	2224	6880 (1640)

Chapter III

POTASSIUM METABOLISM: DIURETIC THERAPY

THE total amount of potassium in the body is about 3000 mmol.

Potassium is predominantly an *intracellular* ion, and only about 2 per cent of the body content is in the extracellular fluid. The plasma potassium level is therefore a poor indicator of the total amount in the body. This apparent disadvantage is more theoretical than real, because it is *plasma potassium concentrations* that are of immediate importance in therapy; either hyperkalaemia or hypokalaemia, if severe, is dangerous, and must be treated whatever the state of the intracellular potassium. However, some assessment of the overall situation should be made to anticipate therapeutic needs.

The low extracellular concentration of potassium ions is important for normal neuromuscular activity and cardiac action; in this respect it resembles calcium and magnesium (p. 231).

FACTORS AFFECTING PLASMA POTASSIUM CONCENTRATIONS

The predominantly intracellular location of potassium provides a reservoir of the ion, and plasma potassium concentration, unlike that of sodium, is relatively little affected by changes in water balance. The hyperkalaemia often associated with dehydration is due more to renal retention (p. 12) than directly to haemoconcentration.

Potassium enters and leaves the extracellular compartment by three main routes:
   the intestine
   the kidney
       the glomerulus
       the tubular cells
   the walls of *all* other body cells

The Intestine

Potassium is absorbed throughout the small intestine. Dietary intake replaces urinary and faecal loss, and amounts to 100 mmol/day or less.

Potassium leaves the extracellular compartment in all intestinal secretions. The approximate concentrations in these secretions are indicated in Table VIII. Although most of them are not very high, the daily volume secreted into the lumen of the gut is large (p. 32) and contains about 100 mmol/day. Most of this potassium is reabsorbed, together with the dietary intake, and only about 10 per cent of this is present in formed faeces. As in the case of sodium, excessive intestinal loss of potassium in diarrhoea stools, in ileostomy fluid or via fistulae is mainly derived from intestinal secretions rather than from dietary intake. However, prolonged starvation can cause potassium deficiency and hypokalaemia.

TABLE VIII

APPROXIMATE POTASSIUM CONCENTRATIONS IN BODY FLUIDS (mmol/l)

Plasma	Gastric	Biliary and pan-creatic	Small intes-tinal	Ileal	Ileostomy (new)	Diarrhoea	Sweat
4	10	5	5	5	15	40	10

The Kidney

The glomerulus.—The concentration of potassium in the glomerular filtrate is the same as in plasma. Because of the very large filtered volume, daily loss by this route would be about a third of the body content (about 800 mmol) if there were no tubular regulation. The net loss, although very variable, is about 10 per cent of this.

The tubules.—Potassium is almost completely reabsorbed in the *proximal tubule* and renal tubular damage can cause potassium depletion.

Potassium is resecreted in the *distal tubule* and *collecting ducts* in exchange for *sodium*. Hydrogen ions compete with potassium for this exchange, which is stimulated by *aldosterone* (p. 38). If the proximal tubule is functioning, potassium loss in the urine depends on three factors:

(i) The amount of *sodium available for exchange*. This depends on the *GFR, sodium load* on the glomerulus, and *sodium reabsorption in the proximal tubule and loop of Henle*. The latter is inhibited by many diuretics (p. 71).

(ii) The relative *amounts of hydrogen and potassium ions* present in the cells of the distal tubule and collecting ducts, and the *ability to secrete H^+ in exchange for Na^+* (inhibited during therapy with carbonate dehydratase inhibitors and in some types of renal tubular acidosis).

(iii) The circulating *aldosterone* level. This is increased following water loss (which usually accompanies intestinal loss of potassium) (p. 46) and in almost all conditions requiring diuretic therapy.

The Cell Wall

Potassium is the predominant intracellular cation and is continuously lost from the cell down a concentration gradient. This loss is opposed by the "sodium pump" situated at the cell surface. This pumps sodium out of the cell in exchange for potassium and hydrogen ions.

In most circumstances the shift of potassium across cell membranes is accompanied by a shift of sodium in the opposite direction, but the *percentage* change in extracellular sodium levels will be much less than that of potassium. A simplified example will demonstrate this. Let us assume extra- and intracellular volumes to be equal, the plasma sodium to be 140 mmol/l and potassium 4 mmol/l, with reversal of these concentrations inside cells. An exchange of 4 mmol/l of sodium for potassium across the cell wall would double the plasma potassium concentration (clinically a very significant change) while only reducing the plasma sodium concentration to 136 mmol/l.

There is *net loss* of potassium from the cell:

if *potassium is lost from the ECF* and replenished from the cell;

if the sodium pump is inefficient, as in diabetic ketoacidosis and in hypoxic states;

in acidosis, when H^+ replaces some of the intracellular potassium.

There is *net gain* of potassium by the cell:

if the *activity of the sodium pump is increased*, as it is after the administration of glucose and insulin: this effect may be used to treat hyperkalaemia. It is the main cause of the change from hyperkalaemia to hypokalaemia during treatment of diabetic coma;

in alkalosis. Induction of alkalosis can be used to treat hyperkalaemia.

Interrelationship Between Hydrogen and Potassium Ions

Extracellular hydrogen ion concentration affects the entry of potassium into all cells: changing the relative proportions of K^+ and H^+ in distal tubular cells affects urinary loss of potassium. In acidosis decreased entry of potassium into the cells from the ECF, coupled with reduced urinary secretion of the ion, causes hyperkalaemia: in alkalosis hypokalaemia is due both to net increased entry of potassium into cells and to increased urinary loss.

As the relationship between K^+ and H^+ is reciprocal, changes in K^+ balance affect the hydrogen ion balance of the body. The sodium pump pumps sodium ions out of the cell in exchange for extracellular potassium and hydrogen ions.

In most cells the pump is situated over the whole cell surface. As K^+ is lost from the ECF, less is available for exchange for sodium, and some of it is replaced by H^+. More H^+ than usual is pumped into the cell causing *intracellular acidosis.*

In absorptive cells, including those in the renal tubule, the pump is probably situated on the luminal surface. In the kidney, the sodium derived from the luminal fluid is pumped into the cell in exchange for K^+ or H^+. If K^+ is lost from the ECF, loss from cells follows, causing a reduction in intracellular K^+, and sodium is reabsorbed in exchange for less K^+ and more H^+ than usual. The *acid urine* is appropriate to the overall intracellular acidosis. H^+ within the tubular cell is formed by the carbonate dehydratase mechanism (p. 85).

$$H_2O + CO_2 \xrightarrow{\text{CD}} H^+ + HCO_3^-$$

As more H^+ is secreted into the urine, the reaction is accelerated and more HCO_3^- is generated and passes into the ECF, accompanied by the reabsorbed sodium. Chronic potassium secretion is therefore accompanied by a *high plasma $[HCO_3^-]$ and extracellular alkalosis*. If other causes for a raised plasma bicarbonate (such as respiratory disease) are absent, and especially if factors known to cause potassium depletion (for instance, diuretic therapy) are present, this finding is a sensitive indicator of potassium depletion, even in the absence of hypokalaemia. *The combination of hypokalaemia and a high plasma $[HCO_3^-]$ is more likely to be due to K^+ depletion, which is common, than primarily to metabolic alkalosis, which is rare.* These are the changes of *chronic* potassium depletion.

In *acute* loss the slight lag in potassium release from cells may result in more severe hypokalaemia for the same degree of depletion than in the chronic state: bicarbonate levels are less likely to be raised, because significant bicarbonate retention by the kidney takes several days.

TABLE IX

INTERRELATIONSHIPS OF PLASMA POTASSIUM AND BICARBONATE LEVELS

Plasma $[K^+]$	Plasma $[HCO_3^-]$	Most likely cause	Examples of Clinical Conditions
Low N or ↓	↑	Chronic K^+ depletion	Diuretic therapy *Chronic diarrhoea (chronic purgative takers)
↓ or ↓↓	N or ↓	Acute K^+ depletion	Severe acute diarrhoea Fistulae etc.
↓	↓	Respiratory alkalosis	Overtreatment on respirator Hysterical overbreathing
↑	↓	Metabolic acidosis	*Renal failure Diabetic ketoacidosis
↑	↑	Respiratory acidosis	Bronchopneumonia
↑	N	Acute K^+ load	Excessive K^+ therapy

* It should be stressed that this is *only a guide*. For instance, in severe diarrhoea bicarbonate loss may be so high that $[HCO_3^-]$ is reduced in spite of potassium depletion: similarly, renal tubular lesions cause depletion of both HCO_3^- and K^+.

Theoretically, potassium excess could cause intracellular alkalosis and extracellular acidosis, with a low plasma bicarbonate. However, *the combination of hyperkalaemia and a low plasma* $[HCO_3{}^-]$ *is more likely to be due to metabolic acidosis, which is common, than primarily to potassium excess, which is extremely rare.* In *respiratory* acidosis the plasma bicarbonate is high (p. 99), but the plasma potassium will also tend to be high.

These various situations are summarised in Table IX.

ABNORMALITIES OF PLASMA POTASSIUM LEVELS

In any clinical situation no single factor entirely accounts for the changes in plasma potassium concentration. For instance, in conditions associated with intestinal potassium loss, concomitant water loss causes secondary hyperaldosteronism; similarly, most conditions requiring diuretic therapy are associated with hyperaldosteronism. This hyperaldosteronism aggravates urinary loss and may also increase entry of potassium into body cells generally. However, if dehydration and sodium depletion are very severe, the reduced amount of filtered sodium means that less is available for exchange with potassium in the distal tubule and aldosterone cannot produce maximal effects: hypokalaemia will then be "unmasked" as the patient is rehydrated. This should be anticipated.

Hypokalaemia

This is usually the result of potassium depletion, although, if the rate of loss of potassium from cells equals or exceeds that from the body, potassium depletion may not cause hypokalaemia. It can occur without depletion if there is a shift into cells, as in alkalotic states and in the rare condition, familial periodic paralysis.

Misleading and temporary hypokalaemia may occur for a few hours after oral administration of diuretics which cause potassium loss. The finding of a higher value on a later specimen does not usually indicate a "laboratory error".

The causes of hypokalaemia may be classified as follows:

1. predominantly due to **loss of potassium from the body.**

   (a) Predominantly due to loss from the ECF in *intestinal secretions.*
      Prolonged vomiting
      Diarrhoea
      Loss through intestinal fistulae

The intestinal loss is usually aggravated by the secondary hyperaldosteronism consequent on water loss. This causes an inappropriately high urinary loss. The important points to be noted are:

(i) Fluid from a *recent ileostomy* and *diarrhoea stools* are particularly rich in potassium (Table VIII), but a prolonged drain of *any* intestinal secretion causes depletion.

(ii) *Habitual purgative takers* may present with hypokalaemia and are often reluctant to admit to the habit.

(iii) The rarely occurring, large *mucus-secreting villous adenomas of the intestine* may cause considerable potassium loss.

(b) Predominantly due to loss from the ECF in *urine.*

(i) Increased activity of sodium : potassium exchange mechanisms in the distal nephron.

Secondary hyperaldosteronism.—This often aggravates other causes of potassium depletion.

Cushing's syndrome and steroid therapy.—Patients secreting excess of, or on prolonged therapy with, glucocorticoids tend to become hypokalaemic due to the mineralocorticoid effect on the distal tubule.

Primary hyperaldosteronism (p. 53).

Synacthen or ACTH therapy and ectopic ACTH (p. 440).

Carbenoxolone therapy.—Carbenoxolone is used to accelerate healing of peptic ulcers. It also potentiates the action of aldosterone, and potassium depletion is sometimes a complication of such therapy.

Liquorice contains glycyrrhizinic acid, which also has an aldosterone-like effect. Subjects fond of liquorice-containing sweets may present with hypokalaemia.

(ii) Excess available sodium for exchange in the distal nephron. *Diuretics inhibiting the "sodium pump" in the loop of Henle* (p. 71). The increased sodium : potassium exchange is aggravated by secondary hyperaldosteronism.

(iii) Decreased renal sodium : hydrogen ion exchange, favouring sodium : potassium exchange.

Carbonate dehydratase inhibitors (p. 72).

Renal tubular acidosis (p. 98).

(iv) Reduced proximal tubular potassium reabsorption. *Renal tubular failure* (e.g. polyuric phase of acute oliguric renal failure).

"Fanconi syndrome" (p. 14).

2. predominantly due to **reduced potassium intake.**

Chronic starvation.—If water and salt intake are reduced, secondary hyperaldosteronism may aggravate the hypokalaemia.

3. predominantly due to **redistribution** in the body. Loss into cells.

Glucose and insulin therapy.—This may be used to treat hyperkalaemia.

Familial periodic paralysis (very rare).—In this condition episodic paralysis occurs associated with entry of potassium into cells.

4. loss from ECF by **more than one route.**
   (a) Into cells and urine
   Alkalosis
   (b) Into cells, urine and intestine
   Pyloric stenosis with alkalosis.—The loss in urine and the loss into cells are probably more important causes of hypokalaemia than the loss in gastric secretion.

HYPERKALAEMIA

This occurs most commonly when the rate of potassium leaving cells is greater than its rate of excretion. The causes of hyperkalaemia may be:

1. predominantly due to **gain of potassium by the body.**
   Gain by ECF from the *intestine or by an intravenous route.*
   Over-enthusiastic potassium therapy.
   Failure to stop potassium therapy when depletion has been corrected.

2. **failure of renal secretion** of potassium.

   (a) Decreased activity of sodium:potassium exchange mechanisms in distal nephron.
   Hypoaldosteronism (as in Addison's disease)
   Diuretics acting on the distal nephron by antagonising aldosterone, or direct inhibition of "sodium pump" (p. 72).

   (b) Too little sodium available for exchange in the distal nephron.
   Renal glomerular failure.—Hyperkalaemia is usually aggravated by the concomitant acidosis and gain of potassium from cells.
   Sodium depletion

3. predominantly due to **redistribution** of potassium in the body.
   Gain of potassium by ECF from cells.
   Severe tissue damage
   Severe acute starvation (as in anorexia nervosa). These cause cell damage and release of K^+ into the ECF.

4. gain by ECF by **more than one route.**
   Reduced renal excretion in spite of gain by ECF from cells.
   Acidosis
   Hypoxia—Failure of sodium pump in all cells.
   Failure in distal tubular cells causes potassium retention. If hypoxia is very severe, lactic acidosis aggravates the hyperkalaemia.

   Diabetic ketoacidosis.—In early untreated diabetes potassium leaves the cells, and in spite of a high urinary loss and consequent body depletion hyperkalaemia is usual. This is due to partial failure of the sodium pump resulting from impaired glucose metabolism because of insulin lack. As the condition becomes more advanced two other factors contribute to hyperkalaemia.

Dehydration with a low GFR.

Acidosis due to ketosis.

All these factors are reversed during insulin and fluid therapy. As potassium enters cells extracellular levels fall and the depletion is revealed. *Plasma potassium levels should be monitored during therapy*, and potassium should be given as soon as concentrations start to fall.

MEASUREMENT OF URINARY AND INTESTINAL LOSSES

Pure urinary or intestinal loss as a cause of hypokalaemia is very rare. Measurement of such losses with a view to quantitative replacement may lead to dangerous errors of therapy. Exchanges across cell walls cannot be measured. If gain of potassium by the ECF from cells is faster than loss from it into urine and intestine, replacement of measured loss could endanger the patient's life by aggravating hyperkalaemia; if loss from ECF into cells is predominant, therapy based on urinary excretion may be inadequate. Moreover, high urinary potassium excretion may be appropriate when, for instance, trauma has damaged many cells, reducing cellular capacity and releasing the ion from cells.

It is plasma potassium levels that are important, and in rapidly changing states frequent estimation of these is the only safe way of assessing therapy. In chronic depletion the plasma bicarbonate level may help to indicate the state of cellular repletion.

DIURETIC THERAPY

In oedematous states the fluid accumulation is accompanied by an excess of sodium in the body, *even if there is hyponatraemia* (p. 56). Diuretics act by inhibiting sodium reabsorption in the renal tubule and secondarily causing water loss. All diuretics tend to affect potassium balance, and this effect should be anticipated.

Diuretics can be divided into two main groups.

1. *Those inhibiting the sodium pump in the loop of Henle* (and perhaps proximal tubular reabsorption of sodium), and therefore water reabsorption: the increased sodium load on the distal tubule and collecting ducts increases sodium: potassium exchange at this site (which is stimulated by the accompanying hyperaldosteronism). In our experience long-term diuretic therapy almost invariably causes significant *potassium depletion*, and sometimes symptomatic hypokalaemia, even if potassium supplements are given (although this has been questioned). *A high plasma bicarbonate* is common in long-continued use of such diuretics. This is because loss of potassium from the ECF results in increased

$Na^+:H^+$ exchange in the distal nephron and at the "sodium pump" in all body cells. In this situation high plasma HCO_3^- levels are a more sensitive indication of K^+ depletion than plasma potassium levels.

Such diuretics are the *thiazide group, frusemide* (furosemide; "Lasix") and *ethacrynic acid* ("Edecrin"). It is claimed that frusemide and ethacrynic acid cause less potassium depletion than the thiazide group.

2. Those either directly *inhibiting aldosterone* or inhibiting the exchange mechanisms in the *distal tubule* and *collecting duct*. These cause *potassium retention* and may lead to hyperkalaemia: potassium supplements should *not* be used. Potassium-retaining diuretics include:

Spironolactone ("Aldactone")—A competitive aldosterone antagonist.
Amiloride ("Midamor") ⎱ Inhibitors of the $Na^+:K^+$ exchange
Triamterene ("Dytac") ⎰ mechanisms in the renal tubule.

This group of diuretics is often used, together with those causing potassium loss, when hypokalaemia cannot be controlled by potassium therapy.

In addition carbonate dehydratase inhibitors such as acetazolamide ("Diamox") (p. 98) inhibit sodium reabsorption and act as diuretics. They are rarely used for this purpose now because of the danger of acidosis.

CLINICAL FEATURES OF DISTURBANCES OF POTASSIUM METABOLISM

The clinical features of disturbances of potassium metabolism (like those of, for example, sodium and calcium) are due to changes in extracellular concentration of the ion.

Hypokalaemia, by interfering with neuromuscular transmission, causes *muscular weakness, hypotonia* and *cardiac arrhythmias*, and may precipitate digitalis toxicity. It may also aggravate paralytic ileus.

Intracellular potassium depletion causes extracellular alkalosis (see p. 67). This reduces ionisation of calcium salts (p. 232) and in long-standing potassium depletion of gradual onset the presenting symptom may be muscle *cramps* and *tetany*. This syndrome is accompanied by high plasma bicarbonate levels.

Prolonged potassium depletion causes lesions in renal tubular cells, and this may complicate the clinical picture.

Severe hyperkalaemia always carries the danger of cardiac arrest. Both hypokalaemia and hyperkalaemia cause characteristic changes in the electrocardiogram.

TREATMENT OF POTASSIUM DISTURBANCES

Abnormalities of plasma potassium should be corrected whatever the state of the total body potassium. However, an attempt should be made to assess the latter so that sudden changes in plasma potassium (for instance, during treatment of diabetic coma) can be anticipated. Treatment should be controlled by frequent plasma potassium estimations.

Hyperkalaemia.—Treatment of hyperkalaemia is based on three principles. In severe hyperkalaemia the first two principles are used.

1. Very severe hyperkalaemia can cause cardiac arrest. Calcium and potassium have opposing actions on heart muscle, and the immediate danger can be minimised by infusion of calcium salts (usually as gluconate) (see Appendix). This allows time to institute measures to lower plasma potassium.

2. Plasma potassium can be reduced rapidly (within an hour) by increasing the rate of entry into cells. Glucose and insulin speed up glucose metabolism and the action of the "sodium pump". Induction of alkalosis by infusion of bicarbonate also increases the rate of entry into cells (see Appendix). For purely practical reasons this treatment (which involves intravenous infusion) cannot be continued indefinitely, but its use allows long-term treatment to be instituted.

3. In moderate hyperkalaemia a slower acting method can be used. Potassium can be removed from the body at a rate higher than, or equal to, that at which it is entering the extracellular fluid by using oral ion exchange resins. These are unabsorbed and exchange potassium for sodium or calcium ions. It will be seen that plasma potassium is lowered at the expense of body depletion. This potassium may have to be replaced later.

Hypokalaemia.—If hypokalaemia is *mild*, potassium supplements should be given *orally* until plasma potassium and bicarbonate levels return to normal. These levels should be monitored regularly. Under-

TABLE X

POTASSIUM CONTENT OF FRUIT AND FRUIT JUICE

	Approximate K+ content (mmol)	Approximate quantity containing 50 mmol	Price for 50 mmol (relative to "Slow-K") in U.K. September, 1977
Tomato juice (Heinz)	82 per litre	610 ml	7
Canned orange juice	42 per litre	1200 ml	8
Fresh orange juice	30 per litre	1700 ml	80
Rose's Orange Cordial (*undiluted*)	16 per litre	3100 ml	37
Bananas	8 per banana	6 bananas	12
"Slow-K" (Ciba)	8 per tablet	6 tablets	1

treatment is more common than overtreatment during oral therapy. The normal subject loses about 60 mmol of potassium daily in the urine, much larger amounts being excreted during diuretic therapy. By the time hypokalaemic alkalosis is present the total deficit is probably several hundred mmol. A patient with hypokalaemia should be given *at least* 80 mmol a day: much more may be needed if plasma levels fail to rise.

In *severe* hypokalaemia, particularly if the patient is unable to take oral supplements, *intravenous potassium* should be given cautiously (see Appendix). Diarrhoea, if present, not only reduces absorption of oral potassium supplements, but may itself be aggravated by them. Such a situation may also be an indication for intravenous therapy.

It is sometimes suggested that hypokalaemia could be treated by a high intake of fruit or of fruit juice. Table X shows that if it were possible to ingest the quantities required, the consequent diarrhoea might well be self-defeating. The relatively high cost is only a minor factor.

SUMMARY

1. Changes in plasma potassium levels are the net result of changes between ECF and cells, kidney and gut.

2. In any clinical situation many factors are involved, and monitoring of plasma potassium levels is the only safe guide to treatment.

3. As hydrogen and potassium ions compete for exchange with sodium across cell walls and in the renal tubule, disturbances of hydrogen ion homeostasis and potassium balance often coexist. A raised plasma bicarbonate may indicate intracellular potassium depletion.

4. Clinical manifestations of disturbances of potassium metabolism are due to its action on neuromuscular transmission and on the heart.

5. Diuretics fall into two main groups:

(*a*) Those inhibiting the "sodium pump" in the loop of Henle, causing potassium depletion.

(*b*) Those antagonising aldosterone either directly, or indirectly by affecting the $Na^+:K^+$ transport mechanism, causing potassium retention.

FURTHER READING

In Chapters I to IV we have used the generally accepted explanation for the handling of potassium ions by the kidney, and for the interrelationship between these and sodium and hydrogen ions. Some work suggests other possible mechanisms. Interested readers are referred to:

SCHULTZE, R. G. (1973). *Arch. intern. Med.*, **131**, 885.

APPENDIX

POTASSIUM-CONTAINING PREPARATIONS

One g of potassium chloride contains 13 mmol of potassium.

FOR INTRAVENOUS USE

For use in serious depletion, or where oral potassium cannot be taken or retained. In most cases oral potassium is preferable. Intravenous potassium should be given with care, especially in the presence of poor renal function and the following rules should be observed:

1. Intravenous potassium should not be given in the presence of oliguria unless the potassium deficit is unequivocal and severe.

2. Potassium in the intravenous fluid should not exceed 40 mmol/l.

3. Intravenous potassium should not usually be given at a rate of more than 20 mmol/hour.

Potassium Chloride Injection B.P.

20 mmol of potassium and chloride in 10 ml.

WARNING. This should *never* be given undiluted. It should be added to a full bottle of other intravenous fluid. (10 ml added to a bottle containing 500 ml of fluid gives a concentration of 40 mmol/l.)

Potassium Chloride and Dextrose Injection B.P.C.

5 per cent dextrose with 40 mmol/l of potassium and chloride. This is hyperosmolal.

FOR ORAL USE

1. Potassium Chloride Tablets B.P.—6·5 mmol K and Cl per tablet.
2. Potassium Effervescent Tablets B.P.—6·5 mmol K per tablet.
3. "Slow-K" (Ciba)—8 mmol K and Cl per tablet.

TREATMENT OF HYPERKALAEMIA

Emergency Treatment

Calcium chloride (or gluconate). A 10 per cent solution is given intravenously with ECG monitoring. This treatment antagonises the effect of hyperkalaemia on heart muscle, but does not alter potassium levels.

WARNING. Calcium should never be added to bicarbonate solutions, because calcium carbonate is insoluble.

Glucose 50 g with 20 units of soluble insulin by intravenous injection lowers plasma potassium rapidly by increasing entry into cells. If the situation is less urgent 10 units of soluble insulin may be added to a litre of 10 per cent dextrose.

If acidosis is present, bicarbonate may be used as an alternative to glucose and insulin injection. 44 mmol (one ampoule) may be injected over 5 minutes.

Long-term Treatment

Sodium polystyrene sulphonate ("Resonium-A"; "Kayexalate") 20–60 g a day by mouth in 20 g doses, *or* 10–40 g in a little water as a retention enema every 4–12 hours. This removes potassium from the body.

INVESTIGATION OF ELECTROLYTE DISTURBANCES

In the last two chapters we have outlined conditions in which sodium or potassium concentrations *may* be abnormal. We have pointed out that an abnormal result, particularly of sodium, may not be clinically significant, while "normal" ones do not guarantee "normal" balance. Before making a request some assessment should be made as to whether the result of an estimation will aid diagnosis or treatment.

Sodium and potassium estimations provide the numerical bulk of the workload of most chemical pathology departments; because of this, they are often estimated simultaneously. However, potassium is more often useful than sodium.

The student should bear the following point in mind:

1. *Plasma sodium* **should** *be estimated regularly*

(*a*) *in the unconscious patient and infants losing fluid,* because of the danger of hypernatraemia;

(*b*) *in the dehydrated patient,* or those with *abnormal losses,* to help diagnosis and to indicate the type of replacement fluid.

2. *Plasma potassium* (*and bicarbonate*) **should** *be estimated regularly* on any patient in whom there is a cause for abnormal levels, because these must be treated:

(*a*) *in patients with abnormal losses* from the gastro-intestinal tract or kidneys (especially due to *diuretic, or steroid or ACTH therapy*);

(*b*) *in patients on potassium therapy*;

(*c*) *in patients in renal failure*;

(*d*) *in patients in diabetic coma or precoma.*

Groups 1 and 2 are numerically small compared to estimations actually requested.

3. **In the conscious, normally hydrated patient, with no abnormal losses plasma sodium estimation rarely helps.** Mild hyponatraemia is common (p. 59), but treatment in such subjects is usually contra-indicated. Unless renal failure is present, potassium estimation is also unhelpful in such subjects.

Chapter IV

HYDROGEN ION HOMEOSTASIS: BLOOD GAS LEVELS

THE pH of the extracellular fluid is normally maintained within about 0·05 units of 7·4. Energy production from metabolism is linked to a series of stepwise reactions in which dehydrogenation is of great importance, the hydrogen being transferred to coenzymes such as NAD. Some of the resulting reduced coenzyme (for example $NADH_2$) supplies hydrogen for synthetic reactions. If oxygen is available most of the rest is dehydrogenated again, the hydrogen ultimately combining with the oxygen to form water during oxidative phosphorylation, the oxidized coenzyme being released for re-use. In the absence of oxygen an adequate supply of coenzyme can only be maintained if the hydrogen is passed on to some intermediate product of metabolism; for instance, pyruvate is reduced by $NADH_2$ to form lactate, which cannot be metabolised until oxygen is available.

Thus, in most tissues, generation of hydrogen ions is linked with the energy production necessary for life, but, if the oxygen supply is adequate, much of the excess is converted to water and pH changes are minimal. However, it has been shown that subjects on a high protein diet pass an acid urine: this seems to be due to the constituent sulphur amino acids. The dehydrogenation step linked to the oxidation of the sulphur to sulphate is not fully understood, but certainly releases hydrogen ions. Excess hydrogen ions (as well as lactate) may also be released during a sudden burst of muscular excercise, and the supply of oxygen may not be adequate to deal immediately with such a load. To prevent a significant change in body pH these hydrogen ions must be inactivated until they can enter aerobic pathways or be eliminated from the body. Under normal circumstances homeostatic mechanisms are so effective that *blood* pH varies very little.

Since hydrogen, and not hydroxyl, ions are produced by metabolism the tendency to acidosis is greater than to alkalosis.

DEFINITIONS

An *acid* is a substance which can dissociate to produce hydrogen ions (protons: H^+): a *base* is one which can accept hydrogen ions. Table XI includes examples of acids and bases of importance in the body.

TABLE XI

Acid		Conjugate Base
Carbonic acid H_2CO_3	$\leftrightharpoons H^+$	$+HCO_3^-$ Bicarbonate ion
Lactic acid $CH_3CHOHCOOH$	$\leftrightharpoons H^+$	$+CH_3CHOHCOO^-$ Lactate ion
Ammonium ion NH_4^+	$\leftrightharpoons H^+$	$+NH_3$ Ammonia
Dihydrogen phosphate $H_2PO_4^-$	$\leftrightharpoons H^+$	$+HPO_4^=$ Monohydrogen phosphate ion
Acetoacetic acid CH_3COCH_2COOH	$\leftrightharpoons H^+$	$+CH_3COCH_2COO^-$ Acetoacetate ion
β-hydroxybutyric acid $CH_3CHOHCH_2COOH$	$\leftrightharpoons H^+$	$+CH_3CHOHCH_2COO^-$ β-hydroxy-butyrate ion.

An *alkali* is a substance which dissociates to produce hydroxyl ions (OH^-). Since OH^- is not a primary product of metabolism alkalis are of relatively little importance in the present discussion.

A *strong acid* is highly dissociated in aqueous solution: in other words it produces many hydrogen ions. Hydrochloric acid is a strong acid, and in solution is almost entirely in the form of H^+Cl^-. However, the examples given in the above list are, chemically speaking, *weak acids*, little dissociated in water and yielding relatively few hydrogen ions. In the body even very small changes of pH are important and result in disturbances of physiology.

Buffering is the term used for the process by which a strong acid (or base) is replaced by a weaker one, with a consequent reduction in the number of free hydrogen ions (H^+); the "shock" of the hydrogen ions is taken up by the buffer with a change of pH smaller than that which would occur in the absence of the buffer.

For example:

$$H^+Cl^- + \qquad NaHCO_3 \rightleftharpoons H_2CO_3 \qquad +NaCl$$
Strong acid   Buffer   Weak acid   Neutral salt

pH is a measure of hydrogen ion activity. It was originally defined as $\log_{10}$ of the reciprocal of the hydrogen ion concentration ($[H^+]$); although it is now known that this definition is not strictly true, it suffices for present purposes. The $\log_{10}$ of a number is the power to which 10 must be raised to produce that number. Thus $\log 100 = \log 10^2 = 2$ and $\log 10^7 = 7$.

Let us suppose $[H^+]$ is 10^{-7} (0·000 000 1) mol/l

Then $\log [H^+] = -7$

But $pH = \log \dfrac{1}{[H^+]} = -\log [H^+] = 7$

For the non-mathematically minded only a few points need be remembered.

Since at pH 6 $[H^+] = 10^{-6}$ (0·000 001) mol/l (1 000 nmol/l)

and at pH 7 $[H^+] = 10^{-7}$ (0·000 000 1) mol/l (100 nmol/l)

a change of *one pH unit* represents a *tenfold change in* $[H^+]$. This is a much larger change than is immediately obvious from the change in pH values. Although changes of this magnitude do not occur in the body during life, in pathological conditions changes of 0·3 of a pH unit can take place. 0·3 is the log of 2. Therefore a *decrease of pH by* 0·3 (e.g. 7·4 to 7·1) represents a *doubling of* $[H^+]$ from 40 nmol/l to 80 nmol/l. Here again the use of pH makes a very significant change in $[H^+]$ appear deceptively small. (Compare the situation if the plasma sodium concentration had changed from 140 to 280 mmol/l). Urinary pH is much more variable than that in the blood: $[H^+]$ can increase 1000-fold (a fall of 3 pH units).

The Henderson–Hasselbalch equation.—We have already seen that a buffer absorbs the "shock" of the addition of H^+ to a system by replacing a strong acid by a weak one. It will be seen that when the bases in column 2 of Table XI buffer H^+, the corresponding acid in column 1 is formed. This weak acid and its conjugate base form a *buffer pair*. In aqueous solution the pH is determined by the ratio of this acid to its conjugate base.

Let us take the bicarbonate pair as an example. Carbonic acid (H_2CO_3) dissociates into H^+ and HCO_3^- until equilibrium is reached (in this case very much in favour of H_2CO_3), and the ratio of the two forms will now remain constant (K). We can therefore write

$$K [H_2CO_3] = [H^+] \times [HCO_3^-]$$

(that is, at equilibrium, the concentration of H_2CO_3 is K times that of the product of $[H^+]$ and $[HCO_3^-]$).

Transposing, $[H^+] = K \dfrac{[H_2CO_3]}{[HCO_3^-]}$

Although some laboratories now express results in terms of $[H^+]$, this practice is not yet widespread. We will use the pH notation in this edition.

$$pH = \log \frac{1}{[H^+]}$$

Taking reciprocals and logarithms in the equation for $[H^+]$ given above (when taking logs, multiplication becomes addition).

$$\text{Log} \frac{1}{[H^+]} = \log \frac{1}{K} + \log \frac{[HCO_3^-]}{[H_2CO_3]}$$

$Log \dfrac{1}{K}$ is called pK

Therefore $pH = pK + \log \dfrac{[HCO_3{}^-]}{[H_2CO_3]}$

This equation (an example of the Henderson–Hasselbalch equation) is valid for any buffer pair. It is important to notice that the pH depends on the *ratio* of the concentrations of base (in this case $[HCO_3{}^-]$) to acid (in this case $[H_2CO_3]$).

In practice it is not possible to measure the very low carbonic acid concentration directly. It is in equilibrium with dissolved CO_2, and if the carbon dioxide concentration is inserted into the equation in place of $[H_2CO_3]$, the overall dissociation constant is now the sum of those of the two reactions

$$K_1 [H_2CO_3] = [H^+] \times [HCO_3{}^-]$$
$$\text{and } K_2 [CO_2] \times [H_2O] = [H_2CO_3]$$

This combined constant is usually written as K' and the pK' is about 6·1. The Henderson–Hasselbalch equation for the bicarbonate system then becomes

$$pH = 6 \cdot 1 + \log \dfrac{[HCO_3{}^-]}{[CO_2]}$$

In practice the partial pressure of CO_2 gas (Pco_2) in blood is measured, and its concentration in solution is derived by multiplying Pco_2 by the solubility constant for carbon dioxide. If the Pco_2 is expressed in kiloPascals (kPa) this constant is 0·225: if it is in mm Hg it is 0·03.

Therefore, if Pco_2 is expressed in kPa, the equation becomes

$$\boxed{pH = 6 \cdot 1 + \log \dfrac{[HCO_3{}^-]}{Pco_2 \times 0 \cdot 225}}$$

We shall use this form in the rest of the chapter.

HYDROGEN ION HOMEOSTASIS

The following points should be noted:
hydrogen ions can be incorporated in water, maintaining normal pH.

This is the normal mechanism during oxidative phosphorylation.

H^+ is incorporated in water during the conversion of H_2CO_3 to CO_2 and water.

$$H^+ + HCO_3{}^- \rightleftarrows H_2CO_3 \rightleftarrows CO_2 + H_2O$$

As this is a reversible reaction H^+ will only continue to be so inactivated if CO_2 is removed. This results in bicarbonate depletion.

hydrogen ions can be lost from the body only through the kidney and the intestine.—This mechanism is coupled, in the kidney, with regeneration of bicarbonate ion (HCO_3^-) and is therefore the ideal method of eliminating any excess H^+.

buffering of hydrogen ions is a temporary measure.—The H^+ is still in the body, and the presence of the weak acid of the buffer pair causes a small change in pH (see the Henderson–Hasselbalch equation). If H^+ is not completely neutralised, or eliminated from the body, and if production continues, buffering power will eventually be used up and the pH will change.

CONTROL SYSTEMS

Carbon dioxide and hydrogen ions are amongst the potentially toxic products of aerobic and anaerobic metabolism respectively. Although most CO_2 is lost through the lungs some is converted to bicarbonate, thus providing a buffering system: inactivating one toxic product provides a means of minimising the effect of the other.

The Henderson–Hasselbalch equation for any buffer pair is:

$$pH = pK + \log \frac{[base]}{[acid]}$$

The buffer pair is most effective at maintaining a pH near its pK. The optimum pH of extracellular fluid is 7·4, but the pK′ of the bicarbonate system is 6·1. Despite this apparent disadvantage bicarbonate is the most important buffer in the body, and accounts for over 60 per cent of blood buffering capacity. Moreover, the system is central to all the other important homeostatic mechanisms for dealing with hydrogen ions, including buffering by haemoglobin (which provides most of the rest of the blood buffering capacity) and secretion of hydrogen ions by the kidney.

Aerobic metabolism provides a plentiful supply of CO_2, the denominator in the equation

$$pH = 6·1 + \log \frac{[HCO_3^-]}{Pco_2 \times 0·225}$$

The Control of CO_2 by the Lungs and Respiratory Centre

The partial pressure of CO_2 in plasma is normally about 5·3 kPa (40 mm Hg). Maintenance of this level depends on the balance between production by metabolism and loss through the pulmonary alveoli.

The sequence of events is as follows:

(a) inspired oxygen is carried from the lungs to tissues by haemoglobin;

(b) the tissue cells utilise the oxygen for aerobic metabolism, and the carbon in organic compounds is oxidised to CO_2;

(c) CO_2 diffuses along a concentration gradient from the cells into the extracellular fluid and is returned by the blood to the lungs, where it is eliminated in expired air;

(d) the rate of respiration, and therefore the rate of CO_2 elimination, is controlled by chemoreceptors in the hypothalamic respiratory centre, which respond to the $[CO_2]$ (or pH) of the circulating blood. If the P_{CO_2} rises much above 5·3 kPa (or if the pH falls) the rate of respiration rises. Normal lungs have a very large reserve capacity for eliminating CO_2.

We now see that not only is there a plentiful supply of CO_2, but that the normal respiratory centre and lungs can control its level within narrow limits, so controlling the denominator in the Henderson–Hasselbalch equation.

Disease of the lungs, or abnormality of respiratory control, will primarily affect the P_{CO_2}.

The Control of Bicarbonate by the Kidneys and Erythrocytes

The renal tubular cells and erythrocytes use some of the CO_2 retained by the lungs to form bicarbonate, and so control the numerator in the Henderson–Hasselbalch equation. Under physiological conditions erythrocytes make fine adjustments to the plasma bicarbonate level in response to changes in P_{CO_2} in the lungs and tissues: the kidney plays the major role in maintaining the circulating bicarbonate concentration.

The Carbonate Dehydratase System

Carbonate dehydratase (carbonic anhydrase; CD) catalyses the first reaction in the chain

$$CO_2 + H_2O \xrightarrow{\text{CD}} H_2CO_3 \rightarrow H^+ + HCO_3^-$$

Not only do erythrocytes and renal tubular cells have a high concentration of CD, but they also have a means of removing one of the products, H^+; thus both reactions continue to the right, and HCO_3^- will be produced. As one of the reactants—water—is freely available, and because one of the products—H^+—is being removed, HCO_3^- production would be accelerated either by a rise in the intracellular concentration of the other reactant—CO_2—or by a fall in the intracellular concentration of the other product—HCO_3^-.

At a plasma P_{CO_2} of 5·3 kPa (a CO_2 concentration of about 1·2 mmol/l —see p. 81) these two tissues maintain the extracellular bicarbonate concentration at about 25 mmol/l in the normal subject. The extra-cellular ratio of $[HCO_3^-]:[CO_2]$ (both in mmol/l) is just over 20:1. It can be calculated from the Henderson–Hasselbalch equation that, with a pK' of 6·1, this ratio represents a pH very near 7·4. An increase of intracellular P_{CO_2}, or a decrease in intracellular $[HCO_3^-]$, accelerates production of HCO_3^-, and minimises changes in the *ratio*, and there-fore changes in pH.

Bicarbonate Generation by Erythrocytes (Fig. 3)

Haemoglobin is an important blood buffer.

$$pH = pK + \log \frac{[Hb^-]}{[HHb]}$$

However, it only works effectively in the body in co-operation with the bicarbonate system.

Because erythrocytes lack aerobic pathways they produce relatively little CO_2. Plasma CO_2 diffuses into the cell along a concentration gradient, where carbonate dehydratase catalyses its reaction with water to form carbonic acid. As H_2CO_3 dissociates, much of the H^+ is buffered by haemoglobin. The concentration of HCO_3^- in the erythro-cyte rises, and it diffuses into the extracellular fluid along a concentra-tion gradient; electrochemical neutrality is maintained by diffusion of chloride into the cell (the "chloride shift").

Under *physiological conditions*, the higher P_{CO_2} in blood leaving tissues stimulates erythrocyte HCO_3^- production, and the lower P_{CO_2} in blood leaving the lungs slows it down; and the arteriovenous differ-

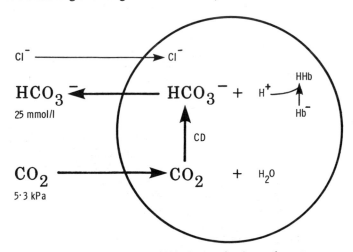

FIG. 3.—Generation of bicarbonate by the erythrocyte.

ence in the ratio $[HCO_3^-]:[CO_2]$, and therefore the pH, is kept relatively constant. There is little effect on the overall HCO_3^- "balance".

In *pathological conditions* affecting PCO_2 the same mechanisms operate: a high intracellular PCO_2 tends to cause a net increase, and a low one a net reduction in extracellular $[HCO_3^-]$. It is important to appreciate these erythrocyte mechanisms to understand the concept of "standard bicarbonate" in *acute respiratory disturbances*. However, because the buffering capacity of haemoglobin is limited the erythrocyte can only make a small contribution to homeostasis in chronic disturbances of H^+ balance (p. 99).

The Kidneys

Carbonate dehydratase is also of central importance in the renal mechanisms involved in hydrogen ion homeostasis. The hydrogen ion is secreted from the renal tubular cell into the tubular lumen, where it is buffered by constituents of the glomerular filtrate. By contrast with the haemoglobin in the erythrocyte, these buffers are constantly being replenished by continuing glomerular filtration. *For this reason, and because H^+ can only be eliminated from the body by the renal route, the kidneys are of major importance in chronic acidosis.* Without them haemoglobin buffering capacity would soon become saturated.

There are two renal mechanisms controlling $[HCO_3^-]$ in the extracellular fluid:

bicarbonate *"reabsorption"*, the predominant mechanism in the maintenance of the steady state. As we shall see, the CO_2 driving the carbonate dehydratase mechanism in the renal tubular cell is derived from filtered bicarbonate in the tubular lumen, and there is no *net* loss of hydrogen ions;

bicarbonate *generation*, a very important mechanism for correcting acidosis, in which the level of CO_2 stimulating the carbonate dehydratase reaction in the renal tubular cell reflects that in the extracellular fluid. There *is* net loss of hydrogen ions. This mechanism is also stimulated by a fall in extracellular $[HCO_3^-]$.

Bicarbonate "reabsorption" (Fig. 4).—Normal urine is almost bicarbonate free.

Bicarbonate cannot be reabsorbed directly, but must first be converted to CO_2 in the tubular lumen, and an equivalent amount of CO_2 is converted to bicarbonate within the tubular cell. The mechanism depends on the action of carbonate dehydratase within the tubular cell, and on H^+ secretion from the cell into the tubular lumen in exchange for the sodium filtered with the bicarbonate. The sequence of events is as shown in Fig. 4.

(a) Bicarbonate passes through the glomerular filter at plasma concentration (about 25 mmol/l).

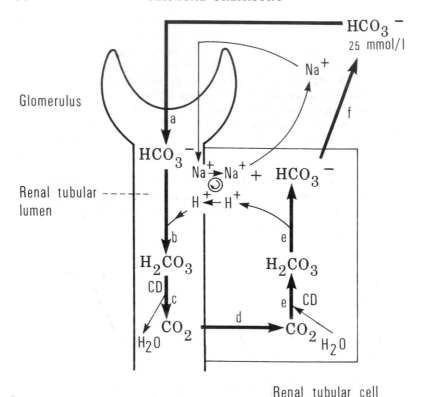

FIG. 4.—"Reabsorption" of filtered bicarbonate by the renal tubular cell.

(*b*) Filtered bicarbonate combines with H^+ secreted by the tubular cell to form H_2CO_3.

(*c*) The H_2CO_3 dissociates to CO_2 and water. In the proximal tubule this reaction is catalysed by carbonate dehydratase on the luminal membrane of the tubular cells: in the distal nephron, where the pH is usually lower, it probably dissociates spontaneously.

(*d*) As the luminal PCO_2 rises, CO_2 diffuses into the tubular cell along a concentration gradient.

(*e*) As the intracellular $[CO_2]$ rises carbonate dehydratase catalyses its combination with water to reform carbonic acid, which dissociates into H^+ and HCO_3^-.

(*f*) As H^+ is secreted (and so starts the reactions from (*b*) again) the intracellular concentration of HCO_3^- rises, and the bicarbonate diffuses into the extracellular fluid accompanied by the sodium reabsorbed in exchange for H^+.

This is a self-perpetuating cycle which reclaims buffering capacity which would otherwise be lost to the body by glomerular filtration. The secreted H^+ is derived from cellular water, and is reincorporated in water in the lumen. Because there is no net change in hydrogen ion balance and no net gain of bicarbonate, this mechanism *cannot correct an acidosis*, but can *maintain a steady state*.

Bicarbonate generation (Fig. 5).—The mechanism in the renal tubular cell for generating bicarbonate is identical to that of bicarbonate reabsorption, but there *is* net loss of H^+ from the body, as well as a net gain of HCO_3^-; this mechanism is therefore ideally suited to correct an acidosis.

The carbonate dehydratase mechanism may be stimulated by a rise in Pco_2 or a fall of $[HCO_3^-]$ within the tubular cell. In this case the rise in $[CO_2]$ is the indirect result of a rise in extracellular Pco_2. The renal

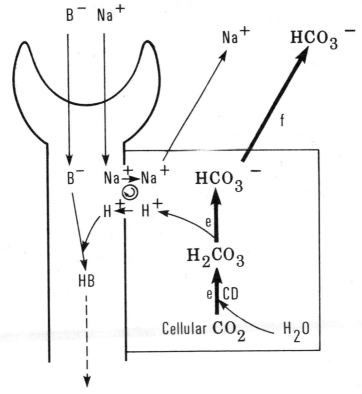

FIG. 5.—Net generation of bicarbonate by the renal tubular cell. (B^- = non-bicarbonate base.)

tubular cell, unlike the erythrocyte, is constantly producing CO_2 by aerobic pathways: this CO_2 diffuses out of the cell into the extracellular fluid down a concentration gradient. An increase in extracellular P_{CO_2}, by reducing the gradient, slows this diffusion, and intracellular P_{CO_2} rises. Conversely, reduction of extracellular $[HCO_3^-]$, by increasing the concentration gradient, increases loss of this anion from the cell.

Normally all filtered bicarbonate is "reabsorbed" by the mechanism described above: once all bicarbonate has been "reabsorbed," net secretion of H^+ and net generation of HCO_3^- depend on the presence of other filtered buffers (B^- in Fig. 5). These buffers, unlike bicarbonate, do not form compounds capable of diffusing into tubular cells; nor is the hydrogen ion incorporated into water. *The H^+ is lost in the urine* as HB. The bicarbonate formed in the cell is derived from cellular CO_2, not luminal bicarbonate, and therefore represents a *net gain in bicarbonate*. As usual, when a mmol of H^+ is secreted into the urine, a mmol of HCO_3^- passes in to the extracellular fluid with sodium.

This mechanism is very similar to that in erythrocytes, but unlike the red cell, the renal tubular cell is not carried between lung and peripheral tissues: it is therefore exposed to a relatively constant P_{CO_2}. Bicarbonate generation coupled with hydrogen ion secretion becomes very important in acidosis, when it is stimulated by a fall in extracellular $[HCO_3^-]$ (metabolic acidosis) or a rise in extracellular P_{CO_2} (respiratory acidosis).

Urinary buffers.—Buffers other than bicarbonate are involved in bicarbonate generation with H^+ secretion. The two most important of these are phosphate and ammonia.

The phosphate buffer pair.—At pH 7·4 most of the phosphate in the glomerular filtrate is in the form of monohydrogen phosphate ($HPO_4^=$), and this can accept H^+ to become dihydrogen phosphate ($H_2PO_4^-$). The pK of this pair is 6·8.

$$pH = 6\cdot8 + \log \frac{[HPO_4^=]}{[H_2PO_4^-]}$$

Even in mild acidosis bone salts are ionised more than at normal pH (p. 232), and the requirement for increased urinary secretion of H^+ is linked with increased buffering capacity in the glomerular filtrate, due to an increase of phosphate liberated from bone.

Buffering by ammonia.—The enzyme glutaminase is present in renal tubular cells and catalyses the hydrolysis of the terminal amino group of glutamine ($GluCONH_2$) to form glutamate ($GluCOO^-$) and the ammonium ion.

$$H_2O + GluCONH_2 \rightarrow GluCOO^- + NH_4^+$$

Ammonia and ammonium ion form a buffer pair

$$pH = pK + \log \frac{[NH_3]}{[NH_4{}^+]}$$

but the pK of the system is about 9·8, and at pH 7·4 the equilibrium is overwhelmingly in favour of $NH_4{}^+$. NH_3 can diffuse out of the cell into the urine much more rapidly than $NH_4{}^+$, and, if the urine is acid, will be prevented from returning by avid combination with secreted H^+ derived from the carbonate dehydratase mechanism. Dissociation of $NH_4{}^+$ in the cell is maintained by removal of NH_3 into the urine. However, this dissociation liberates H^+ in the cell. On the face of it it seems of no advantage to buffer one secreted H^+ in the urine if, at the same time, one is produced in the cell.

A possible explanation is to be found in the fate of the glutamate ($GluCOO^-$) produced at the same time as the ammonium ion. After further deamination to α-oxoglutarate it can be converted to glucose by gluconeogenesis—*a process which utilises an equivalent amount of H^+.* Thus the H^+ liberated into the cell may be incorporated into glucose.

Tubular cell

As usual the net result is a gain of $HCO_3{}^-$.

Both glutaminase activity and gluconeogenesis have been shown to be increased in acidosis.

Acid Secretion by the Stomach

The parietal cells of the stomach secrete H^+ into the gastric lumen. The formation of H^+ depends on the carbonate dehydratase mechanism and generates $HCO_3{}^-$. Electrochemical neutrality is maintained by concomitant secretion of Cl^-, rather than exchange for cation. As H^+Cl^- enters the gastric lumen, $HCO_3{}^-$ diffuses into the extracellular fluid. This mechanism accounts for the post-prandial "alkaline tide", but has little effect on overall hydrogen ion balance in the normal subject because generation of $HCO_3{}^-$ is followed by secretion of $HCO_3{}^-$ into the small intestine. It is of importance in explaining the metabolic alkalosis which may accompany pyloric stenosis (p. 102).

In summary, CO_2 is of central importance in hydrogen ion homeostasis. Despite the apparently unfavourable pK of the bicarbonate buffer system, the tendency of H_2CO_3 to form volatile CO_2, the partial pressure of which can be controlled by the respiratory centre and lungs to about 5·3 kPa, and the ability of the renal tubular cells and erythrocytes to maintain the $[HCO_3^-]$ at 25 mmol/l at this Pco_2, enable the pH to be kept above the pK of the system. The inter-relationship of the controlling tissues is summarised in Fig. 6. Dysfunction of either kidneys

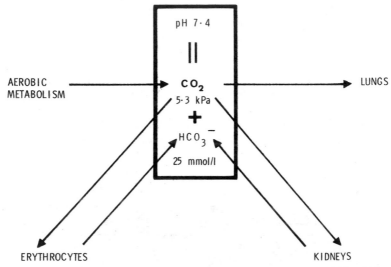

FIG. 6.—Interaction of lungs, kidneys and erythrocytes in control of blood Pco_2, $[HCO_3^-]$ and pH.

or lungs impairs control of extracellular pH. Inhibition of carbonate dehydratase activity (p. 98) affects bicarbonate formation by erythrocytes and renal tubular cells, and bicarbonate reabsorption from the glomerular filtrate, and causes bicarbonate depletion.

Buffers in the body other than bicarbonate and haemoglobin include plasma and tissue proteins; intracellular proteins have an important local role in buffering, but plasma proteins play a very minor part in hydrogen ion homeostasis. Phosphate, with a pK of 6·8, is a very important buffer in the urine, in which concentrations of 25 mmol/l may be reached: its plasma concentration is only about 1 mmol/l and it does not contribute significantly to blood buffering.

DISTURBANCES OF HYDROGEN ION HOMEOSTASIS

Disturbances of hydrogen ion homeostasis involve the bicarbonate buffer pair. In "respiratory" disturbances abnormalities of CO_2 are primary, while in so-called "metabolic" disturbances $[HCO_3{}^-]$ is affected early and changes in CO_2 are secondary.

Measurements Used to Assess Hydrogen Ion Balance

Measurement of blood pH indicates only whether there is overt acidosis or alkalosis. If the pH is abnormal the primary abnormality may be in the control of CO_2 by the lungs and respiratory centre, or in the balance between bicarbonate utilisation in buffering and bicarbonate reabsorption and generation by renal tubular cells and erythrocytes. A normal pH, however, does not exclude a disturbance of these pathways: compensatory mechanisms may be maintaining it. Assessment of these factors is made by measuring components of the bicarbonate buffer system. The concentration of CO_2 is calculated from its measured partial pressure (P_{CO_2}) by multiplying by the solubility constant of the gas (0·225 if P_{CO_2} is in kPa; 0·03 if it is in mm Hg).

pH and P_{CO_2}.—There is a significant arteriovenous difference in pH and P_{CO_2}, and measurement of both must be made on arterial blood. Whole blood (heparinised) must be used because the result of the P_{CO_2} estimation by some methods may depend on the presence of erythrocytes.

Sample collection is important, and details of this are given in the Appendix (p. 109).

Measurement of blood $[HCO_3{}^-]$.—One of two methods may be used to estimate circulating bicarbonate levels. A third value is used to elucidate the cause of the abnormal bicarbonate level.

Plasma total CO_2 (T_{CO_2}): "plasma bicarbonate concentration".— This is probably the most commonly measured index of hydrogen ion homeostasis and is easily performed on plasma. If pH and P_{CO_2} are not required (p. 112) it has the advantage that venous blood can be used. It is an estimate of the sum of plasma bicarbonate, carbonic acid and dissolved CO_2. At pH 7·4 the ratio of $[HCO_3{}^-]$ to the other two components is about 20 to 1 and at pH 7·1 is still 10 to 1. Thus, if the T_{CO_2} were 21 mmol/l, $[HCO_3{}^-]$ would contribute 20 mmol/l at pH 7·4 and just over 19 mmol/l at pH 7·1. Only 1 mmol/l and just under 2 mmol/l respectively would come from $H_2CO_3 + CO_2$. Thus T_{CO_2} is effectively a measure of plasma bicarbonate concentration.

"Actual" bicarbonate.—This estimate is made on whole arterial blood. It is calculated from the Henderson–Hasselbalch equation, using the measured values of pH and P_{CO_2} (in kPa).

$$pH = 6 \cdot 1 + \log \frac{[HCO_3{}^-]}{P_{CO_2} \times 0 \cdot 225}$$

It represents whole blood $[HCO_3{}^-]$ and for reasons discussed above, usually agrees well with plasma T_{CO_2}. It is the estimate of choice if the other two parameters are being measured.

"*Standard*" *bicarbonate*.—The standard bicarbonate is the bicarbonate level which would result from equilibrating whole, arterial blood *in vitro* with a gas with a P_{CO_2} of 5·3 kPa (40 mm Hg). The difference between this value and the "actual" bicarbonate is used to assess the relative contributions to the latter made by abnormalities of the erythrocyte and the renal mechanisms. It is discussed more fully on p. 99.

Acidosis

Acidosis is due to a fall in the ratio $[HCO_3^-]:P_{CO_2}$ in the extracellular fluid.

$$pH = 6·1 + \log \frac{[HCO_3^-]}{P_{CO_2} \times 0·225}$$

In *metabolic* (*non-respiratory*) *acidosis* the primary abnormality in the bicarbonate buffer system is a *reduction in* $[HCO_3^-]$.

In *respiratory acidosis* the primary abnormality in the bicarbonate buffer system is a *rise in* P_{CO_2}.

In either metabolic or respiratory acidosis the ratio of $[HCO_3^-]:P_{CO_2}$, and therefore the pH, can be corrected by a change in concentration of the other member of the buffer pair in the same direction as the primary abnormality. This *compensation* may be partial or complete. The compensatory change in a metabolic acidosis is a reduction in P_{CO_2}: in a respiratory acidosis it is a rise in $[HCO_3^-]$.

In a fully compensated acidosis the pH is normal. However, the levels of the other components of the Henderson–Hasselbalch equation are abnormal. Full correction of all parameters can only occur if the primary abnormality is corrected.

Metabolic Acidosis

The primary abnormality in the bicarbonate buffer system in a metabolic acidosis is a reduction in $[HCO_3^-]$ which, by reducing the ratio $[HCO_3^-]:[CO_2]$ causes a fall in pH. The bicarbonate may be lost in the urine or gastro-intestinal tract, may fail to be generated, or may be utilised in buffering H^+. In the face of a reduction in the number of negatively charged ions electrochemical neutrality might be maintained by replacing it with an equivalent number of other anion(s), or by loss of an equivalent number of cation(s). If we classify metabolic acidosis according to these alternatives many of the associated findings can be explained. The value of chloride estimation can also be assessed.

Because one negative charge balances one positive charge (1 equivalent of cation balances 1 equivalent of anion), and because moles may be multivalent (have more than one charge per mole), milliequivalents (mEq) rather than mmol must be used in any calculation of anion/cation balance. In the normal subject over 80 per cent of extracellular anions is accounted for by chloride and bicarbonate: the remaining 20 per cent or so (sometimes referred to as "unmeasured anion") is made up of proteins, together with the normally low concentrations of, for example, urate, phosphate, sulphate, lactate and other organic anions. The protein concentration remains relatively constant in life, but the levels of the other unmeasured anions can vary considerably in disease.

Sodium and potassium provide over 90 per cent of plasma cation concentration in the normal subject; the balance includes low concentrations of calcium and magnesium, which are not usually measured with sodium and potassium, and which, as we have seen (p. 35), vary very little in life.

The difference between the total concentration of *measured* cations, sodium and potassium, and that of *measured* anions, chloride and bicarbonate, is sometimes called the "anion gap": it is normally about 15 to 20 mEq/l. If we represent the unmeasured anion as A^-, we can express the normal situation as:

$$[Na^+]+[K^+] = [HCO_3^-]+[Cl^-]+[A^-]$$
$$140 + 4 = 25 + 100 + 19 \text{ mEq/l}$$

The anion gap, due to $[A^-]$, is 19 mEq/l.

The changes in anion and cation concentration which coincide with the low $[HCO_3^-]$ of metabolic acidosis are summarised in Table XII.

In the following examples we shall assume a fall in $[HCO_3^-]$ of 10 mmol/l (10 mEq/l).

1. Increase in $[A^-]$.—In **renal glomerular failure** with normal tubular function, bicarbonate generation is impaired by a reduction in the amount of sodium available to exchange for H^+, and in the amount of buffer anion, B^- (Fig. 5) to accept H^+ (p. 88). These buffer anions are components of the unmeasured anion (A^-). For each mEq retained, 1 mEq fewer H^+ can be secreted, and therefore 1 mEq fewer HCO_3^- is generated. The retained A^- therefore replaces HCO_3^-. There is no change in chloride in the uncomplicated case. (Abnormal figures underlined).

$$[Na^+]+[K^+] = [HCO_3^-]+[Cl^-]+[A^-]$$
$$140 + 4 = \underline{15} + 100 + \underline{29} \text{ mEq/l}$$

In this example, as the $[HCO_3^-]$ has fallen from 25 to 15, $[A^-]$ has risen from 19 to 29 mEq/l. The anion gap (entirely due to $[A^-]$) has risen by the same amount.

In generalised renal failure, direct impairment of tubular mechanisms further reduces bicarbonate generation; electrochemical neutrality is maintained by loss of other ions (see below).

TABLE XII

CHANGES IN CONCENTRATION OF IONS BALANCING FALL IN $[HCO_3^-]$ IN METABOLIC ACIDOSIS

Clinical examples	Measured Cation [Na+]	Anion [Cl−]	Unmeasured anion Mixture [A−]	Single [X−]	"Anion Gap"
Glomerular failure	N	N	↑	—	↑
Ketoacidosis and lactic acidosis	N	N	—	↑ (Lactate or ketoacid)	↑
Intestinal loss or renal tubular failure	←————————VARIABLE————————→				
Ureteric transplantation, acetazolamide, renal tubular acidosis, NH_4Cl	N	↑	N	N	N

The last group is the rarest.

IT MUST BE STRESSED THAT THESE ARE THE CHANGES IN UNCOMPLICATED CASES. IN MANY CLINICAL SITUATIONS THERE IS MORE THAN ONE ABNORMALITY.

If renal bicarbonate generation is so impaired that it cannot keep pace with its peripheral utilisation the pH will fall.

$$pH\downarrow = 6{\cdot}1 + \log \frac{[HCO_3^-]\downarrow}{P_{CO_2} \times 0{\cdot}225}$$

Compensation occurs as the respiratory centre responds to the acidosis. CO_2 is lost through the pulmonary alveoli and the pH returns towards normal. *In the compensated case the P_{CO_2} is low.*

The cause of the low $[HCO_3^-]$ is usually obvious if plasma urea levels are estimated.

Correction can only occur when the *GFR increases* (for example, by correction of dehydration). Treatment of the acidosis of irreversible glomerular failure, other than by dialysis, is usually contra-indicated because of the danger of giving sodium salts of alkalis to a patient in whom sodium excretion is impaired; it is, in any case, rarely necessary.

2. **Increase in a single anion (X−) other than chloride.**—Such anions are:
acetoacetate and β-hydroxybutyrate in **ketoacidosis;**
lactate in **lactic acidosis.**

In both these syndromes the increase in $[X^-]$ is due to overproduction rather than to under-excretion: in both there is simultaneous production of H^+. Effectively, therefore, acids (H^+X^-) are being added to the body.

The reduction in $[HCO_3^-]$ is the result of its utilisation in buffering the H^+ which accompanies X^-.

$$HCO_3^- + H^+ \rightarrow H_2CO_3 \rightarrow CO_2 + H_2O$$

and X^- replaces HCO_3^-.

$$[Na^+] + [K^+] = [HCO_3^-] + [Cl^-] + [A^-] + [X^-]$$
$$140 + 4 = \underline{15} + 100 + \underbrace{\underline{19} + \underline{10}}_{\text{"Anion gap"}} \text{ mEq/l}$$

There is no change in chloride in the uncomplicated case.

In this example, as the $[HCO_3^-]$ has fallen from 25 to 15 mEq/l, it has been replaced by 10 mEq/l of X^-. The anion gap is now the sum of $[A^-]$ and $[X^-]$—29 mEq/l.

In the *uncomplicated* case the renal and erythrocyte mechanisms are functioning normally. Bicarbonate reabsorption from the glomerular filtrate is complete.

The falling $[HCO_3^-]$ in the ECF accelerates its generation in the renal tubular cell and erythrocyte (Figs. 3 and 5), and it is only when this fails to keep pace with utilisation that $[HCO_3^-]$ becomes abnormal. The urine will become acid as the rate of H^+ secretion linked to bicarbonate generation increases.

By this ingenious mechanism the additional, and potentially toxic, H^+ is converted to water after it is buffered by HCO_3^-; it is therefore inactivated, near the site of production. Repletion of bicarbonate involves reutilisation of water; this again liberates H^+ but this time in the renal tubular cells—at a site where it is immediately eliminated from the body. Water is being utilised as an inactive carrier for hydrogen from the site of production to the site of secretion (Fig. 7), although it is, of course, not the "same" water nor the "same" CO_2 that is used.

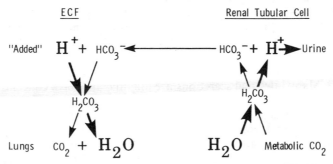

FIG. 7.—Hydrogen ion "shuttle" between site of buffering and kidneys.

Water is freely available and CO_2 is constantly being produced in the tubular cell.

The CO_2 derived from buffering by bicarbonate is lost through the lungs as the respiratory centre is stimulated by the acidosis; because of the large reserve capacity of the lungs, *the blood Pco_2 is never raised if pulmonary function is normal.* The respiratory centre continues to respond to the low pH until the Pco_2 falls low enough to correct the ratio of $[HCO_3^-]:Pco_2$, when the acidosis is *compensated.* Full correction by the kidney is only possible if the rate of H^+ production is reduced to such a level that generation can keep pace with utilisation.

Diagnosis is usually obvious. The commonest cause of *ketoacidosis* is uncontrolled diabetes (p. 192), and the finding of a *high blood glucose concentration* suggests the cause of a low $[HCO_3^-]$; occasionally starvation ketosis can be severe enough to cause acidosis.

The commonest cause of *lactic acidosis* is interference with aerobic metabolism by tissue hypoxia in the *shocked hypotensive patient*; the cause of the acidosis will usually be obvious on clinical grounds. Drugs such as *phenformin* or an overdosage of *salicylates*, may also interfere with lactate metabolism and cause lactic acidosis (pp. 104 and 182). A drug history should be sought if a low $[HCO_3^-]$ is found for no immediately obvious reason.

Lactic acidosis can complicate diabetic ketoacidosis if tissue perfusion is poor. In both lactic acidosis and ketoacidosis, dehydration and hypotension can reduce the GFR so that acidosis is further aggravated by renal factors. It is usually impossible to determine the relative contribution that these different factors make to the fall in $[HCO_3^-]$. However, in practical terms this is not necessary. *Treatment* is directed to improving tissue perfusion by rehydration with the appropriate fluid, and by other measures designed to restore normal blood pressure. Diabetic ketoacidosis should be treated with insulin. If acidosis is very severe intravenous administration of isosmolar, or even hyperosmolar, bicarbonate solution may be indicated: this therapy should be used with caution because of the danger of precipitating plasma hyperosmolality, and hypokalaemia (p. 61). Respiratory dysfunction, which impairs compensatory CO_2 elimination, must also be treated.

3. **Loss of a mixture of anions and cations.**—Loss of intestinal secretions.—Many intestinal secretions are alkaline and have concentrations of HCO_3^- above that of plasma; for instance, the bicarbonate concentration of duodenal juice is about twice that of plasma. Excessive loss of these secretions through fistulae, or in severe diarrhoea, may reduce plasma $[HCO_3^-]$ and cause acidosis. Electrochemical neutrality is maintained by loss of anions and cations in equivalent amounts. The relative proportions of plasma electrolyte concentrations, which depend

on the composition of fluid lost and on that of any replacement fluid, will be very variable. Renal generation of HCO_3^- will be accelerated by the falling concentration, and it is only when intestinal loss is very rapid that $[HCO_3^-]$ becomes abnormal. However, if dehydration is severe enough to reduce the GFR, renal mechanisms may also be impaired and precipitate overt acidosis.

The *diagnosis* is usually obvious on clinical grounds.

Treatment is to restore fluid volume by administering the appropriate solution (p. 61). Usually, if an adequate GFR is maintained, the kidneys will correct the acidosis without bicarbonate administration: in very severe loss it may be necessary to give some bicarbonate.

Renal tubular failure may cause a loss of a similar mixture of ions in the urine. Renal HCO_3^- reabsorption and generation is impaired due to damage to the tubular mechanisms for H^+ secretion. The *cause* of the low $[HCO_3^-]$ is *suggested by polyuria*.

4. Increase in Cl^-.—In groups 1 and 2 the reduction in $[HCO_3^-]$ is compensated for by a rise in unmeasured anion, and $[Cl^-]$ is not affected in uncomplicated cases. In group 3 $[Cl^-]$ is variably affected. The combination of a low $[HCO_3^-]$ and high $[Cl^-]$ is known as "hyperchloraemic acidosis" and is rare; it can usually be predicted on clinical grounds. The "anion gap" in such cases is normal.

$$[Na^+]+[K^+] = [HCO_3^-]+[Cl^-]+[A^-]$$
$$140 + 4 = \underline{15} + \underline{110} + 19 \text{ mEq/l}$$

(*a*) Such a combination would be expected if HCO_3^- were lost in a one-to-one *exchange* for chloride. This occurs if the **ureters are transplanted into the ileum or colon.** The cells of the ileum and colon, like those of the renal tubules, are capable of active transport of ions. Normally reabsorption of water and electrolytes in this part of the intestinal tract is almost complete: however, if fluid which contains chloride enters the lumen, the cells reabsorb some of this chloride in exchange for HCO_3^-. Bicarbonate depletion may therefore occur if urine is delivered into the ileum or colon, as after transplantation of the ureters to this site. This operation may be performed, with total cystectomy, for carcinoma of the bladder. Unless large doses of oral bicarbonate are given, the result is a very low plasma HCO_3^- and very high plasma chloride concentrations.

(*b*) Hydrogen ion secretion, and consequent bicarbonate reabsorption and generation, are affected in tubular disease of any kind. However, if the tubular ability to handle ions other than H^+ is unaffected, hyperchloraemic acidosis results.

Normally 90 per cent of filtered sodium is reabsorbed from the glomerular filtrate isosmotically, accompanied by chloride. Any further sodium returned to the extracellular fluid is exchanged for secreted H^+ or K^+. If H^+ secretion is impaired, and if the same amount of sodium is reabsorbed, then it must either be accompanied by Cl^- or be exchanged for K^+. *This type of hyperchloraemic acidosis is therefore often accompanied by hypokalaemia*—an unusual finding in any other type of acidosis, when hyperkalaemia is the rule.

The two causes of this type of acidosis are:
renal tubular acidosis;
administration of carbonate dehydratase inhibitors.

Renal tubular acidosis.—The group of conditions giving rise to this syndrome is usually the result of acquired tubular lesions; there is a failure to acidify the urine normally, even after ingestion of an acid load such as ammonium chloride (see Appendix, p. 109). There may be either an impairment of the H^+ secreting mechanism itself, or an abnormal permeability of the distal tubular wall to the secreted H^+, allowing its diffusion back into the blood: unlike the situation in generalised tubular failure, the tubular cells retain the ability to form ammonia. Because there is no primary glomerular lesion the plasma urea and creatinine levels are often normal. However, prolonged acidosis increases ionisation of calcium and its release from bone (p. 232): this calcium is often precipitated in the renal tubules and the subject may present with uraemia due to nephrocalcinosis and fibrosis. The increased breakdown of bone salts also partially explains the phosphaturia often accompanying renal tubular acidosis.

Acetazolamide therapy.—Acetazolamide is a drug, used in the treatment of glaucoma, which inhibits the action of carbonate dehydratase. Because H^+ secretion by the renal tubular cell and generation of HCO_3^- by the erythrocyte both depend on the action of carbonate dehydratase (Figs. 3, 4 and 5), bicarbonate reabsorption and generation are impaired with resultant metabolic acidosis.

(*c*) Administration of H^+ with chloride would be expected to cause hyperchloraemic acidosis. A rare cause is administration of **ammonium chloride**: since NH_4^+ forms urea in the liver, with release of H^+, the effect is similar to adding HCl to the extracellular fluid.

To summarise the findings in metabolic acidosis:
$[HCO_3^-]$ *always low*;
P_{CO_2} usually low (compensatory change);
pH low (uncompensated or partially compensated) or normal (fully compensated);
chloride concentration unaffected in most cases; raised after transplantation of the ureters, administration of acetazolamide or ammonium chloride, or in renal tubular acidosis.

Tests which may help to elucidate the cause of a metabolic acidosis are:

plasma urea estimation;
blood glucose estimation;
urine or blood testing for ketones.
The indications for chloride estimations are discussed on p. 112.

Respiratory Acidosis

The plasma concentrations in respiratory acidosis are significantly different from those in non-respiratory disturbances.

The primary defect is retention of CO_2, usually due to generalised respiratory disease (p. 106). The consequent *rise in Pco_2 is the constant finding in respiratory acidosis*. As in the metabolic disturbance, the acidosis is due to the fall in the ratio of $[HCO_3^-]:Pco_2$.

$$pH\downarrow = 6\cdot1 + \log \frac{[HCO_3^-]}{Pco_2\uparrow \times 0\cdot225}$$

Compensatory changes in $[HCO_3^-]$ are initiated by the acceleration of the carbonate dehydratase mechanism in erythrocytes and renal tubular cells by the high Pco_2; $[HCO_3^-]$ generation is speeded up, tending to compensate for the raised Pco_2 (Figs. 3 and 5). The urine becomes acid as more H^+ is secreted.

In **acute respiratory failure** (for instance due to bronchopneumonia or status asthmaticus) the relatively rapid erythrocyte mechanism makes the major contribution to the slight rise in $[HCO_3^-]$. This degree of compensation is often inadequate to prevent a fall in pH.

In **chronic respiratory failure** (for instance due to chronic bronchitis and emphysema) the renal tubular mechanism is of prime importance. Haemoglobin buffering power is of limited capacity, but, so long as the glomerular filtrate provides an adequate supply of sodium and of buffers to maintain H^+ secretion, tubular cells continue to generate bicarbonate until the ratio $[HCO_3^-]:Pco_2$ is normal. Bicarbonate levels as high as twice normal may sometimes be found with a normal pH in stable chronic respiratory failure.

"Standard bicarbonate" (p. 92).—The level of the circulating blood bicarbonate (or "actual bicarbonate") is maintained by three mechanisms, the first of which affects "actual", but not "standard", bicarbonate values.

Generation of H_2CO_3 in erythrocytes from CO_2 and water, which is catalysed by carbonate dehydratase, the hydrogen ion being buffered by haemoglobin (Fig. 3).

A similar mechanism in renal tubular cells, the hydrogen ion being secreted into the urine (Fig. 5).

"Reabsorption" of bicarbonate from the renal tubular fluid as H^+ is secreted into it (Fig. 4). Because, under physiological conditions, urine

contains little HCO_3^-, this mechanism is normally working at near maximum capacity and cannot alter significantly in pathological conditions.

The erythrocyte mechanism is *rapid*, and *depends directly on the circulating* P_{CO_2}. Because of the vast reserve capacity of the normal respiratory pathway to control P_{CO_2}, significant abnormalities of the latter only occur when the lungs or the response of the respiratory centre are abnormal: only in such circumstances will the erythrocyte significantly affect circulating $[HCO_3^-]$. It is of great importance as a short-term response to respiratory disturbances, but is limited in the long term by the buffering capacity of haemoglobin.

This mechanism only requires erythrocytes. *It can therefore occur* in vitro *if whole blood is equilibrated with gases of differing* P_{CO_2}. *In acute respiratory disturbances*, much of the change in blood $[HCO_3^-]$ is due to the erythrocyte mechanism, and can be reversed by equilibrating the blood with a "normal" P_{CO_2} (5·3 kPa; 40 mm Hg). If $[HCO_3^-]$ is now measured again the result is the "standard" bicarbonate.

The renal mechanism increases bicarbonate generation by the kidney in all acidotic states if renal function is normal. It too is affected by blood P_{CO_2} and therefore by respiratory disturbances. If acidosis persists the kidney can continue to influence plasma bicarbonate levels indefinitely. As we have seen, *in the long term* it is the most important mechanism increasing blood buffering capacity, but the cumulative effect takes some time to establish.

This mechanism, which depends entirely on the kidney, does *not* occur *in vitro*. In chronic respiratory disturbances much of the change in circulating $[HCO_3^-]$ is due to this mechanism. Thus, although the "standard" $[HCO_3^-]$ will be nearer normal than the "actual" $[HCO_3^-]$, it will still be abnormal. (In metabolic disturbances, when P_{CO_2} is near normal the "standard" and "actual" bicarbonate are not significantly different.) Measurement of "standard" $[HCO_3^-]$ is most useful in *acute* respiratory disorders, when a metabolic component is suspected (see below).

The findings in the arterial blood in respiratory acidosis are:
P_{CO_2} *always raised*
In *acute* respiratory failure
    pH low
    "Actual" $[HCO_3^-]$ high normal or slightly raised
    "Standard" $[HCO_3^-]$ normal, and lower than "actual" $[HCO_3^-]$
In *chronic* respiratory failure
    pH normal or low depending on severity
    "Actual" $[HCO_3^-]$ raised
    "Standard" $[HCO_3^-]$ raised, but lower than "actual" $[HCO_3^-]$

Mixed Metabolic and Respiratory Acidosis

If there is metabolic acidosis, and CO_2 retention is also present, some of the compensatory increase in blood $[HCO_3^-]$ is used to buffer acid other than H_2CO_3. The rise of actual $[HCO_3^-]$ is impaired, more CO_2 is produced, and the pH falls to lower levels than during CO_2 retention alone. This situation most commonly occurs in the "respiratory distress syndrome" of the newborn, when there is added lactic acidosis due to hypoxia; it may also be due to coexistence of respiratory failure with, for example, ketoacidosis or renal failure.

In such cases the "actual" bicarbonate concentration may be high, normal or low, depending on the relative contributions from the respiratory and metabolic components, and interpretation may be difficult. It is in the *acute* situation of this type that estimation of standard $[HCO_3^-]$ is most useful. If the P_{CO_2} is high and the red cell mechanism is brought into play, the "standard" bicarbonate *must* be nearer normal than the "actual" bicarbonate: if there is no metabolic component it will be within the normal range. However, if there is superadded metabolic acidosis some of the "standard" $[HCO_3^-]$ will have been used in buffering, and this value will be low. If the respiratory disease is chronic, "standard" bicarbonate is less easy to interpret, although a low level with a raised P_{CO_2} would indicate severe superimposed metabolic acidosis.

Alkalosis

Alkalosis is due to a rise in the ratio $[HCO_3^-]:P_{CO_2}$ in the extracellular fluid.

$$pH = 6{\cdot}1 + \log \frac{[HCO_3^-]}{P_{CO_2} \times 0{\cdot}225}$$

In *metabolic alkalosis* the primary abnormality in the bicarbonate buffer system is a *rise in* $[HCO_3^-]$. There is little, if any, compensatory change in P_{CO_2}.

In *respiratory alkalosis* the primary abnormality in the bicarbonate buffer system is a *fall in* P_{CO_2}. The compensatory change is a fall in $[HCO_3^-]$.

The primary products of metabolism are hydrogen ions and CO_2, not hydroxyl ions and HCO_3^-; alkalosis is therefore less common than acidosis.

Alkalosis may present clinically as tetany in spite of normal total plasma calcium levels: this is due to a reduced ionisation of calcium salts in a relatively alkaline medium (p. 232).

Metabolic Alkalosis

A primary rise in plasma $[HCO_3^-]$ may occur in three situations:

Administration of bicarbonate.—Ingestion of large amounts of soluble base (for instance HCO_3^- as sodium bicarbonate in the treatment of "indigestion", or during intravenous bicarbonate infusion). The cause and treatment are both obvious.

Generation of bicarbonate by the kidney in potassium depletion.— This is an extracellular alkalosis only. It is discussed more fully on p. 65.

Generation of bicarbonate by the gastric mucosa when hydrogen and chloride ions are lost in *pyloric stenosis or by gastric aspiration*. This causes *hypochloraemic alkalosis*.

Pyloric stenosis.—Secretion of H^+Cl^- into the gastric lumen is accompanied by generation of equivalent amounts of HCO_3^- by the parietal cells (p. 89). Normally this is followed by secretion of HCO_3^- into the duodenum. Therefore if vomiting occurs with free communication between the stomach and duodenum the loss of secreted HCO_3^- counteracts the effect of its generation by the gastric mucosa. Such vomiting causes relatively little disturbance of hydrogen ion balance: the effects are due to volume and electrolyte loss.

Volume depletion also occurs due to the vomiting of pyloric stenosis. However, the obstruction between the stomach and duodenum reduces loss of HCO_3^-, Na^+ and K^+. Loss of gastric H^+ stimulates further production by the carbonate dehydratase of mucosal cells, with generation of equivalent amounts of HCO_3^-. H^+ continues to be lost, but if duodenal bicarbonate secretion is insufficient to correct the rise in plasma $[HCO_3^-]$ alkalosis may result.

$$pH\uparrow = 6\cdot1 + \log\frac{[HCO_3^-]\uparrow}{P_{CO_2} \times 0\cdot225}$$

Correction of the alkalosis of pyloric stenosis depends on urinary bicarbonate loss. Although acidosis stimulates the respiratory centre, alkalosis does not seem to depress it significantly: CO_2 is not retained in adequate amounts to compensate for the rise in plasma $[HCO_3^-]$. H^+ secretion into the renal tubular lumen is essential for bicarbonate reabsorption from the glomerular filtrate. The rising extracellular $[HCO_3^-]$, derived from gastric mucosal cells, inhibits the formation of H^+ and HCO_3^- by the CD mechanism in renal tubular cells, and diminished renal secretion of H^+ reduces HCO_3^- reabsorption. The consequent loss of HCO_3^- in the urine may compensate for the increased generation by the gastric mucosa.

Two factors may impair this capacity to lose HCO_3^- in the urine, and so may aggravate the alkalosis.

A *reduced GFR* due to volume depletion limits the total amount of HCO_3^- which can be lost.

The *chloride concentration in the glomerular filtrate* may be *reduced*, reflecting hypochloraemia due to gastric loss. Isosmotic sodium re-absorption in the proximal tubule depends on the passive reabsorption of chloride along the electrochemical gradient. A reduction in available chloride limits this isosmotic reabsorption, and more sodium becomes available for exchange with H^+ and K^+. The urine becomes inappropriately acid as H^+ secretion stimulates HCO_3^- reabsorption. Chloride depletion will also increase potassium loss in exchange for sodium, so aggravating the hypokalaemia due to the alkalosis (p. 98).

Thus vomiting due to pyloric stenosis may cause:

hypochloraemic alkalosis;

hypokalaemia;

haemoconcentration and mild uraemia (due to loss of fluid).

The hypokalaemia may be masked while the patient is dehydrated, but should be anticipated during treatment.

Severe hypochloraemic alkalosis is relatively rare because pyloric stenosis is usually treated before this occurs. Nevertheless, when the typical changes in bicarbonate and chloride occur, the diagnosis could (but should not) be made on biochemical grounds alone. The plasma chloride level may be as much as 80 mmol/l lower than that of sodium, rather than the usual 40 mmol/l.

Treatment of the biochemical disorder of pyloric stenosis.—Treatment consists (if renal function is normal) of replacing water and chloride by administering large amounts of saline. The kidney will then correct the alkalosis. The fluid (which, because of the pyloric obstruction, must be given intravenously) should be at least isosmolar saline. If plasma chloride levels are very low hyperosmolar saline may sometimes need to be given. Potassium should be added to the saline if the plasma potassium concentration is low normal or low (see Appendix to Chapter III).

Respiratory Alkalosis

A primary fall in P_{CO_2} is due to abnormally rapid or deep respiration with relatively normal CO_2 transport capacity across the pulmonary alveoli. The causes are:

hysterical overbreathing, which overrides normal respiratory control, with normal lung function;

stimulation of the respiratory centre by **raised intracranial pressure** or **brain stem lesions,** with normal lung function;

stimulation of the respiratory centre by **hypoxia,** with relatively normal overall alveolar diffusion of CO_2 (p. 106);

pulmonary oedema;

lobar pneumonia;

pulmonary collapse or fibrosis;
excessive artificial ventilation.

The fall in P_{CO_2} slows the carbonate dehydratase mechanisms in renal tubular cells and erythrocytes, reducing HCO_3^- generation and reabsorption (Figs. 3, 4 and 5). The *compensatory fall in* $[HCO_3^-]$ tends to correct the pH. It is important not to misinterpret a low $[HCO_3^-]$ with overbreathing as indicating metabolic acidosis, with stimulation of the respiratory centre. In doubtful cases estimation of arterial pH and P_{CO_2} are indicated.

The arterial blood findings in respiratory alkalosis are:
P_{CO_2} *always reduced*;
actual $[HCO_3^-]$ low normal or low;
standard $[HCO_3^-]$ normal in the acute state;
pH raised (uncompensated or partially compensated) or normal (fully compensated).

Salicylates, by stimulating the respiratory centre directly, initially cause respiratory alkalosis. However, in overdosage they uncouple oxidative phosphorylation; consequent impairment of aerobic pathways superimposes a metabolic (lactic) acidosis on the respiratory alkalosis. Both respiratory alkalosis and metabolic acidosis result in low blood HCO_3^- levels, but the pH may be high if respiratory alkalosis is predominant, normal if the two "cancel each other out", or low if metabolic acidosis is predominant. Only by measurement of blood pH can the true state of hydrogen ion balance be assessed.

The possible findings in disturbances of hydrogen ion homeostasis are summarised in Table XIII. The student is urged to read the section on the investigation of hydrogen ion homeostasis (p. 111) to assess realistically how many of these tests are necessary.

BLOOD GAS LEVELS

In respiratory disturbances associated with acidosis a knowledge of the partial pressure of oxygen (P_{O_2}) is as important as that of pH, P_{CO_2} and $[HCO_3^-]$.

Normal gaseous exchange across the pulmonary alveoli involves loss of CO_2 and gain of O_2. However, in pathological conditions a fall in P_{O_2} and rise in P_{CO_2} do not always coexist. The reasons for this are as follows:

CO_2 *is much more soluble than* O_2 *in water*, in which its rate of diffusion is 20 times as high as that of O_2. In *pulmonary oedema*, for example, arterial P_{O_2} falls because its diffusion across the alveolar wall is hindered by oedema fluid. Respiration is stimulated by the hy-

TABLE XIII

SUMMARY OF FINDINGS IN ARTERIAL BLOOD IN DISTURBANCES OF
HYDROGEN ION HOMEOSTASIS

		pH	P_{CO_2}	Actual HCO_3^-	Std HCO_3^-	K^+ levels (plasma)
Acidosis						
Metabolic	Initial state	↓	N	<u>↓</u>	<u>↓</u>	Usually ↑ (↓ in renal tubular acidosis and acetazolamide therapy)
	Compensated state	N	Ⓓ	<u>↓</u>	<u>↓</u>	
Respiratory	Acute change	↓	<u>↑</u>	N or ↑	N	↑
	Compensation	N	<u>↑</u>	ⓐ	↑	
Alkalosis						
Metabolic	Acute state	↑	N	<u>↑</u>	<u>↑</u>	
	Chronic state	↑	N or slightly ↑	<u>↑↑</u>	<u>↑↑</u>	↓
Respiratory	Acute change	↑	<u>↓</u>	N or ↓	N	
	Compensation	N	<u>↓</u>	ⓑ	↓	↓

Arrows underlined = Primary change.
Arrows circled   = Compensatory change.

Note:
1. Generalised potassium depletion can cause extracellular alkalosis.
   Generalised alkalosis can cause hypokalaemia.
   Only the clinical history can differentiate the primary cause of the combination of alkalosis and hypokalaemia.
2. Overbreathing *causes* a low [HCO_3^-] in respiratory alkalosis.
   Metabolic acidosis with a low [HCO_3^-] *causes* overbreathing.
   Only measurement of blood pH and/or P_{CO_2} can differentiate these two.

poxia and by pulmonary distension, and CO_2 is "washed out". However, the rate of transport of O_2 through the fluid cannot be increased enough to restore normal P_{O_2}. The result is a *low or normal arterial* P_{CO_2} and a *low* P_{O_2}. Only in very severe cases is the P_{CO_2} raised; *the haemoglobin of arterial blood is normally 95 per cent saturated with oxygen*, and very little oxygen is carried in simple solution in the plasma: the dissolved O_2 is in equilibrium with the oxyhaemoglobin. Over-breathing air with a normal atmospheric P_{O_2} cannot significantly increase the amount of oxygen carried in the blood leaving normal alveoli: it can, however, reduce the P_{CO_2}. (Breathing pure oxygen by increasing inspired P_{O_2} can increase arterial P_{O_2}, but not the haemoglobin saturation.)

Let us consider the situation in many pulmonary conditions such as *pneumonia, collapse of the lung*, and *pulmonary fibrosis or infiltration*. In such conditions not all alveoli are affected by the disease to the same extent, or in the same way.

(*a*) Some alveoli will be unaffected. The composition of blood leaving these is initially that of normal arterial blood. Increased rate or depth of respiration can lower the P_{CO_2} to very low levels, but not alter either the P_{O_2} or the haemoglobin saturation in this blood.

(*b*) Some alveoli, while having a normal blood supply, may, perhaps because of obstruction of small airways, have little or no air entry. The composition of blood leaving these is near to that of venous blood (it is a right-to-left shunt). Unless increased ventilation can overcome the obstruction it will have no effect on this low P_{O_2} and high P_{CO_2}.

(*c*) Some alveoli may have normal air entry, but little or no blood supply. These are effectively "dead space": increased ventilation will be "wasted", because however much air enters and leaves these alveoli, there is no gas exchange with blood.

Blood from (*a*) and (*b*) mixes in the pulmonary vein before entering the left atrium: the result is mixed venous and arterial blood. The high P_{CO_2} and low P_{O_2} stimulate respiration, and if enough unaffected alveoli (*a*) are present, the reduction of P_{CO_2} in blood leaving these to very low levels may "correct" the high P_{CO_2} from poorly aerated alveoli. For reasons discussed above, neither the P_{O_2} nor the haemoglobin saturation will be significantly altered. The final result is therefore *low or normal arterial* P_{CO_2} with *a low* P_{O_2}.

If the proportion of type (*b*) and (*c*) alveoli is very high, P_{CO_2} cannot be adequately corrected by overventilation, and the result is a *high arterial* P_{CO_2} *and low* P_{O_2}.

In conditions in which almost all the alveoli have a normal blood supply but poor air entry the result will be a *high arterial* P_{CO_2} and *low* P_{O_2}. This may be due to *mechanical or neurological defects in respiratory movement*, or to *obstruction of large or small airways*. However, early in an acute asthmatic attack, stimulation of respiration by alveolar stretching may keep P_{CO_2} normal, or even low.

The two groups of findings in pulmonary disease are summarised in the list below. The conditions marked with* can fall into either group.

Low arterial P_{O_2} *with low or normal* P_{CO_2} can occur in such conditions as:

   pulmonary oedema (diffusion defect)
   pneumonia*
   collapse of the lung*
   pulmonary fibrosis or infiltration*

Low arterial P_{O_2} *with high* P_{CO_2} can occur in such conditions as:

   chest injury, gross obesity, ankylosing spondylitis (impairment of
       movement of the respiratory cage)
   poliomyelitis, lesions of the central nervous system affecting the

respiratory centre (neurological impairment of respiratory drive) laryngeal spasm, severe asthma, chronic bronchitis and emphysema (obstruction to airways)

pneumonia*

collapse of the lung*

pulmonary fibrosis or infiltration*

SUMMARY

HYDROGEN ION HOMEOSTASIS

1. CO_2 is central to hydrogen ion homeostasis. Arterial blood P_{CO_2} is controlled by the respiratory centre and lungs to about 5·3 kPa.

2. At a P_{CO_2} of 5·3 kPa the carbonate dehydratase mechanism in erythrocytes and renal tubular cells maintains the plasma $[HCO_3^-]$ at about 25 mmol/l.

3. In erythrocytes the H^+ produced by the carbonate dehydratase mechanism is buffered by haemoglobin. This has limited capacity, but is of importance in acute disorders of hydrogen ion homeostasis.

4. The renal tubular cell secretes H^+ into the urine in exchange for Na^+. H^+ secretion is essential for HCO_3^- reabsorption: HCO_3^- generation also depends on secretion of H^+ into the urine.

5. Normal urine is HCO_3^- free. Generation of HCO_3^- to replace utilisation depends on the availabilty of urinary buffers, particularly $HPO_4^=$.

6. Renal correction of either acidosis or alkalosis is dependent on a normal glomerular filtration rate.

7. Acidosis results from a decrease in the ratio $[HCO_3^-]:P_{CO_2}$. In a compensated acidosis the ratio is normal, but the levels of HCO_3^- and CO_2 are abnormal.

8. Acidosis may be due to excessive production of hydrogen ion, to failure of the lungs or kidneys, or to excessive loss of bicarbonate.

9. Alkalosis results from an increase in the ratio $[HCO_3^-]:P_{CO_2}$. In a compensated alkalosis the ratio is normal, but the levels of HCO_3^- and CO_2 are abnormal.

BLOOD GAS LEVELS

Low P_{O_2} and Normal or Low P_{CO_2}

1. Carbon dioxide is very much more soluble than oxygen in water. Its level in the blood is therefore less affected than that of oxygen in pulmonary oedema, in which it may even be low due to respiratory stimulation.

2. Arterial blood is 95 per cent saturated with oxygen. Overbreathing therefore cannot increase oxygen carriage in blood from normal alveoli,

but can reduce the P_{CO_2}. In ventilation-perfusion defects, where some alveoli have a normal blood supply, but are ventilated poorly, mixture of the "shunted" blood with blood from normal alveoli results in a low P_{O_2} and normal or low P_{CO_2} in the peripheral arterial blood.

Low P_{O_2} and High P_{CO_2}

3. The P_{O_2} is low and P_{CO_2} is high in total alveolar hypoventilation when neither gas can be adequately exchanged.

APPENDIX

AMMONIUM CHLORIDE LOADING TEST OF URINARY ACIDIFICATION

The ammonium ion (NH_4^+) is acidic because it can dissociate to ammonia and H^+. If ammonium chloride is ingested the kidneys should normally secrete the excess of hydrogen ion.

Procedure

No food or fluid is taken after midnight.

8 a.m. Ammonium chloride 0·1 g/kg body weight is administered orally.

Hourly specimens of urine are collected between 10 a.m. and 4 p.m., and the pH of each specimen measured immediately with a pH meter (pH paper is not very accurate). If the pH of any specimen falls to 5·2 or below, the test can be stopped.

Interpretation

In normal subjects the urinary pH falls to 5·2 or below between 2 and 8 hours after the dose. In *renal tubular acidosis* this degree of acidification fails to occur. In generalised renal failure the response of the functioning nephrons may give normal results.

COLLECTION OF SPECIMENS FOR BLOOD GAS ESTIMATION

1. *Arterial* specimens are preferable to *capillary* ones.

2. The syringe should be moistened with *heparin* and the specimen well mixed. Excess heparin may dilute the specimen and cause haemolysis.

3. Gas exchange with the atmosphere should be minimised by *leaving the specimen in the syringe* and expelling any air bubbles which may be present. The needle may be left on the syringe, and bent over to seal the orifice, or the nozzle stoppered.

4. The rate of erythrocyte metabolism should be minimised by keeping the specimen *on ice* or *in the refrigerator* until the estimation is performed.

5. The estimation should be performed *within an hour*.

Capillary Specimens

In newborn infants arterial puncture may be technically difficult, and it is common to request estimations on capillary specimens (usually heel pricks). The following points should be noted.

1. The area from which the specimen is taken should be warm and pink, because the composition of the capillary blood should be as near arterial as possible. In the presence of peripheral cyanosis results may be dangerously misleading.

2. There should be free flow of blood. Squeezing may cause dilution by ECF.

3. The capillary tube(s) must be heparinised.

4. The tube(s) must be completely filled with blood. Air bubbles will invalidate the results.

5. The ends of the tube(s) should be sealed with Plasticine.

6. Mixing with the heparin should be complete. Usually a small splinter of metal is inserted into the capillary tube, and mixing is performed by passing a magnet up and down the tube.

7. The tube(s) should be put on ice, and the estimation performed at once. *Whenever possible an arterial specimen should be obtained.*

Fetal Specimens

In fetal distress, one indication for termination of labour may be fetal acidosis. If the cervix is dilated, and the fetal head presenting, specimens may be taken by puncturing the scalp and allowing blood to flow into long capillary tubes passed up the vagina. Again, air bubbles invalidate the results.

If is preferable to take an arterial specimen from the mother at the same time, and to compare the fetal and maternal results.

INVESTIGATION OF HYDROGEN ION DISTURBANCES

We have discussed the mechanism behind the abnormal findings in disorders of hydrogen ion balance. Understanding is an essential preliminary to critical assessment of how much investigation is necessary for diagnosis and management. Without such understanding, this assessment cannot be made, and, more dangerously, the results may be misinterpreted.

INTERPRETATION OF PLASMA T_{CO_2}

As explained on p. 91, this commonly performed estimation is effectively a measure of plasma $[HCO_3^-]$. It has the advantage that it can be performed on venous plasma at the same time as, for instance, plasma urea, sodium and potassium.

The measurement of the level of plasma bicarbonate *alone* tells us nothing about the state of hydrogen ion balance. For instance, a low concentration may be associated with compensated or uncompensated non-respiratory ("metabolic") acidosis, or respiratory alkalosis. The pH is determined by the ratio of $[HCO_3^-]$ to $[CO_2]$ according to the Henderson-Hasselbalch equation. Nevertheless, with a little thought T_{CO_2} usually provides adequate information for clinical purposes.

We suggest the following procedure.

If the T_{CO_2} is Low

(a) **Exclude artefactual causes** due to *in vitro* loss of CO_2 from a specimen that has been standing overnight, or from a small specimen with a large dead space above it.

(b) **Reassess the clinical picture.**—Particularly note the presence of the following:

diarrhoea or intestinal fistulae;

any evidence suggesting renal glomerular or tubular failure (Chapter I);

hypotension, dehydration or other evidence of poor tissue perfusion;

a history of ureteric transplantation into the ileum or colon;

a drug history with specific reference to biguanides, such as phenformin, or more rarely, acetazolamide or ammonium chloride.

(c) **Estimate plasma urea and blood glucose** levels (if not already available) and test the urine for ketones.

In the great majority of cases the diagnosis will now be obvious. If it is not, amongst the few remaining possibilities are:

respiratory alkalosis due to overbreathing;

renal tubular acidosis.

In such rare cases estimation of arterial blood pH and P_{CO_2} may help to differentiate respiratory alkalosis from metabolic acidosis. If renal tubular acidosis is suspected, the finding of a high plasma chloride level strengthens the suspicion: an ammonium chloride loading test (p.109) should be performed.

If the Tco_2 is High

(a) **Reassess the clinical picture.**—Particularly note whether:
there is generalised respiratory disease;
there is a cause for potassium depletion (for instance, is the patient taking potassium-losing diuretics?);
bicarbonate has been ingested or infused;
there is severe vomiting which, especially if there is a dyspeptic history, might indicate pyloric stenosis.

(b) **Estimate plasma potassium** (if not already available).

INDICATIONS FOR ESTIMATING ARTERIAL pH AND Pco_2

In most cases of metabolic acidosis, in which the fall in $[HCO_3^-]$ is primary, plasma Tco_2 yields adequate information for clinical purposes. Similarly, a subject with chronic bronchitis and a high plasma concentration of bicarbonate undoubtedly has a respiratory acidosis which may or may not be fully compensated. If there is no possibility of improving air entry into the lungs by use of physiotherapy, expectorants and antibiotics, treatment on a respirator is contra-indicated, because a reduction of Pco_2 will lead to a bicarbonate loss (by reversal of the changes described on p. 99): unless the patient continues to be respired artificially for life, removal from the respirator will result in return of the Pco_2 to its initial high levels, but with a delay of the compensatory increase in bicarbonate for some days. Nothing is to be gained, from the therapeutic point of view, by a knowledge of pH and Pco_2 in such a case.

Arterial blood estimation may be indicated:

1. If there is doubt about the cause of an abnormal Tco_2 (for instance to differentiate metabolic acidosis from respiratory alkalosis).

2. In some conditions in which mixed respiratory and non-respiratory conditions may be present (e.g. cardiac arrest; renal failure complicated by pneumonia; salicylate overdosage). In such conditions, the estimations are indicated only if the results will influence treatment.

3. If there has been an acute exacerbation of the chronic bronchitis, or if the pulmonary disease is of acute onset and potentially reversible (especially as hypoxia may superimpose a lactic acidosis). In such cases vigorous therapy or artificial respiration may tide the patient over until the pulmonary condition improves and more precise information than the plasma Tco_2 concentration is required for monitoring adequate control of treatment.

4. If arterial blood is being taken for estimation of Po_2.

INDICATIONS FOR CHLORIDE ESTIMATION

Chloride levels are consistently affected in only a few conditions. Hyperchloraemia occurs in the metabolic acidosis associated with ureteric transplantation, renal tubular acidosis, and administration of acetazolamide or ammonium chloride. Hypochloraemia occurs in the metabolic alkalosis of pyloric stenosis. If the approach outlined above is followed the level of chloride can usually be predicted, and estimation adds nothing to diagnostic or therapeutic precision. Chloride estimation may help in two situations:

1. If there is a low Tco_2 of obscure origin. The finding of a high chloride (with a normal "anion gap") strengthens a suspicion of renal tubular acidosis. In acidosis due to most other obscure causes the chloride is normal (and "anion gap" increased).

2. If a patient who is vomiting has a high Tco_2. An equivalently low chloride level favours pyloric stenosis.

Chapter V

THE HYPOTHALAMUS AND PITUITARY GLAND

GENERAL PRINCIPLES OF ENDOCRINE DIAGNOSIS

THE principles of the laboratory diagnosis of endocrine disorders are outlined in this and the following three chapters. A few general points may be mentioned here.

Endocrine glands may secrete *excessive* or *deficient* amounts of hormone. This may be caused by a *primary* abnormality of the gland itself, or may be *secondary* to an abnormality in a controlling mechanism, usually located in the hypothalamus or pituitary gland. In the latter case the endocrine gland itself is normal.

The secretion of most hormones is influenced, directly or indirectly, by the final product or metabolic consequence of its secretion. This is usually a *negative feedback* whereby a rising final product (hormonal or non-hormonal) inhibits secretion. This concept is used in many tests of endocrine function.

If results of preliminary tests are near the limits of the "reference range", it is necessary to determine whether they are abnormal or not: if results are definitely abnormal, the abnormality may be primary, or secondary to a disorder of one of the controlling mechanisms. If either of these points is not clear when the results are considered together with the clinical findings, so-called "dynamic" tests should be carried out. In such tests the response of the gland or feedback mechanism is stimulated or suppressed by administration of exogenous hormones.

Suppression tests are used mainly for the differential diagnosis of *excessive hormone secretion*. The substance (or an analogue) normally suppressing hormone secretion by negative feedback is administered and the response measured. *Failure to suppress* implies secretion that is not under feedback control.

Stimulation tests are used mainly for the differential diagnosis of *deficient hormone secretion*. The trophic hormone that normally stimulates secretion is administered and the response measured. A normal response excludes a primary abnormality of the gland whereas *failure to respond* confirms it.

In this Chapter disorders of the pituitary gland and hypothalamus will be discussed. Diseases of their target endocrine organs, the adrenal glands, gonads and thyroid gland will be considered in Chapters VI, VII and VIII.

PITUITARY AND HYPOTHALAMUS

There is a close relationship between the neural and endocrine systems, most obvious in the interaction between the hypothalamus and the two lobes of the pituitary gland. The anterior and posterior lobes are developmentally and functionally distinct, but both depend for normal functioning on hormones synthesised in the hypothalamus. The hypothalamus has extensive neural connections with the rest of the brain, and it is not surprising to find that stress and some psychological disorders affect secretion of pituitary hormones, and hence those of many other endocrine glands.

HYPOTHALAMIC HORMONES

The hypothalamus synthesises two main groups of hormones, which are associated with the function of the posterior and anterior lobes of the pituitary gland respectively.

The Hypothalamus and the Posterior Pituitary Lobe

Three large peptides are synthesised in the hypothalamus, and travel down the nerve fibres of the pituitary stalk to be stored in the posterior lobe of the pituitary gland. Their release into the blood stream is under hypothalamic control.

These peptides are:

Antidiuretic hormone (ADH: vasopressin), which is discussed in Chapter II.

Oxytocin, a hormone which controls ejection of milk from the lactating breast, and which may have some role in the initiation of uterine contractions during labour (although normal labour can proceed in its absence). It can be used in obstetric practice, to induce labour.

Neurophysin, the function of which is doubtful, but which may aid storage of the other two hormones in the posterior pituitary gland.

The Hypothalamus and the Anterior Pituitary Lobe

Nomenclature of hypothalamic and pituitary hormones.—The nomenclature of anterior pituitary hormones and their hypothalamic controlling factors has varied over the years, and different synonyms are still used. We will use the most familiar ones in this edition, but the more recent suggestions are given in the Appendix.

Although all these substances are hormonal, by convention the *name* "hormone" is only used for those of known chemical structure: when their presence is inferred by demonstrable biological activity they are called "factors".

Several small molecules synthesised by the hypothalamus have effects on secretion of specific hormones by the anterior pituitary lobe. The hypothalamus has no direct neural connection with the anterior pituitary, as it does with the posterior lobe. The controlling factors are carried from the site of synthesis to the cells of the anterior lobe through a network of capillary loops which enter the median eminence and are intimately applied to the hypothalamic nerve fibres. This network reforms into vessels that pass down the pituitary stalk into a second network of capillaries in the anterior pituitary. The hypothalamic controlling factors reach the anterior pituitary lobe by this *hypothalamic portal system*; the high local concentration stimulates or inhibits hormonal secretion from the cells of the lobe.

The anterior pituitary gland, in its turn, secretes peptide hormones into the systemic circulation. The cells of the anterior pituitary lobe can be classified histologically by their staining reactions as basophil, acidophil and chromophobe, and more sophisticated techniques can be used to identify the hormones within these cells.

The *basophils* secrete the four hormones which *affect other endocrine glands*. Three of these are glycoproteins of similar chemical structure: they are thyroid-stimulating hormone (TSH), follicle-stimulating hormone (FSH) and luteinising hormone (LH), and they stimulate secretion from the thyroid gland and gonads. A fourth hormone, adrenocorticotrophic hormone (ACTH; corticotrophin) is a simple polypeptide, and acts on the adrenal cortex. Secretion is *stimulated* by the *hypothalamic factors*, thyrotophin-releasing hormone (TRH), FSH/LH-releasing hormone (FSH/LH-RH) and corticotrophin-releasing factor (CRF).

Melanocyte-stimulating hormone (MSH) is identical with part of the ACTH molecule; it stimulates skin pigmentation, and ACTH in excess may have the same effect.

The *acidophils* secrete the two hormones which *affect peripheral tissues directly*—prolactin and growth hormone (GH). They are simple polypeptides with amino acid sequences in common. Secretion of prolactin is controlled by a *hypothalamic* prolactin *inhibitory* factor (PIF)—probably dopamine. GH-controlling factors have not been isolated but the effect of stress in increasing GH secretion provides strong evidence for GH-RF, and a GH-IF may also exist.

Chromophobe cells make up about half the total cells of the anterior pituitary lobe. They contain few stainable granules, and may be precursors of actively secreting cells—although chromophobe adenomas often secrete excessive amounts of prolactin and sometimes of GH and ACTH.

Control of Anterior Pituitary Hormone Secretion

Physiological control.—*Stress*, whether physical or emotional, *stimulates* release of all the simple polypeptide hormones by affecting their hypothalamic controlling factors. Psychiatric illness may mimic endocrine disease. The stress of insulin-induced hypoglycaemia, with assessment of the output of these hormones (ACTH, prolactin and GH) may be used to test anterior pituitary function. Stress does *not* affect the glycoprotein hormones (TSH, FSH, LH).

Negative feedback by a rising level of circulating target cell hormones *reduces* secretion from the basophils. This suppression may be mediated by the hypothalamic centre or may directly affect the anterior pituitary (Fig. 8).

Inherent rhythms.—Many hormones are secreted episodically but, for instance, ACTH (and consequently cortisol) is secreted in a 24-hour

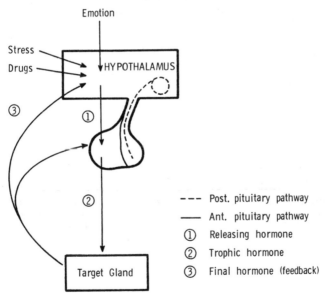

Emotion

Stress

Drugs

HYPOTHALAMUS

③

①

②

Target Gland

- - - Post. pituitary pathway

——— Ant. pituitary pathway

① Releasing hormone

② Trophic hormone

③ Final hormone (feedback)

Fig. 8.—Control of pituitary hormone secretion.

(circadian) rhythm regular enough to be of diagnostic value. The two hormones secreted by acidophils are mainly released during sleep.

Effects of drugs.—Drugs such as *chlorpromazine* interfere with the action of dopamine. This results in reduced GH secretion (reduction of RF) and increased prolactin secretion (reduction of IF). *Bromocriptine* (2-bromo-α-ergocryptine) which may stimulate the attachment of dopamine to pituitary receptor sites, and *L-dopa*, which is converted to dopamine, have the opposite effect in normal subjects. *Bromocriptine* causes a *paradoxical suppression of the excessive GH secretion in acromegalics.*

These effects have been used as pituitary function tests, and in the therapy of hypothalamic-pituitary disorders.

Assay of pituitary hormones.—All the pituitary hormones are peptides. Assay depends on immunological techniques, in which an antibody raised to the substance to be measured is added to the plasma and the antibody-hormone complex formed is quantitated. Specificity depends on the availability of an antibody which reacts *only* with the substance to be measured; because many of the hormones are closely related chemically, there may sometimes be cross-reaction between antibody raised to one hormone and a closely related hormone. The precision of the assay also depends on other factors which are difficult to control. The clinician should remember that the specificity and precision of peptide hormone assays are not comparable with that of, say, sodium estimation, and should interpret results with due caution.

DISORDERS OF PITUITARY HORMONE SECRETION

The main clinical syndromes associated with excessive or deficient pituitary hormone secretion are shown in Table XIV. Excessive secretion usually involves a single hormone, but deficiencies are often multiple.

TABLE XIV

CONSEQUENCES OF A PRIMARY ABNORMALITY OF ANTERIOR PITUITARY HORMONE SECRETION

Hormone	Excess	Deficiency
Growth Hormone	Acromegaly Gigantism	Dwarfism
Prolactin	Amenorrhoea Infertility Galactorrhoea	Lactation failure
ACTH (corticotrophin)	Cushing's disease	Secondary adreno-cortical hypofunction
TSH	Hyperthyroidism (very rare cause)	Secondary hypothyroidism
LH/FSH	Precocious puberty	Secondary hypo-gonadism Infertility

Growth Hormone (GH)

Growth hormone (somatotrophin) is secreted by acidophil cells of the anterior pituitary possibly in response to growth-hormone releasing hormone. GH may be measured in plasma by radioimmunoassay and values are expressed as mU/l of a WHO standard or as μg/l (2 mU/l = 1 μg/l). Secretion is episodic and occurs in relatively infrequent bursts, usually associated with exercise, the onset of deep sleep or in response to the falling blood glucose level about an hour after meals. At other times levels may be very low or undetectable, especially in children.

Physiology.—There are two facets of growth hormone activity.

Growth.—The central role of GH in regulating growth is well illustrated by the dwarfism or gigantism that occur with deficiency or excess of GH in childhood. GH promotes protein synthesis and, in conjunction with insulin, stimulates amino acid uptake by cells. Its effect on skeletal growth however is probably mediated by a peptide, somatomedin, which is synthesised in the liver, and its secretion is dependent on the presence of GH. This mechanism allows a sustained action on skeletal growth despite the marked fluctuations that occur in the plasma levels of GH. Thyroxine is also required for normal growth. The growth spurt at the onset of puberty is probably due to androgens.

Intermediary metabolism.—GH antagonises the insulin-mediated uptake of glucose. In excess, GH may produce carbohydrate intolerance, but this is usually counteracted by increased insulin secretion. GH also stimulates lipolysis and, as the resultant increase in FFA also antagonises insulin release and its action, the effects are difficult to separate. GH secretion is stimulated by hypoglycaemia and suppressed by hyperglycaemia. In adults growth hormone deficiency very rarely produces symptoms.

Control.—In addition to the factors mentioned above GH secretion may be stimulated by stress, starvation, rapidly falling blood glucose levels and by the administration of certain amino acids, or drugs such as L-dopa. Secretion is suppressed by hyperglycaemia and is affected by the levels of other hormones. An impaired response to stimuli may occur in obese patients, in hypothyroidism and hypogonadism, in some cases of Cushing's syndrome and in patients receiving large doses of steroids. Oestrogens potentiate GH response.

GH EXCESS: ACROMEGALY AND GIGANTISM

An adenoma of acidophil cells is the most common cause of GH excess. It is possible that this develops due to prolonged stimulation by the hypothalamus. It is sometimes part of the pluriglandular syndrome

(p. 435). The clinical manifestations depend on whether the condition develops before or after fusion of the bony epiphyses.

In childhood GH excess leads to *gigantism*. In these patients epiphyseal fusion may be delayed by accompanying hypogonadism and extreme heights of over 8 feet may be reached. Mild acromegalic features may develop after bony fusion but these giants frequently die in early adult life from infections or progressive tumour growth.

In adults *acromegaly* develops. Bone and soft tissues increase in bulk and lead to increasing size of the hands and other parts due to soft tissue thickening. There may be excessive hair growth and skin secretion. Changes in facial appearance are often marked, due to the increasing size of the jaw and sinuses. The gradual coarsening of the features may, however, pass unnoticed for many years. An early complaint is that rings and shoes become too small. The viscera also enlarge and cardiomegaly is common. Although the thyroid enlarges, these patients are usually euthyroid. Menstrual disturbances are common.

A different group of symptoms may occur due to the encroachment of the pituitary tumour on surrounding structures such as the optic chiasma. With progressive destruction of the gland, deficiency of the other anterior pituitary hormones may develop. Prolactin levels may however be raised (p. 152).

Non-specific Laboratory Findings

Several biochemical abnormalities may occur, but none is constant or specific enough for diagnosis.

Impaired glucose tolerance may be demonstrated in about 25 per cent of cases. Only about half of these develop symptomatic diabetes. In most cases, however, the insulin response during the GTT is greater than normal due to the antagonism of insulin by GH. It is probable that only those predisposed to diabetes will develop it under these conditions. In others, the pancreas can secrete enough insulin to overcome the antagonism.

Plasma phosphate concentration is *raised* above normal in some severe cases, but may be normal in the later stages, despite activity of the disease.

Hypercalcaemia is a very rare complication.

Diagnosis

The diagnosis of acromegaly is suggested by the clinical features and radiological findings. It is confirmed by the demonstration of a *raised GH level that is not suppressed by glucose*.

Plasma GH levels may also be used to monitor the effects of treatment.

Basal levels.—In a resting subject after an overnight fast GH levels are usually greater than normal and may reach several hundred mU/l. Because of the wide "normal" range, samples from ambulant patients may fail to distinguish acromegalics with only moderately raised levels from normal subjects.

Glucose tolerance test (GTT) with GH levels.—During the course of a GTT on a normal subject, plasma GH falls to less than 10 mU/l (5 μg/l) within 1–2 hours (p. 119). In acromegaly, when secretion is autonomous, this does not occur to the same extent; levels may remain constant or may even increase.

GROWTH HORMONE DEFICIENCY

A small percentage of normally proportioned dwarfed children have growth hormone deficiency. In about half of these there is also deficiency of other hormones. In some patients an organic disorder of the anterior pituitary or hypothalamus may be found, and a rare familial, autosomal recessive form is described, but in many cases the cause is unknown. Emotional deprivation may be associated with growth hormone deficiency indistinguishable by laboratory tests from that due to other causes. Growth failure is progressive and may be noted in the first two years of life. GH is the only means of treatment, and as supplies of this are limited it is essential that GH deficiency be proved.

Diagnosis

Basal plasma GH levels in normal children are usually low and rarely exclude the diagnosis. Blood taken at a time when physiologically high levels are expected (p. 119) may avoid the more unpleasant stimulation tests. Such times are 60 to 90 minutes after the onset of sleep and about 20 minutes after vigorous exercise.

Stimulation of GH secretion.—If GH deficiency is not excluded by the above measurements, one or more stimulation tests are necessary. The simplest is the measurement of plasma GH after oral administration of Bovril (p. 129) (stimulation by the amino acids). If this is negative, intravenous infusion of arginine, or an insulin hypoglycaemia test, is necessary. A clearly normal response excludes the diagnosis, and a clearly deficient one confirms it. Intermediate responses do not usually indicate GH deficiency, but need evaluating further. Test procedures, and diagnostic criteria are usually provided by the authority supplying GH for therapy: both of these may vary in different countries. Once GH deficiency is established a cause should be sought by appropriate radiological and clinical means.

HYPOPITUITARISM

Causes of Hypopituitarism

Destruction of, or damage to, the gland or hypothalamus by a primary or secondary tumour. Secondary tumours are commonly of mammary origin.

Infarction, most commonly postpartum (Sheehan's syndrome) or, rarely, other vascular catastrophes.

Pituitary surgery or irradiation.

Rarer causes include:
head injury;
haemochromatosis.

Isolated deficiencies may be idiopathic.

Established panhypopituitarism presents a distinct clinical picture. It is, however, not common and the question of hypofunction of the anterior pituitary usually arises in one of the following circumstances:

in a patient with clinical and radiological evidence of a pituitary or local brain *tumour*;

in a patient presenting with *hypogonadism, hypothyroidism* or *adrenocortical insufficiency* which is shown by preliminary testing to be *secondary* in origin;

in a patient with *dwarfism* shown to be *due to growth hormone deficiency*.

While isolated hormone deficiency, particularly of growth hormone or gonadotrophins, may sometimes occur, usually several hormones are involved. If a deficiency of one hormone is demonstrated it is therefore important to establish if the secretion of the others is normal or impaired. This is needed, not only to guide substitution therapy, but also to assess the ability of the pituitary to respond to stress such as surgery.

The anterior pituitary has considerable functional reserve and clinical features of deficiency are usually absent until destruction of about 70 per cent of the gland has occurred. Amenorrhoea and infertility may, however, develop early if destruction is associated with raised prolactin levels (p. 152).

Consequences of Hormone Deficiencies

In general, with progressive pituitary damage, gonadotrophin and growth hormone deficiencies occur first. ACTH and/or TSH deficiency may follow even years later or not at all. The clinical and biochemical consequences are generally those of target gland failure and are summarised below.

Secondary hypogonadism, due to gonadotrophin deficiency, presents

as amenorrhoea, infertility, atrophy of secondary sex characteristics and impotence or loss of libido. Characteristically axillary
and pubic hair disappears. In children puberty is delayed.

Retarded growth in children due to growth hormone and thyroid
insufficiency (TSH).

Secondary hypothyroidism (TSH deficiency) may be clinically indistinguishable from primary hypothyroidism.

Secondary adrenocortical hypofunction (ACTH deficiency) may
resemble the primary form (Addison's disease) except that hyperpigmentation (due to excessive ACTH secretion) does not occur. In
contrast there is often striking pallor, possibly due to MSH deficiency. Aldosterone secretion (which is controlled by angiotensin,
not ACTH) is normal and the sodium deficiency so typical of
Addison's disease does not occur. However, there may be marked
dilutional hyponatraemia due to deficiency of cortisol, which is
necessary for normal water excretion: cortisol is also necessary for
maintenance of normal blood pressure, and hypotension is associated with ACTH deficiency. Hyperkalaemia is unusual. Because
of cortisol and/or growth hormone deficiency, these patients are
very sensitive to insulin and fasting hypoglycaemia may be a
feature.

Lactation failure may occur after postpartum pituitary infarction
(Sheehan's syndrome), due to prolactin deficiency. In other cases,
however, prolactin levels are often raised. *Galactorrhoea* may then
be present.

The patient with hypopituitarism, like the patient with Addison's
disease, may die because he cannot respond to the stress caused by,
for example, infection or surgery with an adequate increase in cortisol
secretion. Other life-threatening complications are hypoglycaemia, water
intoxication and hypothermia.

Evaluation of Anterior Pituitary Function

Usually plasma thyroxine and plasma or urinary steroids are requested
first. However, low values, although suggestive of hypopituitarism, are
by no means diagnostic. Furthermore "normal" values do not exclude
pituitary disease. Not only does the pituitary have a large functional
reserve but TSH and ACTH secretion are usually affected only late in
the disease. If other pituitary hormones are measured directly problems
of interpretation may still arise as basal values may "normally" be very
low. Once suspected therefore, the diagnosis of hypopituitarism is best
confirmed or excluded by *direct* measurement of pituitary hormones
after *stimulation.*

Combined pituitary stimulation test (p. 128).—The most useful and
certainly the least time-consuming test from the patient's point of view,

is the simultaneous intravenous administration of insulin, LH/FSH-RH and TRH.

Stress (insulin-induced *hypoglycaemia*) stimulates GH, ACTH and prolactin (p. 117).

LH/FSH-RH stimulates LH and FSH.

TRH stimulates TSH and prolactin (p. 152).

The response is monitored by measuring these hormones in blood samples taken over the subsequent 1–2 hours. The ACTH response is usually assessed by measuring plasma cortisol; this however assumes a normally functioning adrenal cortex. A reduced cortisol response must be further investigated by adrenal cortex stimulation tests (p. 147).

If the above hormone determinations are not available, *indirect* evidence should be sought by demonstrating target gland hypofunction that responds to administration of the relevant trophic hormone, although results are often difficult to interpret.

Secondary adrenocortical hypofunction (p. 141) is diagnosed by the finding of abnormally low plasma or urinary steroid levels that respond to ACTH stimulation.

Secondary thyroid hypofunction is diagnosed by low plasma thyroxine levels or by a low radioactive iodine neck uptake that responds to TSH stimulation.

In both instances it must be remembered that *prolonged* hypopituitarism may lead to secondary atrophy of the target gland with a consequent diminished response to stimulation.

Note that these tests establish only the presence or absence of hypopituitarism. The *cause* must be sought by clinical and radiological means.

Hypothalamus or Pituitary?

In the above discussion we have not distinguished between hypothalamic and pituitary causes of pituitary hormone deficiency or, more correctly, between deficient releasing factor and deficient pituitary hormone secretion. Isolated hormone deficiencies are more likely to be of hypothalamic than of pituitary origin.

Some tests evaluate both hypothalamic and pituitary function, and some only the latter. It is sometimes possible to distinguish the level of the lesion. The TSH response to TRH (p. 169) for example, differs in hypothalamic and pituitary causes of secondary hypothyroidism. In hypogonadism a hypothalamic cause can be postulated if the basal gonadotrophin levels are low, and there is a delayed, but quantitatively normal, response to LH/FSH-RH.

When other synthetic releasing hormones become available it may be possible to make the distinction in cases of GH and ACTH deficiency by comparing the response to stress (both components) and to releasing factors (pituitary only).

PITUITARY TUMOURS

The clinical presentation of pituitary tumours depends on the types of cell involved and on the size of the tumour.

1. Tumours of secretory cells produce the clinical effects of excessive hormone secretion.

GH—Acromegaly or gigantism
ACTH—Cushing's syndrome
Prolactin (p. 152)—Infertility, amenorrhoea, and, occasionally, galactorrhoea.

2. Large tumours may present:

with such symptoms as visual disturbances and headache due to pressure on the optic chiasma or to raised intracranial pressure;

with deficiency of some or all of the anterior pituitary hormones.

Tumours of non-secretory cells may cause no endocrine effects and, for example, radiological investigations may be more helpful than chemical ones. A combined pituitary function test may indicate subclinical hormone deficiency, but is difficult to interpret. Plasma prolactin estimation may help: there is often hyperprolactinaemia, either because the tumour secretes prolactin, or because it impairs normal hypothalamic inhibition (but see p. 153 for the problems of interpreting hyperprolactinaemia).

SUMMARY

1. The anterior pituitary secretes growth hormone (GH), prolactin, adrenocorticotrophic hormone (ACTH; corticotrophin), melanocyte-stimulating hormone (MSH), thyroid-stimulating hormone (TSH) and the two pituitary gonadotrophins, follicle-stimulating hormone (FSH) and luteinising hormone (LH). The posterior pituitary secretes anti-diuretic hormone (ADH) and oxytocin.

2. The secretion of anterior pituitary hormones is controlled by regulating factors from the hypothalamus. These, in turn, are controlled by circulating hormone levels (feedback), or respond to stimuli from higher cerebral centres.

3. GH controls growth and has a number of effects on intermediary metabolism. *Excessive GH* secretion causes gigantism or acromegaly. Laboratory evidence of autonomous GH secretion is obtained by failure of suppression of plasma levels of GH during a glucose tolerance test. *GH deficiency* in childhood leads to dwarfism. Diagnosis of such deficiency is made by demonstrating a subnormal GH response to appropriate stimuli.

4. Anterior pituitary hormone deficiency usually involves several

hormones. Less commonly, isolated deficiency occurs. The clinical features of hypopituitarism are those of gonadotrophin and sex hormone deficiency and of secondary hypofunction of the adrenal cortex and thyroid. Diagnosis is made by demonstrating reduced anterior pituitary reserve after stimulation.

5. Pituitary tumours may produce GH, ACTH or prolactin excess, or may cause hypopituitarism. In patients without obvious endocrine disturbance hyperprolactinaemia may be found.

FURTHER READING

General Endocrinology

HALL, R., ANDERSON, J., SMART, G. A. and BESSER, M. (1974). *Fundamentals of Clinical Endocrinology*, 2nd edit. London: Pitman Medical Publishing Co.

Medicine (2nd Series, 1975). (A monthly add-on series of practical general medicine), Nos. **7, 8** and **9**—Endocrine Diseases. Medical Education (International) Ltd.

Hypothalamus and Pituitary

BESSER, M. (1975). The Hypothalamus and Pituitary Gland. *Medicine* (2nd Series) No. 7, 324. Medical Education (International) Ltd.

BESSER, G. M. (ed.) (1977). The hypothalamus and pituitary. *Clinics in Endocr. and Metab.*, **6**, 1–276.

This issue is devoted to the normal and abnormal function of the hypothalamus and pituitary.

APPENDIX

Nomenclature of Pituitary Hormones

A recently proposed nomenclature, not yet in general use, should be mentioned for reference. The pituitary hormone is usually identified by the suffix "*-tropin*", and the corresponding hypothalamic hormone by "*-liberin*" (releasing) or "*-statin*" (inhibiting). For example, the release of *somatotropin* (growth hormone) may be promoted by *somatoliberin* or inhibited by *somatostatin*. The other pituitary hormones are *prolactin* (unchanged), *corticotropin* (ACTH), *melanotropin* (MSH), *thyrotropin* (TSH), *lutropin* (LH), and *follitropin* (FSH).

Specimen Collection

Many hormones are stable in drawn blood for several hours or even days. Others, such as insulin, renin and ACTH, are unstable and samples should be kept cool and plasma separated from cells without delay. Your laboratory should be consulted for details of specimen collection and handling.

Procedure for Repeated Sampling

Many "dynamic" tests of endocrine gland function require several blood samples over a short period of time. Repeated venepuncture is unpleasant for the patient and is also undesirable because it introduces an element of stress which may interfere with the results. An indwelling needle helps to avoid these problems. The following is a suitable procedure:

(i) An indwelling needle (a "Braunula", a "Butterfly" or similar needle of at least 19G) is introduced into a suitable forearm vein. It is secured in position with adhesive strapping.

(ii) Isotonic saline is infused *slowly* to keep the needle open.

(iii) At least 30 minutes is allowed between insertion of the needle and sampling to allow any stress-induced elevation of hormones to return to normal.

(iv) All samples may be taken through this needle:

    (*a*) disconnect the saline infusion (this can be avoided by the use of a disposable 3-way tap between the infusion set and the needle);

    (*b*) aspirate and *discard* 1–2 ml of saline-blood mixture;

    (*c*) *using a separate syringe*, aspirate the required volume of blood;

    (*d*) re-establish the saline flow.

If there is an untoward reaction such as severe hypoglycaemia after insulin administration, intravenous therapy can be given without delay through the indwelling needle.

TESTS OF PITUITARY FUNCTION

The tests outlined below are those in use in our laboratories. The references on which these descriptions are based should be consulted for further details of procedure and interpretation.

Insulin Stress Test

This test is used to assess the ability of the anterior pituitary to secrete growth hormone, and indirectly ACTH, in response to the stress of hypoglycaemia. As plasma cortisol is measured, the whole hypothalamic-pituitary-adrenal cortex axis is tested.

After an *overnight fast*:

1. Insert indwelling intravenous cannula (see above).
2. Wait at least 30 minutes for hormone levels to return to basal levels.
3. Take *basal blood* samples.
4. Inject soluble insulin intravenously (for dose see below).
5. Take *further blood samples* at 30, 45, 60, 90 and 120 minutes after insulin administration.

Notes.—1. Glucose, cortisol and GH are estimated on all blood samples. The samples for hormone estimation should be kept at 4°C until delivered to the laboratory.

2. *The dose of soluble insulin* must be sufficient to lower blood glucose levels to less than 2·2 mmol/l (40 mg/dl). The usual dose is 0·15 U/kg body weight. If pituitary or adrenocortical hypofunction is probable, or if a low fasting blood glucose has previously been found, the dose should be reduced to 0·1 or 0·05 U/kg. Conversely if there is resistance to the action of insulin (Cushing's syndrome, acromegaly or obesity) 0·2 or 0·3 U/kg may be required.

3. The test is potentially dangerous and should be done only under *direct medical supervision. Glucose* for intravenous administration should be *immediately available* in case severe hypoglycaemia develops. At the conclusion of the test the patient should be given something to eat.

4. If it is necessary to administer glucose, *continue with the blood sampling*. The stress has certainly been adequate.

Interpretation.—*Provided the blood glucose has fallen below 2·2 mmol/l (40 mg/dl)*, preferably with clinical signs of hypoglycaemia, a normal response is a rise in plasma GH of more than 26 mU/l (about 13 μg/l) to exceed 40 mU/l (about 20 μg/l) and in plasma cortisol of more than 200 nmol/l (7 μg/dl) to exceed 550 nmol/l (20 μg/dl). Failure to do so indicates hypothalamic or anterior pituitary hypofunction. Primary adrenocortical hypofunction as a cause of failure of plasma cortisol to rise must be excluded by a tetracosactrin stimulation test (p. 147). In Cushing's syndrome neither cortisol nor GH levels rise.

Combined Anterior Pituitary Function Test

(Harsoulis, P., Marshall, J. C., Kuku, S. F., Burke, C. W., London, D. R., and Fraser, T. R. *Brit. med. J.*, 1973, **4**, 326; and Mortimer, C. H., Besser, G. M., McNeilly, A. S., Tunbridge, W. M. G., Gomez-Pan, A., and Hall, R. *Clin. Endocr.*, 1973, **2**, 317).

Thyrotrophin-releasing hormone (TRH) and LH/FSH-RH can be administered with insulin to provide a single test assessing anterior pituitary reserve.

The procedure is identical with the standard insulin stress test, except that 100 μg of LH/FSH-RH and 200 μg of TRH are injected intravenously immediately after the insulin. The responses to the combined stimuli are the same as when they are applied separately.

Bovril Test of Growth Hormone Secretion

Bovril stimulates growth hormone secretion in children and most women, but not usually in adult males. It is an ideal screening test for possible growth hormone deficiency in children who have low sleeping or post-exercise plasma GH levels. It is safe, and does not require intravenous infusion or special supervision.

1. After an overnight fast, blood is taken for estimation of basal GH levels.

2. The child drinks Bovril ($13 \cdot 3$ g/m^2 body surface or $0 \cdot 33$ g/kg body weight) dissolved in warm water.

3. Further blood samples are taken 30, 60, 90 and 120 minutes after the Bovril.

As little as $0 \cdot 5$ ml of blood may be needed for the estimation. If this is so in your laboratory, the specimen may be obtained from a finger prick, obviating the need for repeated venepuncture.

Interpretation.—A plasma GH level of 40 mU/l (20 µg/l) or more excludes GH deficiency. Poor responses must be confirmed by another test, such as the insulin stress test.

LH/FSH-releasing Hormone Test

(Mortimer, C. H., Besser, G. M., McNeilly, A. S., Marshall, J. C., Harsoulis, P., Tunbridge, W. M. G., Gomez-Pan, A., and Hall, R. *Brit. med. J.*, 1973, **4**, 73).

The synthetic LH/FSH-releasing hormone stimulates the release of gonadotrophins from the normal anterior pituitary. 100 µg of LH/FSH-RH is given by rapid intravenous injection. *Plasma LH and FSH levels are measured in blood drawn before and at 20 and 60 minutes after* the injection. No side-effects have been described.

In general, patients with *primary gonadal* failure have an *exaggerated* response, and those with *hypopituitarism* an *impaired* response. Most patients with low basal gonadotrophin levels show a variable but definite response.

Chapter VI

ADRENAL CORTEX: ACTH

THE adrenal glands are divided into the functionally distinct cortex and medulla. The *cortex* is part of the hypothalamic-pituitary-adrenal endocrine system. Histologically the adult adrenal cortex has three layers. The outer thin layer (*zona glomerulosa*) is functionally distinct and secretes aldosterone. The inner two layers, the *zona fasciculata* and the *zona reticularis*, form a functional unit and secrete the bulk of the adrenocortical hormones. A wider fourth layer is present in the fetal adrenal gland, but disappears after birth. Although the *medulla* is part of the sympathetic nervous system, glucocorticoids are probably needed for the final stage of adrenaline (epinephrine) synthesis.

CHEMISTRY AND BIOSYNTHESIS OF STEROIDS

Steroid hormones, bile salts, and the "vitamin Ds" are all derived from cholesterol, and steroid-producing tissues are rich in cholesterol. Figure 9 shows the internationally agreed numbering of the 27 carbon atoms, and the lettering of the four rings of the cholesterol molecule: the products of cholesterol are also indicated. The bile acids and vitamin Ds are dealt with elsewhere in the book (pp. 299 and 235).

As shown in Fig. 9, the side chain on the 17-position is the main determinant of the type of hormonal activity, but substitutions in other positions modify activity within a group.

The first hormonal product of cholesterol is pregnenolone. Several major synthetic pathways diverge from pregnenolone (Fig. 10). The final product depends on the tissue and the enzymes it contains.

Adrenal Cortex

Zonae fasciculata and reticularis.—*Cortisol* (the most important C_{21} steroid) is formed by progressive addition of hydroxyl groups at positions 17, 21 and 11.

Androgens (e.g. androstenedione) are formed after removal of the side chain to produce C_{19} steroids.

Synthesis of these two groups is stimulated by ACTH from the anterior pituitary (p. 134). ACTH secretion is influenced by plasma cortisol levels via negative feedback. Impaired cortisol synthesis, due to an inherited deficiency of C_{21} or C_{11} hydroxylase (congenital adrenal

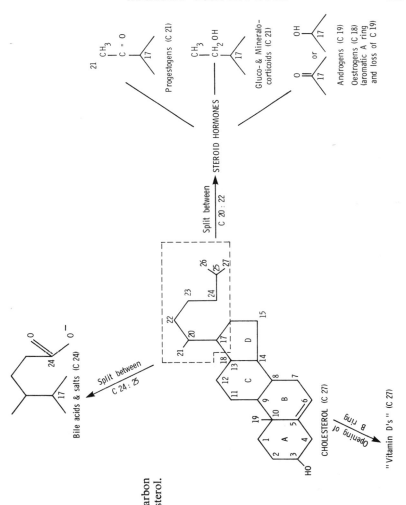

Fig. 9.—Numbering of steroid carbon atoms and products of cholesterol.

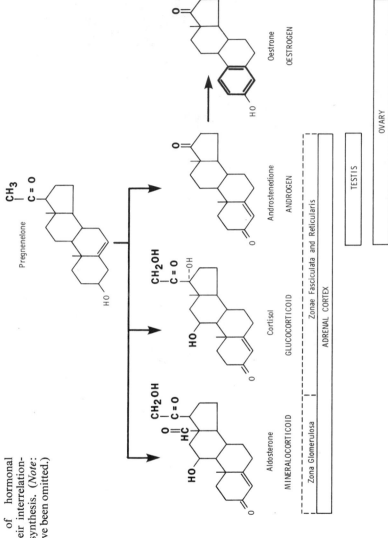

Fig. 10.—Examples of hormonal steroids, showing their interrelationships and sites of synthesis. (*Note*: intermediate steps have been omitted.)

hyperplasia) results in increased ACTH stimulation with increased activity of *both* pathways. The resulting excessive androgen production produces virilisation (p. 143).

Zona glomerulosa.—A different series of hydroxylations produces *aldosterone*. Synthesis of this steroid is controlled by the renin-angiotensin system (p. 38) and not normally by ACTH. In ACTH deficiency, therefore, aldosterone secretion is unaffected.

Glucocorticoids such as cortisol stimulate gluconeogenesis (p. 176) and help maintain blood pressure and renal water excretion. The **mineralocorticoid** effect of aldosterone was discussed on p. 38. **Adrenal androgens** stimulate protein synthesis and, at physiological levels, have little androgenic effect. Small chemical differences thus produce a large difference in biological effect and, for example, dexamethasone may be synthesised chemically to accentuate one physiological action.

Testicular androgens are a product of androstenedione; *oestrogens* are derived from androgens by removal of C_{19} and aromatisation of the A ring. In this chapter only adrenal steroids are considered.

CONTROL OF ADRENAL STEROID SECRETION

The hypothalamus, anterior pituitary and adrenal cortex form a functional unit (the hypothalamic-pituitary-adrenal axis—Fig. 11).

Cortisol is secreted in response to *adrenocorticotrophic hormone* (*ACTH*; corticotrophin) from the anterior pituitary. ACTH secretion is in turn dependent on a hypothalamic *releasing hormone* (corticotrophin releasing factor; CRF). At least three mechanisms influence CRF secretion.

Inherent rhythm.—Sampling at frequent intervals has shown that ACTH is secreted *episodically*, each burst of secretion being followed 5 to 10 minutes later by cortisol secretion. These episodes are most frequent in the early morning (between the 6th and 8th hour of sleep) and least frequent in the few hours prior to sleep; in general, plasma cortisol levels are *highest* between about 07.00 and 09.00 hours and *lowest* between about 23.00 and 04.00 hours.

Loss of this circadian rhythm is one of the earliest features of Cushing's syndrome.

Feedback mechanism.—Plasma cortisol levels cannot be the only factor influencing ACTH secretion, as shown by the parallel circadian rhythm in ACTH and cortisol levels. Nevertheless sustained high levels of cortisol do suppress ACTH secretion and deficient cortisol production is associated with high ACTH levels, providing evidence of *negative feedback* control. Several dynamic tests are based on this phenomenon.

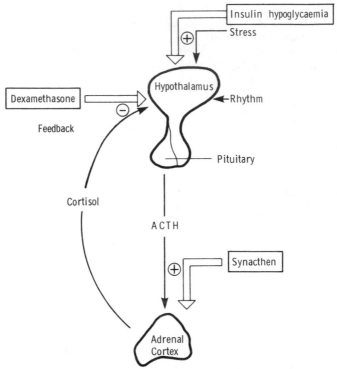

Fig. 11.—Control of cortisol secretion (+ = stimulates; − = inhibits). Dynamic tests shown in boxes.

Stress.—Stress, physical or mental, can override the first two mechanisms and cause sustained ACTH secretion. The consequent diagnostic difficulties are discussed on p. 137. The stress of insulin-induced hypoglycaemia can be used to test the axis.

ACTH (Corticotrophin)

ACTH is a single chain polypeptide made up of 39 amino acids. Only the 24 N-terminal amino acids are required for biological activity and this peptide has been synthesised (*tetracosactrin*; "Synacthen") and can be used in diagnosis and therapy in place of ACTH. ACTH *stimulates cortisol* synthesis and secretion by the adrenal cortex. It has much less effect on adrenal androgen production and, at physiological levels, virtually no effect on aldosterone production. ACTH also has some melanocyte-stimulating activity.

ESTIMATIONS USED IN THE DIAGNOSIS OF ADRENOCORTICAL DISORDERS

The most useful estimation is that of *cortisol*, which may be measured in plasma or in urine by competitive protein-binding or radioimmunoassay. Alternatively, *11-hydroxycorticosteroids* (11-OHCS) may be measured by a fluorimetric method, but this is subject to non-specific interference.

Whichever method is used, it is important to note that, like thyroxine, cortisol is largely protein-bound. Primary changes in cortisol-binding globulin levels will alter values (p. 162) and should not be misinterpreted as indicating adrenocortical dysfunction.

Plasma ACTH estimation by radioimmunoassay is less generally available. It is most useful in the differential diagnosis of Cushing's syndrome (p. 138).

Two chemical estimations of groups of steroids in the urine are:

total 17-oxogenic steroids (T17-OGS) which include the metabolites of the steroids on the cortisol synthetic pathway. It reflects cortisol secretion. The term "oxogenic" is used because, during estimation, the steroids are converted to the more stable and easily measured 17-oxo form;

17-oxosteroids which include the metabolites of the androgen pathway. In the male 10–20 per cent of the total is derived from metabolites of testicular testosterone (not itself a 17-oxosteroid). This estimation therefore reflects *adrenal androgen secretion*.

The selection of tests is discussed in the following pages. Blood is easily collected but plasma levels reflect adrenocortical activity only at that moment. Because of the episodic nature of cortisol secretion, isolated values may be misleading. The measurement of steroids in a 24-hour urine sample reflect the *overall daily secretion*, but it is difficult to ensure accurately timed collections (p. 459).

DISORDERS OF THE ADRENAL CORTEX

The main disorders of adrenocortical function are:

Hyperfunction
Excessive secretion of
cortisol—Cushing's syndrome;
aldosterone—primary aldosteronism (Conn's syndrome);
androgens—adrenocortical carcinoma
          congenital adrenal hyperplasia.

Hypofunction
primary—Addison's disease
          congenital adrenal hyperplasia (deficient cortisol and aldosterone);
secondary—ACTH deficiency (deficient cortisol).

CUSHING'S SYNDROME

Cushing's syndrome is produced by an excess of circulating cortisol. The *causes* are:

bilateral adrenal hyperplasia due to:
excessive ACTH secretion by the pituitary (with or without a pituitary tumour);
ectopic ACTH production by a non-endocrine tumour. The most frequent cause of this is bronchial carcinoma (see Chapter XXII);
adenoma or carcinoma of the adrenal cortex;
administration of cortisone or its analogues. This should not present a diagnostic problem.

In adults pituitary-dependent adrenal hyperplasia is the commonest non-iatrogenic cause.

Clinical features.—The commonest form of Cushing's syndrome, pituitary-dependent adrenal hyperplasia, usually occurs in women of reproductive age.

The commonest presenting feature is obesity, which may be limited to the trunk. Other clinical features are a round plethoric face (moon-face), purple striae on the abdomen, hypertension, muscular weakness, osteoporosis and frequently hirsutism. Menstrual irregularities, especially amenorrhoea, are common. Psychological disturbances may occur.

The most prominent metabolic abnormalities in Cushing's syndrome reflect the *glucocorticoid* action of cortisol. About two-thirds of the patients have *impaired glucose tolerance* as shown by a diabetic glucose tolerance test and of these many have hyperglycaemia and consequent glycosuria. As cortisol has the opposite action to insulin the resultant picture resembles, in certain aspects, that of diabetes mellitus. Protein breakdown is accelerated and the liberated amino acids are either converted to glucose (gluconeogenesis) or lost in the urine leading to a negative nitrogen balance. This is evident clinically as muscle wasting and osteoporosis. The mechanism underlying the characteristic distribution of adipose tissue is not known.

The *mineralocorticoid* effect of excess cortisol is seen in some, but not all, cases of Cushing's syndrome. *Sodium retention* by the renal tubules causes parallel water retention, and plasma sodium concentration is usually normal. *Potassium loss* in the urine may cause severe depletion, with hypokalaemic alkalosis.

Androgen secretion may be increased and may be a factor in the hirsutism, but virilisation is uncommon.

Investigation of Suspected Cushing's Syndrome

Because untreated Cushing's syndrome has a high morbidity and mortality rate, any suspected case must be adequately investigated. Such cases are usually obese, red-faced, hairy, hypertensive women, of whom most do *not* have Cushing's syndrome. There is no test, or group of tests, that is specific for the disorder. To avoid the equally serious error of diagnosing non-existent Cushing's syndrome, it is necessary to approach the problem in a logical sequence. At all stages results must be interpreted with a full knowledge of the clinical state.

What we require to know is:

(*a*) is there excessive cortisol secretion?

(*b*) if so, does the patient have any other condition that may cause it?

(*c*) if it is Cushing's syndrome, what is the cause?

(a) **Is there excessive cortisol secretion?**—The first step is to measure *cortisol levels in plasma*. Morning values may be unequivocally raised, but in many cases overlap with those of normal subjects. One of the earliest features of Cushing's syndrome is diminution or loss of the circadian variation of plasma cortisol. High levels of cortisol are found in a patient with Cushing's syndrome in late evening, when cortisol secretion and plasma levels are normally at a minimum.

Measurement of *cortisol in urine* is a sensitive index of secretion. Only non-protein-bound ("free") cortisol is filtered and excreted. Normally however, CBG is almost fully saturated, and increased cortisol secretion produces a disproportionate rise in the free fraction, with a corresponding increase in urinary free cortisol. Because of the loss of the circadian rhythm, these raised plasma values are present for a longer than normal time and daily urinary cortisol excretion is further increased.

If both plasma and urinary levels are normal, the diagnosis of Cushing's syndrome is highly unlikely. One problem is that cortisol secretion in Cushing's syndrome may fluctuate considerably. If values are normal and the condition is strongly suspected on clinical grounds the tests should be repeated on another occasion. *There is no point in applying further tests at this time.*

(b) **Is there another cause for excessive cortisol secretion?**—Excessive cortisol secretion, as demonstrated by the above tests, may be due to causes other than Cushing's syndrome. The following are of particular importance:

Stress overrides the other controlling mechanisms of ACTH secretion and leads to loss of the normal circadian variation of plasma cortisol. As urinary free cortisol is the most sensitive index of cortisol secretion, values may be markedly increased. This may occur in even relatively minor physical illness as well as in psychiatric disturbances, particularly endogenous depression.

Severe alcohol abuse can produce hypersecretion of cortisol that mimics Cushing's syndrome clinically and biochemically. The abnormal findings revert to normal on withdrawal of alcohol.

Simple obesity, often accompanied by hirsutism and hypertension, may be associated with a shortened half-life, and increased turnover of cortisol. Urinary cortisol metabolites are often increased and if T 17-OGS are measured, elevated values are found. Plasma and urinary cortisol levels are usually normal.

If there is still doubt about the diagnosis the *short dexamethasone suppression test* may be of value. The basis of this test is that in all forms of Cushing's syndrome the normal ACTH-cortisol feedback relationship is disturbed. In *pituitary-dependent adrenal hyperplasia*, the feedback centre is relatively insensitive to feedback and the anterior pituitary continues to secrete ACTH in the presence of raised plasma cortisol levels. The synthetic steroid dexamethasone has a feedback effect similar to cortisol, and a dose which suppresses ACTH secretion in normal subjects will not do so in these patients. In Cushing's syndrome due to an *adrenal tumour* or to *ectopic ACTH* secretion, ACTH secretion from the pituitary is already suppressed and dexamethasone will have no effect. Elevated cortisol levels due to any of these types of Cushing's syndrome are not suppressed by dexamethasone. As dexamethasone is not measured by the usual methods of estimating cortisol, the response may be monitored by plasma or urine cortisol levels.

While normal suppression is a strong point against the presence of Cushing's syndrome failure to suppress is not diagnostic. Severely stressed or obese patients and some alcoholics may also fail to suppress. In doubtful cases the tests should be repeated after correction of the primary condition.

(c) **What is the cause of the Cushing's syndrome?**—The test of choice is *estimation of plasma ACTH*. Levels are moderately raised in pituitary-dependent adrenal hyperplasia, markedly raised in that due to ectopic ACTH secretion and very low in cases of adrenocortical tumour (feedback suppression by high cortisol levels).

If ACTH estimation is not available, a *high-dose dexamethasone test* may help. The principle of the test is as described earlier except that a high enough dose of dexamethasone is given to suppress the relatively insensitive feedback centre in pituitary-dependent disease. In the other two categories, when pituitary ACTH is already suppressed, even this high dose will have no effect.

It must be emphasised that these tests can only be interpreted if the diagnosis of Cushing's syndrome has already been made. In an ill patient, with excessive (but appropriate) cortisol secretion, the plasma ACTH levels will be raised but the high dose of dexamethasone may suppress it. The erroneous diagnosis of pituitary-dependent Cushing's syndrome may be made.

Other findings may indicate the cause. Very high steroid levels are uncommon with pituitary-dependent hyperplasia or adrenocortical adenoma and suggest carcinoma or ectopic ACTH secretion. Markedly

TABLE XV

TEST RESULTS IN CUSHING'S SYNDROME

	Adrenocortical hyperplasia		Adrenocortical tumour	
	Pituitary dependent	Ectopic ACTH	Carcinoma	Adenoma
DIAGNOSIS				
Plasma cortisol				
morning	raised, may be N	raised	raised	raised, may be N
evening	raised			
after 2 mg dexamethasone	no suppression			
Urinary cortisol	raised	usually very high		raised
AETIOLOGY				
Plasma ACTH	high normal or moderately raised	raised	very low	
8 mg dexamethasone	suppression	no suppression		
17-oxosteroids	usually N	often raised	usually very high	normal

increased 17-oxosteroid excretion may occur with adrenocortical carcinoma and distinguish this condition from adenoma. In children adrenocortical carcinoma is a relatively common cause.

Table XV summarises the results of tests in Cushing's syndrome.

ADRENOCORTICAL HYPOFUNCTION

Primary Adrenocortical Hypofunction (Addison's Disease)

Addison's disease is caused by destruction of all zones of the adrenal cortex. Tuberculosis was once the commonest cause but it has now been superseded in many countries by idiopathic atrophy of the gland, probably due to an autoimmune process. Rare causes of destruction (which must be bilateral) include amyloidosis, mycotic infections and secondary malignancy.

Clinical features.—The presentation of Addison's disease depends on the degree of adrenal destruction. If there is extensive adrenal destruction the patient may be shocked or dehydrated (Addisonian crisis): this is a medical emergency requiring immediate treatment. If the destruction is less extensive the disease may be diagnosed only after a prolonged period of ill health. The clinical features of this form—tiredness, pigmentation of the skin and buccal mucosa, weight loss and hypotension—are common to many severe chronic diseases.

The most serious consequences are due to the *mineralocorticoid deficiency*. *Sodium depletion* is the cardinal biochemical abnormality of the Addisonian state. Loss of sodium through the kidneys is accompanied by loss of water and while these parallel one another the plasma sodium concentration remains in the normal range (see Fig. 2, p. 41). The dehydration and consequent haemoconcentration are nevertheless evident from the clinical state and usually from the raised haematocrit and total protein concentration. During a crisis, however, there is almost invariably hyponatraemia, as sodium loss exceeds water loss. As the loss is due to inability of the tubules to reabsorb sodium adequately in the absence of aldosterone the urine will contain an inappropriately high sodium concentration for the degree of volume depletion (p. 51). (Estimation of urinary sodium is, however, not a useful diagnostic test.) The Addisonian crisis is therefore the result of massive sodium depletion. Other abnormalities usually include a raised plasma potassium concentration and metabolic acidosis (compare the reverse in mineralocorticoid excess). The fluid depletion leads to a reduced circulating blood volume and renal circulatory insufficiency with a reduced glomerular filtration rate and a moderately raised plasma urea. In the more chronic form of Addison's disease only some of these features may be present.

The absence of *glucocorticoids* is shown by the marked insulin sensitivity of the hypoadrenal state. The glucose tolerance curve is flat and there may be fasting hypoglycaemia, but this only occasionally gives rise to symptoms.

Androgen deficiency is not clinically evident because testosterone production by the testis is unimpaired.

The pigmentation that develops in Addison's disease is probably due to high circulating ACTH and MSH levels resulting from the lack of cortisol suppression of the feed-back mechanism.

Secondary Adrenal Hypofunction

ACTH release may be impaired by disease of the hypothalamus or of the anterior pituitary, most commonly due to tumour or infarction. Corticosteroid therapy suppresses ACTH release and after such therapy, especially if prolonged, the ACTH-releasing mechanism may be slow to recover, and there may be temporary adrenal atrophy after prolonged lack of stimulation.

Complete anterior pituitary destruction results in the picture of hypopituitarism (p. 122) but if destruction is only partial there may be sufficient ACTH for basal requirements. As in partial adrenal cortical destruction (p. 140) the deficiency may only become evident under conditions of stress. In these patients, as in those on corticosteroid therapy, stress may precipitate acute adrenal insufficiency. The most usual sources of stress are infections and surgery.

The adrenal crisis due to ACTH deficiency is due to lack of glucocorticoids only. It differs from an Addisonian crisis in that aldosterone secretion (not under the influence of ACTH) is normal; consequently there is not the characteristic sodium depletion and dehydration. The condition is nevertheless potentially fatal and is ushered in by mental disturbance, nausea, abdominal pain and usually hypotension (cortisol deficiency). As in primary glucocorticoid deficiency there may be hypoglycaemia and marked insulin sensitivity. Plasma sodium levels may be low, not because of sodium loss as in Addisonian crisis, but because of dilution due to the delayed water excretion of cortisol deficiency.

If the patient presents in this state blood may be taken for plasma cortisol estimation but treatment should be instituted at once. This single estimation may distinguish acute adrenal insufficiency (low levels) from clinically similar conditions where normal adrenal response to stress produces raised cortisol. It does not distinguish primary from secondary adrenal insufficiency.

Investigation of Suspected Adrenocortical Hypofunction

If the patient presents in probable acute adrenal failure immediate treatment is necessary and there is no time to perform diagnostic tests. Blood should be taken for immediate electrolyte and glucose measurement and later plasma cortisol estimation. Further tests can await recovery from the crisis: the underlying abnormality will not be affected by therapy. This initial sample should distinguish between adrenal failure and clinically similar crises where plasma cortisol levels are markedly raised due to stress.

In the less acutely ill patient, morning plasma cortisol is measured, and, if unequivocally high, no further tests are necessary. In most, however, levels will be within the "normal" range, and further tests are necessary. *The essential clinical abnormality in adrenocortical hypofunction is that the patient is unable to respond to stress by increasing cortisol secretion normally.* As this may be due to adrenal (primary) or hypothalamic-pituitary (secondary) pathology, it may be necessary to test the whole axis. Lesions of the adrenal cortex are the commonest cause of adrenal hypofunction, and its integrity should be tested first: furthermore, interpretation of pituitary function tests depends on knowing that the adrenal cortex can respond to ACTH. The synthetic ACTH, tetracosactrin (for example, Synacthen) which has the same biological action as ACTH should be used: it lacks the antigenic part of the molecule, and there is much less danger of an allergic reaction. If it is thought that withdrawal of steroids might precipitate an adrenal crisis, it may be safer to perform these stimulation tests under cover of a cortisol analogue, such as dexamethasone, that does not interfere with plasma cortisol measurement.

The following approach is recommended:

1. A *short stimulation test* is done. In Addison's disease and most cases of secondary insufficiency (in which there is adrenal atrophy), plasma cortisol levels fail to rise normally. If ACTH deficiency is suspected despite a normal response, proceed to step 3.

2. Any subnormal response is further evaluated by giving *repeated daily injections of Synacthen*. The atrophic adrenal cortex due to prolonged lack of stimulation by ACTH, will respond by progressively increasing cortisol secretion. This will not occur in Addison's disease.

3. In suspected ACTH deficiency, the full stress pathway must be tested by an *insulin stress test* (p. 128). Unless there is reason to suspect isolated ACTH deficiency (e.g. after prolonged steroid therapy) a full pituitary function test should be done, to evaluate the reserve of other hormones.

Plasma ACTH estimation is of value in some cases. When inappropriately low cortisol values have been found, a raised ACTH level indicates primary, and a low level secondary, insufficiency. This will distinguish between true adrenocortical disease (where mineralocorticoid replacement is required) and reversible atrophy due to prolonged lack of ACTH.

Urinary steroid levels are frequently low in many chronic diseases and the estimation has *no place* in the investigation of adrenocortical hypofunction.

Metyrapone test.—The drug metyrapone inhibits cortisol synthesis. The falling cortisol level stimulates ACTH secretion and results in an increased secretion of cortisol precursors which may be measured in blood or urine. It has been used as a test of ACTH secretion but suffers from several major disadvantages.

Metyrapone produces cortisol deficiency which is undesirable, especially in a patient whose cortisol secretion is already embarrassed. As the test depends on negative feedback, it *cannot* be done under steroid cover.

It tests only the feedback mechanism and a normal result does not guarantee that the patient will respond normally to stress.

False negative results are not uncommon and the drug may have unpleasant side-effects.

Corticosteroid Therapy

There is the risk of adrenocortical hypofunction when long-term corticosteroid therapy is stopped. This may be due either to secondary adrenal atrophy or to impairment of ACTH-releasing mechanisms.

A simple means of testing the ACTH-releasing mechanisms is to estimate the morning plasma cortisol levels two or three days after cessation of steroid therapy. A level within the normal range indicates a functioning pituitary and feedback centre. It must be emphasised, however, that this does not test the all-important stress pathway (p. 134). Other tests are considered above.

CONGENITAL ADRENAL HYPERPLASIA

Rarely there may be an inherited deficiency of one of the enzymes involved in the biosynthesis of cortisol, most commonly C_{21} hydroxylase. As a result, plasma cortisol levels tend to be low and the feed-back centre is stimulated to secrete greatly increased amounts of ACTH. This, in turn, increases steroid synthesis by the main pathways shown in Fig. 12, and cortisol levels may remain normal if the block is not complete. There will, in any case, be increased production of androgens and of cortisol precursors behind the block. As in Addison's disease a "normal" cortisol level may merely represent a gland working at full capacity but with no reserve to meet stress. In addition, about one-third of the cases seen in infancy have associated impairment of aldosterone synthesis.

With this background of deficient cortisol production leading to excessive ACTH secretion and therefore to androgen overproduction, the clinical manifestations can be explained.

The *increased androgen secretion* may:

affect fetal development, causing female pseudohermaphroditism which presents at birth;

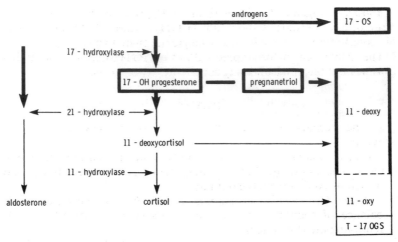

FIG. 12.—The abnormalities occurring in congenital adrenal hyperplasia due to 21-hydroxylase deficiency. Substances of diagnostic importance are shown in boxes.

cause virilisation and precocious puberty in the male at the age of 2 to 3 years;

cause milder virilisation in females at or after puberty.

Aldosterone deficiency (due to impairment of either 21-hydroxylase or 18-hydroxylase) is even rarer and may result in an Addisonian crisis in the first few days of life. All female babies with ambiguous genitalia should have plasma electrolytes estimated. In male babies with no obvious physical abnormalities the diagnosis may not be suspected.

Diagnosis.—The immediate precursor to the block in 21-hydroxylase deficiency is *17-hydroxyprogesterone*. *Plasma levels* of this steroid are raised and may be measured directly, or its metabolite, *pregnanetriol*, may be measured in the urine. It should be remembered that pregnane-triol levels may be raised as a result of stress.

The increased production of androgens and cortisol precursors results in raised urinary 17-oxosteroids and total 17-oxogenic steroids. All these abnormalities can be corrected by administration of cortisone or dexamethasone, which cuts off the excessive ACTH secretion.

The major difficulty in diagnosis using the urine tests is the necessity for a 24-hour collection. Not only is the accuracy of a timed collection difficult to ensure in infants, but, in the patient presenting in a salt-losing crisis, therapy should not be withheld for 24 hours.

A quicker and more accurate method of diagnosis is to determine the *11-oxygenation index* using a *random* specimen of urine. Because the final stage of cortisol synthesis is 11-hydroxylation, any block in the pathway (as in congenital adrenal hyperplasia) results in a decreased proportion of total 17-oxogenic steroids with an –OH group at position 11. The ratio of the steroids without an 11-OH to those with 11-OH (the 11-oxygenation index) will usually be greater than normal (Fig. 12). The urinary chromatographic pattern of corticosteroids and androgens will be abnormal, and may contribute to diagnosis.

SUMMARY

1. The steroids of the adrenal cortex may be classified into three groups.

Glucocorticoids, for example cortisol, which influence intermediary metabolism, body fluid balance and blood pressure. Secretion is controlled by ACTH from the pituitary.

Mineralocorticoids, for example aldosterone, which influence the balance of sodium and potassium. Secretion is controlled by the renin-angiotensin system.

Androgens, for example androstenedione, which probably have an anabolic action.

2. ACTH (and therefore cortisol) secretion is stimulated by CRF from the hypothalamus. The important controlling factors are:

an *inherent rhythm* which produces an overall circadian variation in plasma cortisol levels. These are lowest around midnight and highest in the morning.

negative feedback by circulating *cortisol*;

stress which may override both the above controls.

3. The most useful estimation for the diagnosis of adrenocortical disorders is that of cortisol, in plasma or urine. Plasma ACTH measurement is of value in certain circumstances. Groups of urinary steroids, e.g. total 17-oxogenic steroids (cortisol metabolites) and 17-oxosteroids (androgen metabolites), may be measured but are of limited usefulness.

4. *Cushing's syndrome* is due to excessive circulating cortisol. The causes are hyperplasia or tumour of the adrenal cortex. Hyperplasia is due to excess ACTH either from the pituitary (pituitary-dependent) or from a non-endocrine tumour.

There are three stages in investigation.

(*a*) Is there excessive cortisol secretion?

(*b*) If so, can it be accounted for by a condition other than Cushing's syndrome?

(*c*) If not, what is the cause of the Cushing's syndrome?

5. Primary adrenocortical hypofunction (*Addison's disease*) is due to destruction of the adrenal cortex with loss of all its hormones. Aldosterone deficiency causes sodium and consequent water depletion.

Diagnosis is made by demonstrating that the adrenal cortex cannot respond to stimulation by exogenous ACTH (or an analogue).

6. *Secondary adrenal insufficiency* is caused by diminished ACTH secretion by the pituitary. This may be due to disease of the hypothalamus or pituitary or it may be a result of corticosteroid therapy. It is usually diagnosed by demonstrating a definite, but impaired, response to tetracosactrin (synthetic ACTH) stimulation. In cases without adrenal atrophy the hypothalamic-pituitary-adrenal axis may be assessed by the insulin stress test.

7. *Congenital adrenal hyperplasia* is due to an inherited enzyme deficiency in the biosynthesis of cortisol. Symptoms are due to a deficiency of cortisol and to excessive androgen secretion. Diagnosis is made most rapidly by the 11-oxygenation index, or by estimation of plasma 17-hydroxyprogesterone. Urinary 17-oxosteroids are raised.

FURTHER READING

NABARRO, J., and BROOK, C. (1975). Diseases of the adrenal cortex. *Medicine* (2nd series) No. 8, 351.

HANKIN, M. E., THEILE, H. M., and STEINBECK, A. W. (1977). An evaluation of laboratory tests for the detection and differential diagnosis of Cushing's syndrome. *Clin. Endocr.*, 6, 185.

STEINBECK, A. W., and THEILE, H. M. (1974). The adrenal cortex. *Clinics in Endocr. & Metab.*, 3, 557.

BINDER, C., and HALL, P. E. (eds.) (1972). *Cushing's Syndrome: Diagnosis and Treatment*. London: Wm. Heinemann Medical Books.

See also references on p. 126.

APPENDIX

Plasma Cortisol

Competitive protein binding and radioimmunoassay methods are most commonly used. Samples for these require no special precautions other than avoidance of stasis, but the time of sampling should be recorded on the request form. The test may also measure related steroids so, if the patient is taking steroids, check with your laboratory before interpreting results.

11-OHCS estimation is influenced by non-steroidal fluorescence. As this increases steadily in drawn blood, the sample should be sent to the laboratory on ice as soon as possible. Some drugs, notably *spironolactone*, fluoresce and produce falsely high results in blood and urine.

Tetracosactrin Stimulation Tests

Tetracosactrin is also called Synacthen (Ciba) or Cortrosyn (Organon).

1. 30-minute stimulation test

The patient should be resting quietly.

(*a*) Blood is taken for basal cortisol (or 11-OHCS) levels.

(*b*) 250 μg of tetracosactrin, dissolved in about 1 ml of sterile water or isotonic saline, is given by intramuscular injection.

(*c*) 30 minutes later blood is taken for cortisol estimation.

Normally plasma cortisol increases by at least 190 nmol/l (7 μg/dl), to a level of at least 540 nmol/l (20 μg/dl).

2. 5-hour stimulation test

(*a*) Blood is taken for basal cortisol.

(*b*) 1 mg depot tetracosactrin is injected intramuscularly.

(*c*) Further blood samples are taken 1 hour and 5 hours after (*b*).

Normally plasma cortisol rises to a level of between 600 and 1300 nmol/l (22 and 46 μg/dl) at 1 hour, and to between 1000 and 1800 nmol/l (37 and 66 μg/dl) at 5 hours. Such a response excludes primary (but not secondary) adrenocortical hypofunction. The cause of an impaired response should be investigated by prolonged stimulation.

3. 3-day stimulation test

(*a*) 1 mg depot tetracosactrin is given daily by intramuscular injection.

(*b*) Plasma cortisol is estimated in blood drawn 5 hours after each injection.

No further response after 3 days indicates *primary adrenocortical hypofunction*. An increasing response over this period indicates *secondary adrenal atrophy*.

If there is danger of an adrenal crisis if steroids are withdrawn, dexamethasone, which does not contribute significantly to the cortisol estimation, may be given. Repeated injections of depot tetracosactrin are painful and may, if there is an adrenocortical response, lead to sodium and water retention: *this test is contra-indicated in patients in whom such retention may be dangerous.*

DEXAMETHASONE SUPPRESSION TESTS

Overnight test (for diagnosing Cushing's syndrome).

Dexamethasone (2 mg) is given as a single oral dose at 23.00 hours. Plasma cortisol levels are measured at 08.00 hours the next morning. Suppression is defined as a plasma cortisol level of less than 200 nmol/l.

High-dose test (for deciding the cause of diagnosed Cushing's syndrome).

Dexamethasone (2 mg) is given orally every six hours for two days, starting at 08.00 hours. At 08.00 hours on the third day plasma cortisol is measured. Suppression is defined as levels less than 50 per cent of previously measured basal values.

Interpretation is discussed on p. 138.

The insulin stress test is described on p. 128.

Chapter VII

THE GONADS: GONADOTROPHINS: PROLACTIN

THE study of the endocrinology of the male and female reproductive systems is a specialised and rapidly growing field. Nevertheless many disorders can be assessed using the general principles of endocrine diagnosis, based on a knowledge of normal function. This section will deal only with such principles and for more detail the reader should consult the references at the end of the chapter.

THE HYPOTHALAMIC-PITUITARY-GONADAL AXIS

Secretion of gonadal hormones (like those of the adrenal and thyroid) is stimulated by the pituitary hormones (in this case luteinising hormone (LH) or follicle-stimulating hormone (FSH)), which in turn are controlled by their hypothalamic-releasing factors. One such factor, LH/FSH-releasing hormone, has been identified and influences both LH and FSH secretion. It has been synthesised and may be used to test pituitary function.

THE FEMALE

The ovary secretes oestrogens and progesterone under the influence of LH and FSH. *Slowly rising or sustained high levels of both hormones inhibit pituitary gonadotrophin secretion* by negative feedback, but a *rapid increase in oestrogens*, as occurs just before ovulation, seems to *stimulate LH* secretion (by positive feedback). The ovary also secretes small amounts of androstenedione, an androgen which is converted by peripheral tissues to testosterone; plasma testosterone levels are about a tenth of those in males.

Oestrogens are *secreted by ovarian follicular cells* in the first half of the menstrual cycle, and by the *corpus luteum* during the luteal phase and pregnancy. They differ chemically from androgens only in that they have an aromatic A ring (Figs. 9 and 10); the consequent loss of the C_{19} methyl group means that they have only 18 carbon atoms. This minor chemical difference has far-reaching biological effects, and accounts for "la différence". Oestrogens are essential for the development of female secondary sex characteristics and for normal menstruation; levels are usually undetectable in children. *Oestradiol* and *oestrone* are the most important oestrogens, and are metabolised to the relatively inactive *oestriol.*

Progesterone is secreted by the *corpus luteum* and is chemically similar to the adrenal progestogens (p. 131). It prepares the endometrium to receive the fertilised ovum, and is necessary to maintain early pregnancy.

THE NORMAL MENSTRUAL CYCLE (Fig. 13)

The probable mechanism is as follows:

Follicular (Pre-ovulatory) Phase

At the beginning of this phase the ovarian follicles are undeveloped. Because ovarian hormone levels are low the pituitary secretes FSH and LH.

1. *FSH and LH together* cause *growth of some follicles and maturation of the follicular cells.* LH also stimulates *secretion of hormones* from these cells and circulating oestrogen levels rise steadily. This stimulates regeneration of the previously shed endometrium.

2. The rising oestrogen levels cause a slight fall in FSH secretion by negative feedback to the pituitary gland.

Ovulation

Just before ovulation there is a surge of oestradiol secretion, possibly because the follicular cells have developed to a stage when they become

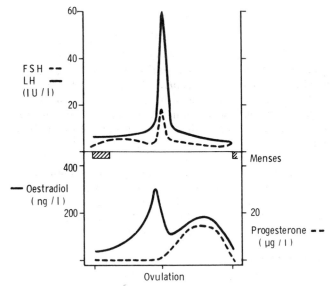

FIG. 13.—Plasma hormone levels during a typical menstrual cycle.

autonomous, and secrete independently of stimulation by FSH. The *rapid rise* in oestrogen concentration *stimulates LH secretion*, and therefore ovulation from a developed follicle. This follicle will develop into the corpus luteum.

Luteal (Postovulatory or Secretory) Phase

This phase is characterised by the rise and fall of the corpus luteum. The corpus luteum takes over ovarian hormone secretion.

1. *LH stimulates the development of the corpus luteum*, and the secretion of *progesterone and oestrogens* from it.

2. *Progesterone prepares the endometrium* to receive a fertilised ovum.

The subsequent events depend on whether the released ovum is fertilised. If it is not, the menstrual cycle takes its course; if it is, pregnancy may supervene.

3. Falling levels of ovarian hormones cause endometrial sloughing and menstrual bleeding.

4. As ovarian hormone levels fall, FSH and LH levels begin to rise. The cycle recommences.

Interpretation of hormone levels must be made in relation to the stage of the cycle. However, as variations in cycle length are due to differences in duration of the follicular phase, this can be a problem. The length of the luteal phase is constant.

PREGNANCY

If the ovum is fertilised it may implant in the endometrium which has been prepared by progesterone (see 2 of "Luteal Phase"). Gonadotrophin secretion is soon taken over by the chorion and developing placenta. This *chorionic gonadotrophin (HCG) is similar in structure and action to LH*, and prevents the involution of the corpus luteum as pituitary gonadotrophin levels fall. *Oestrogen and progesterone levels therefore continue to rise*, and the events in 3 and 4 of "Luteal Phase" are prevented.

LH and HCG are both made up of two polypeptide subunits. The α subunits are identical, and the rapid, non-specific immunological *assay of urinary HCG, used as a pregnancy test*, may give false positive results with high levels of LH, such as occur post-menopausally. The β subunits differ and an assay directed at that of HCG may be useful in the detection and monitoring of some trophoblastic tumours (β-HCG assay).

Lactogenic Hormones

During the later months of pregnancy the placenta and the pituitary secrete hormones which, together, prepare the breast for lactation.

Levels of the placental hormone, *human placental lactogen* (*HPL*), may be used to assess placental function. The pituitary hormone, *prolactin* may be secreted in excess by pituitary tumours, and is therefore of pathological importance.

Prolactin differs from other pituitary hormones because the hypothalamic control is predominantly inhibitory (PIF, p. 116). Secretion can be stimulated by TRH, but this is unlikely to be a physiological control mechanism. As with GH, levels *increase during sleep*, and *during physical and psychological stress*. The circulating concentration increases progressively after the eighth week of pregnancy and at term may be 10 to 20 times higher than that of the non-pregnant woman. Apart from the effect on the breast, the high concentration interferes with gonadotrophin-gonadal function, producing a period of relative infertility during lactation. Despite continuing lactation, non-pregnant levels are reached 2 or 3 months postpartum.

THE MENOPAUSE

Post-menopausally plasma oestrogen levels fall due to ovarian "failure". Removal of normal premenopausal feedback to the pituitary causes increased FSH and, to a lesser extent, LH secretion. The findings are identical with those of primary gonadal failure (p. 154).

DISORDERS OF GONADAL FUNCTION IN THE FEMALE

Measurement and Interpretation of Gonadal Hormone Levels in the Female

Plasma oestrogens and prolactin may be measured by *radioimmunoassay*. As discussed on p. 118, such methods are relatively imprecise, and may sometimes not be completely specific. These factors, and the physiological variations already discussed must be taken into account when interpreting results. *Total urinary oestrogens*, and *urinary pregnanediol* (the most important metabolite of progesterone) may be measured by *chemical methods*.

Hyperprolactinaemia

The patient with hyperprolactinaemia presents most commonly with amenorrhoea and infertility, occasionally accompanied by galactorrhoea.

The most important causes are *pituitary tumours*, which may be so small that they are undetectable by other means. *Idiopathic hyperprolactinaemia* has recently been recognised in anovulatory infertility, and may follow the use of oral contraceptive preparations. Patients with no apparent cause for hyperprolactinaemia should be watched for the development of an overt pituitary adenoma. The higher the prolactin level, the greater the likelihood that a tumour is the cause.

Hyperprolactinaemia should be interpreted with caution. It is particularly important to take samples for prolactin estimation at least 2–3 hours after waking to eliminate misleadingly elevated levels during sleep, and to remember that even the minor stress of venepuncture may cause prolactin secretion. Furthermore, drugs—in particular chlorpromazine, reserpine and methyldopa (p. 117)—may produce this effect, and a drug history should always be taken.

Hyperprolactinaemia is occasionally found, rarely with amenorrhoea or galactorrhoea, in:

severe primary hypothyroidism (probably due to stimulation of secretion by excessive TRH). A pituitary pathology should not be diagnosed on the basis of this finding;

after surgical section of the pituitary stalk (lack of PIF). This is unlikely to pose a diagnostic problem;

about 20 per cent of cases of chronic renal failure (cause unknown).

Hypoprolactinaemia may be the result of anterior pituitary destruction, such as occurs in postpartum infarction of the gland (Sheehan's syndrome). The other findings of hypopituitarism (p. 122) are of more importance.

Investigation of Gonadal Disturbances in the Female

Women with possible gonadal dysfunction may present with any or all of the following complaints:

delayed puberty and primary amenorrhoea;

infertility, with or without amenorrhoea;

hirsutism;

virilisation.

Primary amenorrhoea is the term used when the patient has *never menstruated*. The commonest cause is delayed puberty. The age of the menarche is very variable; unless other signs of endocrine disturbances (such as hirsutism or virilisation) are present extensive investigation should probably be postponed until after the age of 18 years.

Secondary amenorrhoea occurs when the patient who *has previously menstruated* has now stopped. By far the commonest causes are *pregnancy* and the *menopause* (the timing of which is also extremely variable). These should always be thought of before extensive, expensive and sometimes uncomfortable and dangerous investigations are started.

(Note that "primary" and "secondary" when used to describe amenorrhoea have different meanings from "primary" and "secondary" endocrine disease).

The investigation of these disturbances cannot be comprehensively covered here, and the reader is referred to the references given at the end of the chapter for further details. The following is a guide to the more obvious points.

The patient with primary amenorrhoea should be carefully examined, for instance, for the stigmata of Turner's syndrome (congenital gonadal failure) or of intersexuality or other endocrine disease. The presence of hirsutism or virilism should be noted. *It is essential to distinguish between hirsutism and virilism.*

Hirsutism (excessive growth of facial and body hair) is a common complaint and is often associated with menstrual irregularities. It is often familial. It is probably due to slightly excessive androgen secretion from the adrenal cortex, the ovary or both. A number of stimulation and suppression tests have been used in an attempt to establish the source of androgen. In many cases, particularly those due to polycystic ovaries, the plasma testosterone and LH levels are slightly raised.

Virilism is uncommon but much more serious. The patient presents with *other evidence of excessive androgen secretion* such as enlargement of the clitoris, increased hair growth of male distribution, receding temporal hair, deepening of the voice and breast atrophy. The main causes are:

Ovarian tumours (such as arrhenoblastoma and hilus cell tumour) secreting androgens.

Adrenocortical pathology
    Tumours, usually carcinoma.
    Hyperplasia—pituitary-dependent Cushing's syndrome (rarely);
            congenital adrenal hyperplasia.

Other signs of hypopituitarism suggest a hypothalamic or pituitary cause for amenorrhoea and infertility.

Estimations should usually include gonadotrophins and plasma or urinary oestrogen levels.

If plasma or urinary *oestrogen is low* and *gonadotrophins are high* the cause is *primary gonadal failure,* and no other chemical estimations are indicated.

If the results of *both estimations are low* the failure is *secondary* to disease in the pituitary or hypothalamus.

In secondary hypogonadism ("hypogonadotrophic hypogonadism") the deficient secretion of gonadotrophins must be confirmed and distinguished from low "normal" levels by stimulation tests. LH/FSH-releasing hormone can be used to assess anterior pituitary function. Other pituitary function tests must always be performed if secondary failure is suspected. The results of these tests do not usually distinguish between hypothalamic and pituitary causes of reduced gonadotrophin secretion.

A raised LH with normal FSH is characteristic of, although not invariable in, the *polycystic ovary syndrome.*

Plasma prolactin should be estimated if a patient presents with amenorrhoea and/or infertility for no obvious reason.

Detection of ovulation.—It may be important to establish whether a patient complaining of infertility has ovulated, either spontaneously or as a result of therapy to induce ovulation. Plasma progesterone should be measured on a blood sample taken during the second half of the menstrual cycle. A value within the "normal range" for the time of the cycle is good presumptive evidence of ovulation while a value in the range expected in the follicular phase (Fig. 13) indicates the absence of a corpus luteum, and therefore of ovulation.

Effects of Drugs on the Female Hypothalamic-pituitary-gonadal Axis

Oral contraceptives contain synthetic oestrogens and progestogens. By suppressing pituitary gonadotrophin secretion they *prevent ovulation*. Withdrawal mimics involution of the corpus luteum and results in menstrual bleeding.

Clomiphene, by displacing steroids from hypothalamic binding sites prevents negative feedback and may initiate gonadotrophin release, even when circulating levels of oestrogen and progesterone are high: it may be used therapeutically to *induce ovulation* in subjects with secondary hypogonadism ("fertility drug").

Gonadotrophin therapy may be used if clomiphene fails to induce ovulation in women with secondary hypogonadism. Human menopausal gonadotrophin (predominantly FSH, with some LH) is given to mimic the follicular phase. It carries the risk of "hyperstimulation" and *must be monitored* by frequent plasma or urinary oestrogen estimation until levels reach those of the normal pre-ovulatory peak. HCG is then given to induce ovulation by mimicking the LH peak. Success may be assessed by demonstrating rising progesterone levels.

Bromocriptine (p. 117) rapidly *reduces pathologically high levels of prolactin*, whatever the cause: the accompanying amenorrhoea and infertility are cured.

THE MALE

LH stimulates the interstitial (Leydig) cells of the testis to secrete the powerful androgen, testosterone. Prepubertal levels of testosterone are low, and the increase at puberty is responsible for the development of male characteristics. *Testosterone inhibits gonadotrophin secretion* by a negative feedback similar to that of the ovarian hormones in the female.

Testosterone is usually estimated by radioimmunoassay in plasma.

FSH stimulates spermatogenesis.—There is some evidence that a substance ("inhibin") produced during sperm maturation provides a negative feedback to FSH secretion.

Basal prolactin levels are *lower in males* than in females.

GONADAL DISTURBANCES IN THE MALE

Delayed puberty and infertility are the usual presenting features of gonadal disturbances in the male. As in the female, evidence of other endocrine disease and of Klinefelter's syndrome (primary hypogonadism) should be sought (p. 154). Infertility, and very rarely, galactorrhoea may be due to hyperprolactinaemia but *there is no association between hyperprolactinaemia and gynaecomastia*. The method of distinguishing primary from secondary testicular failure is similar to that for distinguishing primary from secondary ovarian failure in the female, but testosterone is estimated instead of oestrogen.

If male infertility is due to *predominant seminiferous tubular failure* (failure of spermatogenesis), Leydig cell function and testosterone secretion may be normal; LH levels are therefore also normal. FSH levels, however, are raised, possibly due to interruption of a spermatogenesis-hypothalamic feedback inhibition.

Urinary 17-oxosteroids are largely of adrenal origin (p. 135) and their excretion has been used as an index of androgen secretion. This is an insensitive index of testicular function.

INVESTIGATION OF PRECOCIOUS PUBERTY IN EITHER SEX

Hormone estimations may help to distinguish *true precocious puberty*, due to excessive gonadotrophin secretion, from *pseudoprecocious puberty*. In the latter, sex hormone production by an ovarian or testicular tumour will lead to the development of the secondary sexual characteristics, but with undeveloped gonads. Plasma gonadotrophin levels will be suppressed. In boys congenital adrenal hyperplasia must be considered (p. 143).

DETECTION AND FOLLOW-UP OF TROPHOBLASTIC TUMOURS

Trophoblastic tumours (hydatidiform mole, chorionepithelioma), and some teratomas secrete HCG. For diagnostic purposes the ordinary pregnancy tests are usually adequate but they are not sensitive enough for follow-up studies. HCG estimation in plasma or urine by radioimmunoassay is at least 20 times more sensitive and, performed at regular intervals, allows early detection and treatment of recurrence.

β-HCG (p. 151) assay distinguishes HCG at low concentration from LH, which may cross-react in the less specific assay. It will *not* differentiate between pregnancy and recurrence of a tumour, because in both cases β-HCG levels are high.

SUMMARY

1. The secretion of gonadal hormones is controlled by the pituitary gonadotrophins LH and FSH. These, in turn, are influenced by hypothalamic-releasing factors and, via negative feedback, by circulating gonadal hormone levels.

2. In the female, cyclical hormonal changes prepare the endometrium for implantation of a fertilised ovum. In the first half of the cycle LH and FSH stimulate ovarian follicle development and oestrogen secretion. A mid-cycle LH surge induces ovulation and converts a follicle into a corpus luteum which secretes oestrogens and progesterone. In the absence of successful implantation the corpus luteum involutes and, as hormone levels fall, endometrial breakdown and shedding occurs (menstruation). If implantation occurs the developing placenta produces HCG which maintains the corpus luteum and prevents menstruation.

3. In the female, gonadal hormone levels must be interpreted in relation to the stage of the menstrual cycle.

4. In primary ovarian failure (and at the menopause) oestrogen levels are low and gonadotrophin levels high (negative feedback). In secondary ovarian failure both oestrogen and gonadotrophin levels are low.

5. In the male LH stimulates testosterone secretion from the testicular Leydig cells, and FSH stimulates spermatogenesis. Testicular failure may be confined to failure of spermatogenesis (raised FSH only) or also involve the Leydig cells when testosterone levels are low and those of both LH and FSH raised.

6. Hyperprolactinaemia is commonly due to a pituitary adenoma. In the female it produces amenorrhoea and infertility, sometimes accompanied by galactorrhoea. Reducing prolactin secretion by bromocriptine restores fertility.

7. Gonadal hormone and gonadotrophin estimation is of value in:
   detection of ovulation
   monitoring of therapy designed to induce ovulation
   assessment of amenorrhoea
   evaluation of hypogonadism and infertility
   evaluation of delayed puberty
   evaluation of virilism
   follow-up of trophoblastic tumours.

FURTHER READING

RYAN, K. (ed.) (1973). Gynecologic endocrinology. *Clin. Obstet. Gynec.*, **16**, 167.
This contains articles on induction of ovulation and menstrual abnormalities.

LONDON, D. (1975). The testis. *Medicine* (2nd series) No. **8**, 370. Medical Education (International) Ltd.

LONDON, D. (1975). The ovary. *Medicine* (2nd series) No. **9**, 381. Medical Education (International) Ltd.

Chapter VIII

THYROID FUNCTION: TSH

THE three hormones known to be produced by the thyroid are thyroxine, tri-iodothyronine and calcitonin. Thyroxine and tri-iodothyronine are products of the thyroid follicular cell and influence metabolism throughout the body. Calcitonin is produced by a specialised cell (the C cell) and influences calcium metabolism. This is an anatomical rather than a functional relationship and in lower animals the calcitonin-secreting cells may be completely separated from the thyroid. Calcitonin is considered briefly on p. 236.

PHYSIOLOGY

Chemistry

The thyroid hormones are synthesised in the thyroid gland by iodination and coupling of two molecules of the amino acid tyrosine. Thyroxine has four iodine atoms and is referred to as T_4; tri-iodothyronine has three and is therefore often known as T_3. The chemical structures are shown in Fig. 14.

Iodine Metabolism

Iodide in the diet is absorbed rapidly from the small bowel. Most natural foods contain adequate amounts of iodide except in regions where the iodide content of the soil is very low. In these areas there used to be a high incidence of goitre, but general use of artificially iodised salt has made this a less common occurrence. Sea foods have a high iodide content and fish and iodised salt are the main dietary sources of the element.

Normally about one-third of the absorbed iodide is taken up by the thyroid and the other two-thirds is excreted via the kidneys.

Biosynthesis of Thyroid Hormones

The steps in the biosynthesis of the thyroid hormones are outlined below and in Fig 15. An understanding of the sequence of events is helpful in appreciating both the mechanism of action of the drugs used in the treatment of thyroid disorders, and the results of the several congenital enzyme deficiencies (p. 169).

(a) *Iodide is actively taken up* by the thyroid gland. The concentration of iodine in the gland is normally about twenty times that in plasma but

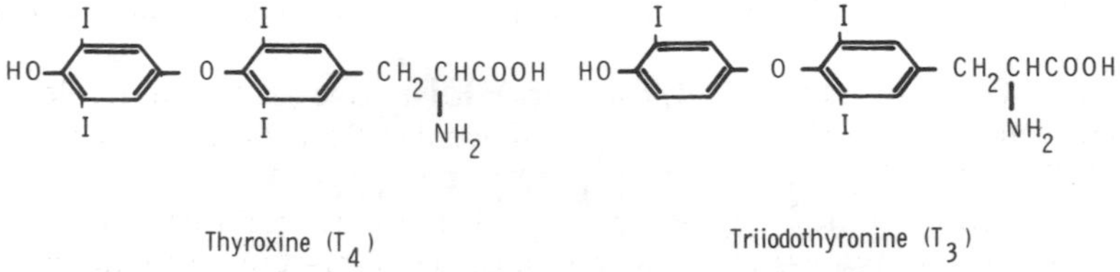

Thyroxine (T_4) Triiodothyronine (T_3)

FIG. 14.—Chemical structure of the thyroid hormones.

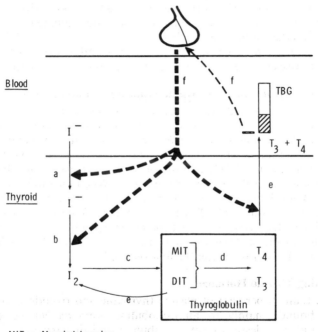

MIT = Monoiodotyrosine

DIT = Diiodotyrosine

TBG = Thyroxine – Binding Globulin

FIG. 15.—Synthesis and regulation of thyroid hormones.
(See text for explanation of small letters.)

may exceed it by a hundred times or more. The salivary glands, gastric mucosa and mammary glands are also capable of concentrating iodine. Uptake is *blocked by thiocyanate or perchlorate.*

(*b*) Trapped *iodide is rapidly converted to iodine.*

(*c*) It is then *incorporated* into mono- and di-iodotyrosine (MIT and

DIT). This step is *inhibited by carbimazole, propylthiouracil and related drugs.*

(d) MIT and DIT are *coupled* to form thyroxine (T_4) (2 molecules of DIT) and tri-iodothyronine (T_3) (1 molecule of DIT and 1 molecule of MIT).

The iodotyrosines, and T_3 and T_4, do not exist in a free state in the thyroid gland, but are incorporated in the large protein molecule *thyroglobulin*. This iodoprotein is the main component of the colloid of the thyroid follicle.

(e) *Release* of T_3 and T_4 occurs by breakdown of thyroglobulin by proteolytic enzymes. The hormones pass into the blood. Mono- and di-iodotyrosine released at the same time are de-iodinated by another enzyme and the iodine re-utilised.

Each step is controlled by specific enzymes and congenital deficiency of any of these enzymes can lead to goitre and, if severe, hypothyroidism.

Control of Thyroid Hormone Synthesis and Secretion (Fig. 15)

The uptake of iodide (a), conversion to iodine (b) and release of thyroid hormones (e) are promoted by the action of thyroid-stimulating hormone (thyrotrophin; TSH) from the anterior pituitary gland.

TSH secretion is stimulated by the hypothalamic thyrotrophin-releasing hormone (TRH) and suppressed (by negative feedback) by high circulating levels of thyroid hormones. When the latter are pathologically high the stimulating action of TRH is overridden and even a large intravenous dose of TRH fails to increase TSH levels. This is the basis of the TRH stimulation test for hyperthyroidism. In infants TRH (and consequently TSH) is secreted on exposure to cold. This factor does not seem to be operative after the first year of life.

Circulating Thyroid Hormone

More than 99 per cent of plasma thyroxine and tri-iodothyronine is protein-bound, mainly to an α-globulin, *thyroxine-binding globulin* (*TBG*), and, to a lesser extent, to albumin and thyroxine-binding pre-albumin. It is the small free fraction which is physiologically active, and which regulates pituitary TSH secretion (compare calcium, p. 232, and cortisol, p. 135).

The level of plasma T_4 is usually about 100 nmol/l (8 μg/dl) but is higher in the first month of life and in old age (particularly in women). Plasma tri-iodothyronine levels are much lower (about 1·5 nmol/l) and tend to be lower in old age. Despite this difference in total concentrations, the free fractions of T_4 and T_3 are almost equal and T_3 is metabolically the more active. Some of the plasma T_3 is secreted directly but most is derived from the peripheral deiodination of T_4.

"Reverse T_3", differing from T_3 only in the position of the iodine atoms on the thyronine molecule, may be another peripheral product: its biological significance is uncertain, but it is probably inactive.

In iodine deficiency or after treatment of hyperthyroidism, the ratio of T_3 to T_4 secreted is increased. As a result, the patient may be clinically euthyroid despite low T_4 levels.

Actions of Thyroid Hormones

Thyroid hormones influence and speed up many metabolic processes in the body. They are essential for normal growth, mental development and sexual maturation. They also increase the sensitivity of the cardiovascular and central nervous systems to catecholamines and so influence cardiac output and heart rate. The mechanism of thyroid hormone action is unknown.

THYROID FUNCTION TESTS

It is essential to confirm a clinical diagnosis of thyroid dysfunction by laboratory tests. Treatment may be prolonged and, in the case of hypothyroidism, life-long. During treatment the original clinical picture disappears and it is not an uncommon problem to be faced with a patient who has taken thyroxine for many years for probable "thyroid trouble". Without an adequately established diagnosis it may be difficult to assess the past history.

The tests described in this section determine only the *functional state* of the thyroid gland. The aetiology must be established by other means, some of which will be mentioned, briefly.

Plasma Thyroxine and Free Thyroxine Index

Plasma thyroxine.—Most techniques in common use assess levels of protein-bound plus free thyroxine: more than 99 per cent of that circulating is protein-bound, and it is predominantly this fraction that is being measured. Clinical effects of changes in thyroxine secretion by the thyroid gland are, on the other hand, due to changes in the *free* fraction. Despite this, the protein-bound and free fractions parallel each other *provided that the level of the binding-protein stays nearly constant*, and the assay is then a valid reflection of the hormone state [(*a*) and (*b*) in Fig. 16].

If, on the other hand, the primary change is in the TBG level there is no change in the level of *free* hormone, but there is a change in measured T_4. If secretion of hormone from the thyroid gland is normal, the *proportion* of the protein saturated with T_4 remains at about 30 per cent of the total, whatever the level of TBG. An *increase in TBG* results in a proportional *increase* in *both protein-bound T_4 and in free binding sites* [(*c*) Fig. 16], and a *decrease in TBG decreases both T_4 and free binding*

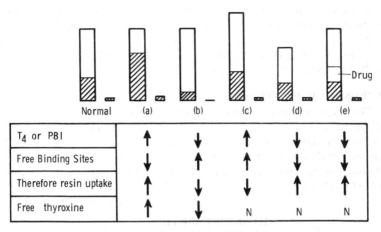

	Normal	(a)	(b)	(c)	(d)	(e)
T$_4$ or PBI		↑	↓	↑	↓	↓
Free Binding Sites		↓	↑	↑	↓	↓
Therefore resin uptake		↑	↓	↓	↑	↑
Free thyroxine		↑	↓	N	N	N

FIG. 16.—Interpretation of tests of circulating hormone (see text).

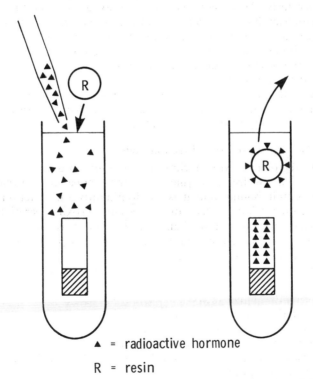

▲ = radioactive hormone

R = resin

FIG. 17.—Assessment of free binding sites on thyroxine-binding globulin.

sites [(*d*) Fig. 16]. These might be misinterpreted as hyperthyroidism and hypothyroidism if only T_4 were measured. If we compare (*a*) and (*b*) with (*c*) and (*d*) we will see that, by contrast, if the primary change is in the amount of thyroid hormone, the number of free binding sites varies inversely with the level of measured T_4. If we could assess the number of these sites the cause of the changes in T_4 should become apparent.

Assessment of free binding sites.—Methods of assessing the level of free binding sites all depend on *in vitro* addition of radioactive thyroid hormone to the patient's plasma in amounts which exceed the capacity of TBG to bind it. The more free sites there are on the TBG in the specimen, the more the added radioactive hormone will be bound by the protein. A resin may now be added to take up the *unbound* hormone: the more free sites there were on the protein the less radioactive hormone will be left to bind to the resin (Fig. 17). If the resin is separated from the plasma, the amount of radioactivity *either* left on the TBG, *or* bound to the resin can be determined. The more left in the plasma, *or*, the less on the resin, the more free sites there were on the TBG.

Therefore:

in primary disease of thyroid function measured T_4 and free sites vary inversely;

in primary disorders of TBG levels, measured T_4 and free sites change in the same direction.

There are many modifications of this technique, and many formulae have been devised, using these two measurements, to "correct" the T_4 and therefore to give an assessment of free T_4 (free thyroxine index or FTI). The student should consult his own laboratory for details. It is worth noting that any formula which employs the results of two assays, both with some degree of analytical variation, may increase the likelihood of yielding a misleading answer. It may be better to inspect the results of the two assays separately. If a formula is used, it should be noted that "correction" is often incomplete when levels of TBG are very abnormal.

Causes of abnormal TBG concentration.—An increase in TBG concentration occurs due to high levels of circulating oestrogens:

in pregnancy, and in the newborn infant;

during oestrogen therapy or while taking oral contraceptives.

A decrease in TBG concentration occurs:

due to protein loss, as in the nephrotic syndrome;

in the rare congenital TBG deficiency.

TBG binding sites occupied by drugs.—Some drugs, such as salicylates and diphenyl hydantoin ("Epanutin"), occupy binding sites on TBG and displace thyroxine. The resultant low T_4 may or may not be accompanied by a raised resin uptake. Low T_4 values can only be interpreted with a knowledge of what drugs the patient is taking.

"Normalised" thyroxine levels (for example, *effective thyroxine ratio*).—A number of tests are becoming available in which the T_4 value is corrected for abnormalities of protein binding in a single procedure and is expressed as a ratio of a standard serum. This is analogous to the FTI and may be used as a screening test. Borderline results, especially if low normal, should be confirmed by other tests.

Plasma TSH may be measured by radioimmunoassay. Estimation is of particular value in the diagnosis of primary hypothyroidism. Unfortunately most currently available techniques are unable to distinguish between low normal and subnormal levels. Dynamic tests, however, overcome this problem (p. 168).

Plasma Tri-iodothyronine (T_3)

T_3 can be measured in plasma by radioimmunoassay, and the estimation is of particular value in certain situations (p. 167). T_3, like T_4, is bound to protein, and it may be necessary to assess binding sites to interpret the results.

Protein-bound Iodine

An alternative assessment of plasma T_4, still used in some laboratories, is to estimate hormonal iodine, usually as the protein-bound iodine (PBI). These estimations suffer from the major disadvantage that non-hormonal iodine compounds may be included. This produces *falsely raised values* which may either *mimic hyperthyroidism* or, more seriously, *mask hypothyroidism* by elevating low values to within the normal range. The causes of iodine contamination are considered on p. 173.

Radio-iodine Uptake Tests

A group of thyroid function tests utilise tracer doses of radioactive iodine (usually ^{131}I or ^{132}I) which is metabolised by the body in the same way as the natural element. Most commonly the *uptake over the neck* (thyroid) is measured 6 or 24 hours after oral administration. It is used as an index of thyroid activity, although uptake reflects only the first stage of thyroid hormone synthesis. In *hyperthyroidism* the neck uptake is usually *above normal* and in *hypothyroidism* it is *low*.

These tests suffer from one major disadvantage—results are largely dependent on the size of the body iodine pool. The radioactive iodine enters and is diluted by the body iodine pool. If, for example, the pool is greater than normal, the tracer dose will be more diluted than normal. The iodine taken up by the thyroid will therefore contain a lower proportion of radioactive isotope and the *measured neck uptake will be low*, even though absolute iodine uptake and thyroid activity are normal.

The opposite situation, a *falsely high measured neck uptake*, will apply if there is less iodine than normal in the body, as in iodine deficiency (p. 173).

Other Tests Used in the Assessment of Thyroid Disease

Thyroid "scans".—After giving a dose of ^{131}I larger than that used in the neck uptake the areas of activity are "mapped" by a gamma camera or by scanning the gland with a counter. In this way overactive "hot" or underactive "cold" areas can be detected.

Thyroid antibodies.—A group of diseases, the autoimmune thyroid diseases, are associated with circulating thyroid antibodies. The two antibodies most useful clinically are:

complement fixing microsomal antibody;

antibody to thyroglobulin.

Antibodies may be detected in high titre in most cases of Hashimoto's thyroiditis and many cases of idiopathic hypothyroidism. Their role, if any, in the aetiology of these conditions is uncertain. A high incidence of antibodies in thyrotoxicosis reflects the focal thyroiditis often seen histologically in the gland and correlates with development of postoperative hypothyroidism in these cases.

DISORDERS OF THE THYROID GLAND

The commonest presenting features in patients with thyroid disease are:

hyperthyroidism—due to excessive thyroid hormone secretion;

hypothyroidism—due to deficient thyroid hormone secretion;

goitre (thyroid enlargement), with or without disturbance of secretion.

Excess or deficiency of circulating thyroid hormone produces characteristic clinical changes. Thyroid disease may exist without hyper- or hypofunction. These changes, and the causes of the abnormal thyroid activity, will be considered only in so far as they influence the results of thyroid function tests.

HYPERTHYROIDISM (THYROTOXICOSIS)

The syndrome produced by sustained excess of thyroid hormone may be easily recognised or may remain unsuspected for a long time. The key feature is a speeding up of metabolism. Clinical features include tremor, tachycardia (and sometimes arrhythmias), weight loss, tiredness, sweating and diarrhoea. Anxiety and emotional symptoms may be prominent. In some cases a single feature predominates, such as weight loss, diarrhoea, tachycardia or atrial fibrillation.

The *causes* of hyperthyroidism are:
Graves' disease (commonest form);
toxic multinodular goitre or single functioning nodule (occasionally an adenoma);
ingestion of thyroid hormones.
Rare causes:
secretion of a thyrotrophin by tumours of trophoblastic origin;
struma ovarii (thyroid tissue in ovarian teratoma);
administration of iodine to a subject with iodine-deficiency goitre;
TSH-secreting tumour of the pituitary.

Graves' disease occurs at any age and is commoner in females. It is characterised by one or more of the following:
hyperthyroidism due to diffuse hyperplasia of the thyroid;
exophthalmos;
localised (pretibial) myxoedema.

Graves' disease is one of the autoimmune thyroid diseases. Thyroid antibodies (p. 166) are detectable in some cases in addition to *thyroid-stimulating immunoglobulins*, which probably cause hyperfunction.

Toxic nodules, single or multiple, in a nodular goitre may secrete thyroid hormones *autonomously*. *TSH is suppressed* by negative feedback and the rest of the thyroid tissue is inactive. Such "hot" areas can be detected by a thyroid scan. Toxic nodules are found more commonly in the older age groups, and the patient may present with only one of the features of hyperactivity, most commonly cardiovascular symptoms.

The *feature common to all forms of hyperthyroidism* (with the exception of the *very* rare TSH-secreting pituitary tumour) is that *secretion is not dependent on TSH from the pituitary*.

Diagnosis of Hyperthyroidism

As the commonest abnormality in hyperthyroidism is a raised plasma T_3 concentration, it is logical to measure this hormone first. However T_4 (and FTI) estimation is more generally available and if these are unequivocally raised in a clinically thyrotoxic patient no further tests are necessary.

If the results are equivocal or if there is clinical doubt plasma T_3 must be measured (and interpreted in conjunction with a measurement of binding sites). In a significant proportion of patients with hyperthyroidism T_4 levels may be normal, at least in the early stages of the disease.

Rarely, both T_4 and T_3 levels are within "normal limits" or only mildly raised. What we require now is proof that these levels are in fact abnormally high for that particular patient. If they are, pituitary secretion of TSH will be suppressed. Unfortunately current TSH methods cannot distinguish "low normal" from "abnormally low" levels. Either of the following tests may help. The *TRH test* will show that plasma TSH levels *fail to rise normally* after TRH administration, because there is negative feedback from the abnormal T_4 levels. Alternatively a T_3-*suppression test* may be done. The radioactive iodine neck uptake is measured before and after a short course of tri-iodothyronine in doses about twice that of normal thyroid hormone output (that is 120 μg/day for six days). This will suppress any TSH secretion. If the neck uptake does not fall it indicates that TSH secretion was already suppressed, confirming abnormal T_4 levels.

The response of a patient being treated for hyperthyroidism may be monitored by plasma T_4 or T_3 estimation.

HYPOTHYROIDISM

Hypothyroidism is due to suboptimal circulating levels of one or both thyroid hormones. The condition develops insidiously and, in its early stages, symptoms are non-specific. Many of the features of frank hypothyroidism are, not unexpectedly, the opposite of those in hyper-thyroidism. There is a generalised slowing down of metabolism, with mental dullness, physical slowness and weight gain. The skin is dry, hair falls out and the voice becomes hoarse. The face is puffy and the subcutaneous tissues thickened (myxoedema).

In the most severe form myxoedema coma, with profound hypo-thermia, may develop. In children, growth may be impaired and con-genital thyroid hormone deficiency leads to cretinism. The fully de-veloped case of myxoedema is easily recognised, but lesser degrees of hypothyroidism with slow onset may be commoner than previously thought.

The causes of hypothyroidism are:

primary.—Disease of the gland (autoimmune thyroiditis):
   idiopathic hypothyroidism
   Hashimoto's disease.
The result of treatment:
   post-thyroidectomy
   post-^{131}I therapy for hyperthyroidism.
Uncommon causes:
   dyshormonogenesis
   exogenous goitrogens and drugs.

secondary to TSH deficiency, due to anterior pituitary or hypothalamic disease. This is much less common than primary hypothyroidism.

The essential difference between primary and secondary hypothy-roidism is that TSH levels are raised in the former and low in the latter.

Hashimoto's disease and "primary" hypothyroidism are now consid-ered to be different manifestations of the same basic disorder. There is progressive destruction of thyroid tissue and circulating thyroid antibodies are present. The term "dyshormonogenesis" includes the congenital deficiencies of the enzymes involved in thyroxine synthesis. Deficiencies have been described at each of the stages shown in Fig. 15 (p. 160), with differing biochemical features. The end result in each case is reduced thyroxine synthesis and hypothyroidism. Goitre is almost invariable in dyshormonogenesis, due to continuous TSH stimulation. The commonest form is failure to incorporate iodine into tyrosine (stage (c) of Fig 15).

If secondary hypothyroidism is long-standing, irreversible atrophic changes occur in the thyroid gland.

Diagnosis of Hypothyroidism

Again the first step is to measure plasma T_4 and perhaps to assess free binding sites. If hypothyroidism is likely, despite a T_4 that is low "normal" or only moderately reduced, it is desirable to estimate plasma TSH.

(a) If the T_4 is unequivocally low it is necessary to establish whether primary or secondary hypothyroidism exists.

(b) The lower range of "normal" is poorly defined. Levels within the "normal" range may in fact be suboptimal for that particular patient. Because of the T_4-pituitary feedback however, this fact will be established by the resultant increased TSH secretion. Early hypothyroidism may be a difficult clinical diagnosis and plasma TSH should be measured in all suspect cases.

These two estimations should confirm the diagnosis in most patients with primary hypothyroidism. In doubtful cases and in suspected secondary hypothyroidism a TRH test should be done. The possible responses are:

an exaggerated rise of plasma TSH with levels *higher at 20 than at 60 minutes* is seen in *primary hypothyroidism*;

an exaggerated rise with TSH levels *higher at 60 than at 20 minutes* is seen in cases of *secondary hypothyroidism* due to *hypothalamic* dysfunction. This delayed response probably indicates that the TSH-secreting cells of the pituitary have been without normal TRH stimulation;

a subnormal response is seen in cases of *secondary hypothyroidism* of *pituitary* origin.

Exceptions to these patterns may occur.

The response of a patient being treated with thyroid hormones is followed by plasma T_4 and TSH estimations.

Special problems in the diagnosis of hypothyroidism.—Two situations warrant special consideration.

Neonatal hypothyroidism.—The incidence of neonatal hypothyroidism varies in different populations from 1/5 000 to 1/10 000. It is therefore commoner than many inborn errors (p. 356) for which routine screening is advocated. The signs of hypothyroidism in the newborn are minimal but if treatment is not started within the first few months permanent brain damage results. Because of the high but variable plasma T_4 values in the first month of life (due to the high maternal oestrogen levels in late pregnancy, which increase TBG levels in both mother and infant) interpretation of this test is difficult. It is rare for the FTI to correct for an extreme rise in TBG. Plasma TSH levels are usually raised however and this test should be done in all suspected cases at about a week after delivery.

A patient already on treatment.—No patient should be given thyroid replacement therapy until the clinical diagnosis has been confirmed and documented by laboratory tests. If it is necessary to confirm the diagnosis in a patient already on treatment but without adequate laboratory confirmation of the disease, plasma T_4 and TSH levels should be measured and, unless hypothyroidism is confirmed by a low T_4 and raised TSH, treatment stopped. The tests should then be repeated at intervals but the diagnosis cannot be excluded until at least 6 weeks after cessation of therapy.

The results of thyroid function tests are summarised in Table XVI.

TABLE XVI

RESULTS OF THYROID FUNCTION TESTS

	True T_4 abnormalities			TBG abnormalities	
		Hypothyroidism		Raised levels	Low levels
	Hyperthyroidism	Primary	Secondary		
Plasma T_4	↑	↓	↓	↑	↓
Resin-uptake test	↑	↓	↓	↓	↑
Free-thyroxine index	↑	↓	↓	N	N
Plasma TSH levels	↓	↑	↓	N	N
Response to TRH	Absent	Exaggerated	See text	Normal	Normal

EUTHYROID GOITRE

Thyroxine synthesis may be impaired by iodine deficiency, by drugs such as para-aminosalicylic acid, or possibly by minor degrees of enzyme deficiency. The consequent slight reduction of circulating thyroxine level results, by the feedback mechanism, in increased TSH secretion. This stimulates thyroxine synthesis and levels are therefore maintained in the normal range. As a result, the thyroid enlarges (goitre), but hypothyroidism is avoided. In areas with low iodine content of the soil, iodine deficiency used to be common (endemic goitre).

Plasma T_4 levels are usually normal. Low levels in clinically euthyroid patients are often accompanied by raised plasma T_3 levels (see p. 162). TSH levels are often normal.

Inflammation of the thyroid (thyroiditis), whether acute or subacute, may produce marked but temporary aberrations of thyroid function tests. These conditions are relatively uncommon.

Other Biochemical Findings in Thyroid Disease

Cholesterol level.—In *hypothyroidism* the synthesis of cholesterol is impaired but its catabolism is even more impaired and plasma cholesterol levels are high. This may help both in diagnosis and in following the effect of treatment but many other factors cause raised cholesterol levels. Although, statistically, low cholesterol levels occur with *hyperthyroidism*, plasma cholesterol estimation is of little value in the individual case because of overlap with the normal range.

Several clinical tests of thyroid function, such as Achilles tendon reflex duration and ECG, depend on the peripheral action of thyroxine (on muscle contraction and on cardiac muscle).

Hypercalcaemia is very rarely found with severe thyrotoxicosis. There is an increased turnover of bone, probably due to direct action of thyroid hormone (p. 236).

Glucose tolerance tests may show a "lag" curve in thyrotoxicosis probably due to rapid absorption. In hypothyroidism, by contrast, the glucose tolerance curve may be flat. This test is of little use in diagnosis.

Plasma creatine kinase (CK) levels are often raised in hypothyroidism but, again, the estimation does not help diagnosis. CK levels may also be raised in thyrotoxic myopathy.

Urinary total 17-oxogenic steroid and **17-oxosteroid** values are low in untreated hypothyroidism, and this finding does not necessarily indicate a pituitary origin of the disease.

SUMMARY

1. There are two circulating thyroid hormones (apart from calcitonin) —thyroxine and tri-iodothyronine. Their synthesis depends on an adequate supply of iodine, and is controlled by circulating thyroid hormone levels via TSH from the anterior pituitary gland.

2. Circulating thyroid hormone levels may be assessed by T_4 or PBI estimation and assessment of free binding sites. Changes in TBG may give abnormal results. The free thyroxine index is an expression of thyroid status regardless of TBG changes.

3. An excess of circulating thyroid hormone produces the syndrome of hyperthyroidism. This may occur in Graves' disease or may be due to a functioning nodule of the thyroid.

The diagnosis of hyperthyroidism is confirmed by measuring plasma T_4 (and, if necessary, binding sites) and, if the result is equivocal, plasma T_3. In doubtful cases a TRH test in which there is a subnormal rise in plasma TSH *in the presence of high levels of thyroid hormones* may clinch the diagnosis.

4. A decreased circulating thyroid hormone level produces the syndrome of hypothyroidism. This may be primary, due to disease of the thyroid gland, or be secondary to pituitary or hypothalamic disease. TSH levels are raised in the first group and low in the second.

The diagnosis of hypothyroidism is made by measuring the plasma T_4 and TSH. Doubtful cases are confirmed by the TRH test, which may also distinguish pituitary from hypothalamic causes of secondary hypothyroidism.

5. Euthyroid goitre represents compensated thyroid disease. Thyroid function tests may be normal.

6. Radioiodine neck uptake tests measure the activity of the thyroid gland. Results are influenced by a variety of factors other than thyroid activity, and the estimation may be unreliable for the primary diagnosis of thyroid disease.

FURTHER READING

IRVINE, W. J., and TOFT, A. D. (1976). The diagnosis and treatment of thyrotoxicosis. *Clin. Endocr.*, 5, 687.

STEPANAS, A. V., MASHITER, G., and MAISEY, M. N. (1977). Serum triiodothyronine: clinical experience with a new radioimmunoassay kit. *Clin. Endocr.*, 6, 171.

BURR, W. A., RAMSDEN, D. B., EVANS, S. E., HOGAN, T., and HOFFENBERG, R. (1977). Concentration of thyroxine-binding globulin: value of direct assay. *Brit. med. J.*, 1, 485.

BAYLISS, R. I. S., and HALL, R. (1975). The thyroid gland. *Medicine* (2nd Series) 7, 297.

HALL, R., SMITH, B. R., and MUKHTAR, E. D. (1975). Thyroid stimulators in health and disease. *Clin. Endocr.*, 4, 213.

BURGER, H. G., and PATEL, Y. C. (1977). Thyrotropin releasing hormone—TSH. *Clin. Endocr. Metab.*, 6, 83.

APPENDIX

Many factors alter thyroid function tests so that the results no longer reflect the thyroid status of the patient. The commoner (and avoidable) ones are listed below with the time periods that should elapse before measuring the PBI or radioactive iodine uptake. In the case of drugs used in the treatment of thyroid disease the PBI reflects the resultant thyroid status of the patient whereas the neck uptake does not. The times given are approximate—for greater detail on individual drugs or radio-opaque media see:

DAVIS, P. J. (1966). *Amer. J. Med.*, **40**, 918 (for PBI).

MAGALOTTI, M. F., HUMMON, I. F., and HIERSCHBIEL, E. (1959). *Amer. J. Roentgenol.*, **81**, 47 (for radio-iodine uptake).

TABLE XVII

SUBSTANCES INTERFERING WITH THYROID FUNCTION TESTS

Iodine Interference	PBI	Neck uptake
Dietary iodide (fish, iodised salt, etc.)	—	3 days
Drug iodide such as KI (suspect in cough mixtures and amoebicides such as "Enterovioform")	2 weeks	8 weeks
Iodine-containing contrast media		
(a) Rapidly cleared (for example, IVP, angiogram, cholecystogram, salpingogram, bronchogram)	8 weeks	8 weeks
(b) Slowly cleared (for example, myelogram)	1 year + (may be indefinite)	1 year +
Interference with Thyroxine Metabolism		
(a) Blocking uptake (for example, perchlorate, thiocyanate, nitrate)	—	2 weeks
(b) Inhibiting organic binding—thioureas, thiouracils	—	6 months
(c) Thyroid hormones, PAS	—	8 weeks

TRH test (Hall, R., Ormston, B. J., Besser, G. M., Cryer, R. J., and McKendrick, M. *Lancet*, 1972, **1**, 759).

(a) A basal blood sample is taken.

(b) 200 µg of TRH in 2 ml saline is injected rapidly intravenously.

(c) Further blood samples are taken 20 and 60 minutes after the TRH injection.

TSH is measured on all samples.

The maximum response occurs at 20 minutes. Published normal values are:

Males   5·9–18·3 mU/l
Females 6·8–20·0 mU/l

Chapter IX

CARBOHYDRATE METABOLISM
AND ITS INTERRELATIONSHIPS

IN most parts of the world carbohydrate is the major source of energy intake. Under normal circumstances starch is the main dietary carbohydrate, disaccharides contribute significantly and monosaccharides are a minor component of the diet.

CHEMISTRY

The basic units of carbohydrate are *monosaccharides*. Physiologically the most important groups are the *hexoses* (six carbon atoms) and *pentoses* (five carbon atoms).

The main hexoses of physiological importance, *glucose, fructose* and *galactose*, are all reducing sugars and therefore reduce Benedict's solution (and Clinitest tablets). Some important pentoses are *ribose* (in RNA) and *deoxyribose* (in DNA).

The common *disaccharides* are *sucrose* (fructose + glucose), *lactose* (galactose + glucose) and *maltose* (glucose + glucose). Lactose and maltose are reducing sugars, sucrose is not.

Polysaccharides are long-chain carbohydrates. *Starch*, found in plants, is a mixture of amylose (straight chains) and amylopectin (branched chains). *Glycogen*, found in animal tissue, has a highly branched structure. Both these polysaccharides are composed of glucose subunits.

BIOCHEMISTRY

THE IMPORTANCE OF EXTRACELLULAR GLUCOSE LEVELS

The brain is the tissue most vulnerable to hypoglycaemia. Cerebral cells derive their energy from aerobic metabolism of glucose. They *cannot*:

store glucose in significant amounts;

synthesise glucose;

metabolise substrate other than glucose and ketones. The latter usually provide very little of the energy requirements of the brain because normal plasma ketone levels are very low;

extract enough glucose for their needs from the extracellular fluid at low concentrations, because entry of glucose into the brain is not facilitated by hormones.

It is therefore clear that the brain is very dependent on extracellular concentrations of glucose for its energy supply, and that hypoglycaemia is likely to impair cerebral function. As we saw on p. 35, hyperglycaemia —especially of rapid onset—can also cause cerebral dysfunction by its effect on extracellular osmolality. In normal subjects the plasma (extracellular) glucose concentration usually remains between 4·5 and 10 mmol/1, despite the intermittent load entering the body from the gastro-intestinal tract.

The relatively low upper level of plasma glucose is also important in minimising loss of this important energy source from the body. The renal tubule reabsorbs almost all glucose from the glomerular filtrate up to a concentration of about 10 mmol/1, so that normal urine is nearly glucose free, even after a carbohydrate meal. As we shall see, this retained glucose can be stored until required.

MAINTENANCE OF EXTRACELLULAR GLUCOSE LEVELS

Plasma glucose concentrations depend on the balance of glucose entering and leaving the extracellular compartment. As little is normally lost unchanged from the body, maintenance of the relatively narrow range of 4·5 to 10 mmol/1 in the face of widely varying input from the gastro-intestinal tract is most likely to depend on exchange with cells. If we understand the interaction between tissues which effects this control we can also explain pathological disturbances of carbohydrate metabolism (including ketosis and lactic acidosis).

The Liver

The liver is the most important single organ in ensuring a constant supply of fuel for other tissues, including the brain, under a wide variety of conditions. It is also of importance in helping to control plasma glucose concentration post-prandially. It is well-adapted to these roles for many reasons:

1. Portal venous blood leaving the absorptive area of the intestinal wall reaches the liver first. The hepatic cells are in a key position to buffer the hyperglycaemic effect of a high carbohydrate meal (Fig. 18).

2. The liver cell can store some of a temporary excess of glucose as glycogen. The rate of glycogen synthesis from glucose-6-phosphate (G-6-P) may be stimulated by insulin (Fig. 18) which is secreted by the β cells of the pancreas in response to systemic hyperglycaemia.

3. The liver can convert some of a temporary excess of glucose to fatty acids (a process also stimulated by insulin): these are ultimately stored as triglyceride in adipose tissue (Fig. 18 and p. 178).

4. The entry of glucose into hepatic cells (as into cerebral cells) is not affected directly by insulin, but depends on extracellular concen-

trations. The conversion of glucose to G-6-P— the first step in glucose metabolism in all cells—can be catalysed in the liver by the enzyme glucokinase, with a low affinity for glucose relative to that of hexokinase found in most tissues. Glucokinase is induced by the insulin secreted in response to systemic hyperglycaemia. For these reasons proportionately less glucose is extracted by hepatic cells during fasting, when portal levels are low, than after a high carbohydrate meal. This helps to maintain a fasting supply to vulnerable tissues such as brain.

5. Under aerobic conditions the liver can synthesise glucose and glycogen by *gluconeogenesis* and *glycogenesis*, using glycerol, lactate or the carbon chains resulting from deamination of most amino acids (most of which are products of other tissues). The liver is therefore an important organ for storing energy as glycogen.

6. The liver contains the enzyme, glucose-6-phosphatase, which, by hydrolysing the G-6-P yielded by glycogen breakdown (glycogenolysis) or by gluconeogenesis, releases glucose and helps to maintain extracellular fasting levels. Hepatic glycogenolysis is stimulated by the hormone glucagon, secreted by the α-cells of the pancreas in response to hypoglycaemia.

7. During fasting the liver can convert fatty acids released from adipose tissue to ketones which can be used by other tissues, including brain, as fuel when glucose is in short supply.

This combination of factors is unique to the liver. The renal cortex is the only other tissue capable of gluconeogenesis, and of converting G-6-P to glucose: its glycogen storage capacity is small. The gluconeogenic capacity of the kidney is probably mainly of importance in hydrogen ion homeostasis (p. 89); the ability of the cells to release the resultant glucose is an obvious necessity if it cannot be stored, but is probably no more than a minor "spin-off" in the overall economy of the body.

Other tissues can store glycogen to a greater or lesser extent, but can only use it locally as they do not contain glucose-6-phosphatase; this glycogen plays no part in maintaining plasma glucose levels.

Systemic Effects of a Glucose Load (Fig. 18)

We have seen that the liver modifies the potential hyperglycaemic effect of a high carbohydrate meal by extracting relatively more glucose from the portal blood than in the fasting state. Some glucose, however, passes the liver unchanged and the rise in systemic concentration stimulates the β-cells of the pancreas to secrete insulin, which may further stimulate hepatic and muscle glycogenesis. More importantly, *entry of glucose into adipose tissue and perhaps into muscle cells*, unlike that into liver and brain, *is stimulated by insulin*, and plasma glucose falls rapidly to near fasting levels. Even after oral ingestion of 50 g of

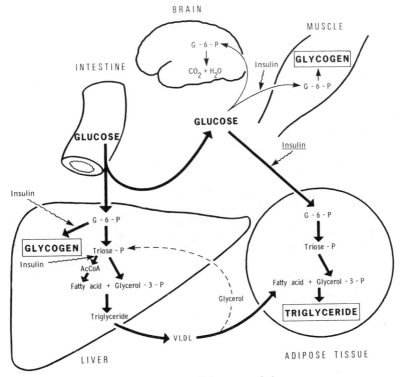

FIG. 18.—Postprandial storage of glucose.

pure glucose, as in the glucose tolerance test, the level 2 hours later is below 6·7 mmol/l (120 mg/dl): failure to fall to this level in 2 hours indicates relative or absolute insulin deficiency (diabetes mellitus, p. 187). Conversion of intracellular glucose into G-6-P in adipose and muscle cells is catalysed by the enzyme *hexokinase which, because its affinity for glucose is greater than that of hepatic glucokinase*, ensures that glucose enters the metabolic pathways at systemic concentrations which are lower than those in portal blood.

Both muscle and adipose tissue store the excess post-prandial glucose, but the mode of storage and the function of the two types of cell is very different: by examining the interrelationship of each of these tissues with the liver many of the disturbances of carbohydrate metabolism can be explained.

KETOSIS

Adipose Tissue and the Liver

Adipose tissue contains the most important long-term energy store in the body. Excessive utilisation of fat stores is associated with ketosis.

Adipose tissue, in conjunction with the liver, converts excess glucose to triglyceride and stores it in this form rather than as glycogen. The component fatty acids are derived from glucose entering the liver, and the component glycerol from glucose entering adipose tissue cells.

In the *liver* glycolysis yields:

glycerol-3-phosphate from triose phosphates, which are intermediates on the glycolytic pathway;

fatty acids from acetyl CoA.

The triglyceride formed after condensation of glycerol-3-phosphate and fatty acid is transported to adipose tissue in VLDL, where it is hydrolysed by lipoprotein lipase (p. 218). The released fatty acids (of hepatic origin) condense with glycerol-3-phosphate derived from glucose entering *adipose tissue* under the influence of insulin, and the resultant triglyceride is stored. Far more energy can be stored as triglyceride than as glycogen.

During *fasting*, when exogenous glucose is unavailable, endogenous adipose tissue triglyceride is reconverted to free fatty acids (FFA) and glycerol by lipolysis (Fig. 19). Both are transported to the liver where glycerol enters the gluconeogenic pathway at the triose phosphate stage, and the resultant glucose can be released into the blood stream at a time when plasma glucose concentration would otherwise tend to fall. The liver converts the FFA to acetyl CoA, and can also form *acetoacetic acid* by enzymatic conversion of two moles of acetyl CoA; acetoacetic acid can be reduced to *β-hydroxybutyric acid* and decarboxylated to *acetone*. These "ketones" can be used as an energy source by brain and other tissues at a time when glucose is in relatively short supply.

Ketosis therefore occurs when fat stores are the main energy source and may result from *fasting*, or from reduced nutrient absorption due to vomiting. Mild ketosis may occur after as little as the 12 hours' preparation for a glucose tolerance test (and should not then be misinterpreted as diabetic ketosis). After short fasts acidosis is not usually detectable, but after longer periods more hydrogen ions may be produced than can be dealt with by homeostatic mechanisms and plasma bicarbonate levels fall. For some time the plasma glucose concentration is maintained by hepatic mechanisms (p. 176), but in prolonged starvation (as in anorexia nervosa) or in ketotic hypoglycaemia of infancy (p. 203) hypoglycaemia may occur. The brain may suffer less from ketotic than from the same degree of insulin-induced hypoglycaemia: in the former

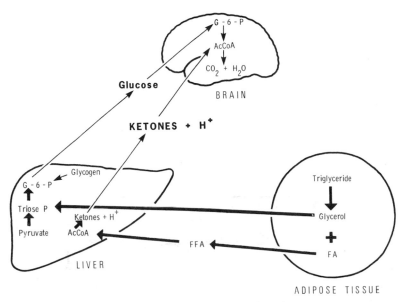

FIG. 19.—Fasting pathways: ketosis.

the brain adapts to ketone metabolism, while in the latter ketone levels are low, depriving the brain of its only non-glucose energy source.

Diabetic ketoacidosis is more severe. Hyperglycaemia is invariable, differentiating it from the ketosis of fasting. The mechanism of the ketosis is the same. In starvation ketosis the supply of glucose to cells of adipose tissue is insufficient for normal glycolysis and lipogenesis. In insulin deficiency the intracellular glucose deficiency is due to impaired entry of high extracellular concentrations of glucose into these cells: the high extracellular levels are misleading as an index of intracellular events. Ketosis reflects the fact that lipolysis is the predominant pathway.

LACTATE PRODUCTION AND LACTIC ACIDOSIS

Striated Muscle and the Liver

Glucose enters the muscle post-prandially under the influence of insulin and is stored as glycogen. Because of the absence of glucose-6-phosphatase this glycogen can only supply local needs. Quantitatively muscle glycogen stores are second only to those in liver.

Muscular contraction (Fig. 20).—During muscular activity glycogenolysis is stimulated by adrenaline, and the resultant G-6-P is drawn upon by rapid glycolysis and by oxidation in the tricarboxylic acid cycle

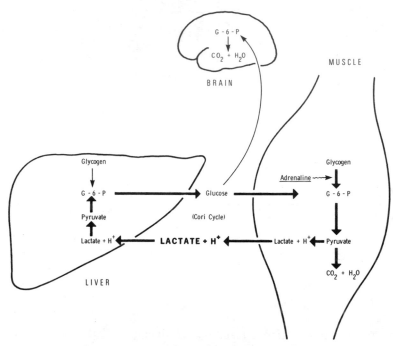

FIG. 20.—Pathways during muscular contraction.

to supply the necessary energy. Under these conditions the rate of glycolysis may outstrip the availability of oxygen, and glycolytic products then exceed the immediate aerobic capacity to oxidise them.

The overall reaction for anaerobic glycolysis is

$$\text{Glucose} \rightarrow 2\ \text{Lactate}^- + 2\text{H}^+$$

The lactate is carried in the blood stream to the liver where it can be used for gluconeogenesis, providing further glucose for the muscle (Cori cycle). During gluconeogenesis H^+ is also reutilised. Under aerobic conditions the liver is predominantly a lactate-consuming organ.

This physiological accumulation of lactic acid during muscular contraction is a temporary phenomenon, and rapidly disappears at rest, when slowing of glycolysis allows aerobic processes to "catch up".

Pathological Lactic Acidosis

Lactic acid produced by anaerobic glycolysis may be oxidised to CO_2 and water in the TCA cycle, or be reconverted to glucose by gluconeogenesis in the liver. Both the TCA cycle and gluconeogenesis require oxygen: *glycolysis is the only pathway that does not need oxygen.*

Pathological accumulation of lactate might be due to increased production, or to decreased utilisation.

Production may be increased by:
increased rate of anaerobic glycolysis.

Utilisation may be decreased by:
impairment of the TCA cycle;
impairment of gluconeogenesis.

The *clinical syndromes* associated with lactic acidosis usually involve more than one of these factors.

Tissue hypoxia, due to the poor tissue perfusion of the "shock syndrome", is the commonest and most important cause of lactic acidosis (Fig. 21). Under these circumstances tissue hypoxia causes hyperlactataemia because:

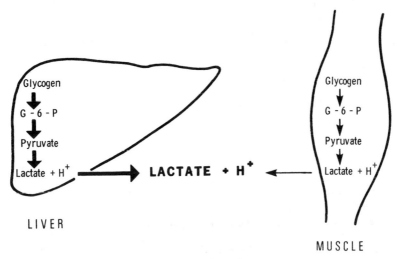

LIVER

MUSCLE

FIG. 21.—Hypoxic pathways.

the TCA cycle cannot function anaerobically and oxidation of pyruvate and lactate to CO_2 and water is impaired;

hepatic and renal gluconeogenesis from lactate cannot occur anaerobically;

anaerobic glycolysis is stimulated because the falling ATP levels cannot be regenerated by the TCA cycle, as they are in aerobic conditions.

The combination of impaired gluconeogenesis and increased anaerobic glycolysis converts the liver from a lactate and H^+ consuming organ to one generating large amounts of lactic acid.

In severe hypoxia (such as after cardiac arrest) acidosis is severe.

This hypoxic syndrome may also complicate diabetic ketoacidosis with dehydration.

Other causes of lactic acidosis are outlined below. In all of them there is imbalance between pyruvate production and utilisation.

Phenformin, an antidiabetic drug, can cause severe lactic acidosis. Phenformin inhibits both the TCA cycle and gluconeogenesis.

Severe illnesses such as leukaemias may be associated with lactic acidosis. A variety of factors may be involved, including poor tissue perfusion and increased anaerobic glycolysis by malignant tissue.

Infusion of fructose may cause lactic acidosis. Unlike glucose it enters hepatic cells at low blood concentration, and the rate of glycolysis may exceed the capacity of the TCA cycle in even a mildly hypoxic liver.

In *glucose-6-phosphatase deficiency* (von Gierke's disease, p. 202) the rate of glycolysis increases when G-6-P cannot be converted to glucose in liver and kidney.

The treatment of lactic acidosis is that of the cause, and of the acidosis. Measurement of blood lactate levels is only rarely necessary, because it is the acidosis that is dangerous: lactate itself is harmless.

HORMONES CONCERNED WITH GLUCOSE HOMEOSTASIS

Many of the important effects of these hormones have already been described. The more detailed actions are summarised in Table XVIII.

Insulin is synthesised in the β-cells of the pancreas as the protein proinsulin. Release into the ECF follows hydrolysis to a smaller, physiologically active, double-chain peptide (51 amino acids), and the most important stimulus to this is *hyperglycaemia*. Amino acids such as leucine have a less important effect (but see leucine sensitivity, p. 203). Its important *hypoglycaemic action* is due to stimulation of entry of glucose into cells of adipose tissue (p. 178), and is contributed to by its actions on intermediary metabolism.

Growth hormone (GH), glucocorticoids, adrenaline and glucagon all have a *hyperglycaemic action*. GH, *glucocorticoids and adrenaline* are discussed in more detail on pp. 119 and 133 respectively. Physiologically secretion is *stimulated by stress*, and pathological excess accounts for the hyperglycaemia of the clinical syndromes acromegaly (GH) (p. 119), Cushing's syndrome (glucocorticoids) (p. 136) and phaeochromocytoma (adrenaline and noradrenaline) (p. 430). *Glucagon* is a single-chain polypeptide synthesised by the α-cells of the pancreatic islets and secretion is *stimulated by hypoglycaemia*.

Some of the principles that we have discussed are applied in parenteral feeding. A brief discussion of this subject will be found in the Appendix, p. 212.

TABLE XVIII

ACTIONS OF HORMONES ON INTERMEDIARY METABOLISM

	Insulin	Glucagon	Growth Hormone	Glucocorticoids	Adrenaline
Carbohydrate metabolism **(a) in liver**					
—glycolysis	+ +				
—glycogenesis	+ +				
—glycogenolysis	—				+
—gluconeogenesis		+ +		+	+
(b) in muscle					
—glucose uptake	+ +		—	—	
—glycogenesis	+		+		+
—glycogenolysis	—				
Protein —synthesis	+		+		
—breakdown	—			+	
Fat —synthesis	+				
—lipolysis	—		+	+	+
Secretion—stimulated by	Hyperglycaemia Amino acids Glucagon GIT hormones Adrenaline Fasting	Hypoglycaemia Amino acids Fasting	Hypoglycaemia Other	Hypoglycaemia Stress	Stress
—inhibited by		Insulin			
Results	Uses and stores available glucose	Provides glucose	Spares glucose	Provides glucose	
			Provide FFA as alternative fuel		
Blood FFA levels	Fall		Rise		
Blood glucose levels	Fall		Rise		

+ stimulates   — inhibits

INTERPRETATION OF BLOOD AND URINE GLUCOSE LEVELS

Glucose in Urine

If glucose is detectable in the urine by the usual qualitative tests (see Appendix) there is, by definition, glycosuria. Because it is almost completely reabsorbed in the proximal tubule, urinary glucose is normally

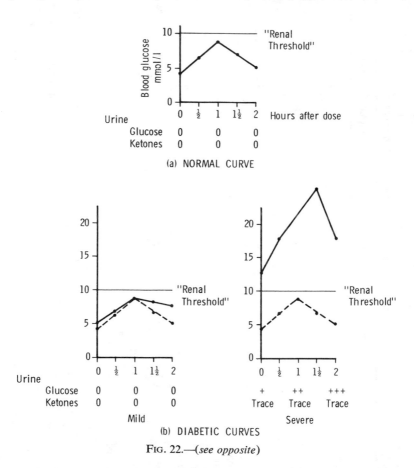

FIG. 22.—(*see opposite*)

undetectable. As blood, and therefore glomerular filtrate, levels increase, progressively more escapes absorption and glycosuria usually occurs at blood levels above 10 mmol/1. This figure is often called the "renal threshold". As blood glucose does not normally exceed this level at any time of day, glycosuria often, but not always, indicates hyperglycaemia.

However, as with most substances, the proximal tubule absorbs a certain proportion of glucose presented to it per unit time and the above "threshold" is only valid if the glomerular filtration rate is normal. A low GFR, whether due to glomerular damage or to circulatory insufficiency, reduces the rate of fluid delivered to the tubules and provided tubular function is normal, higher concentrations of glucose (but the same amount/minute) may be almost completely reabsorbed:

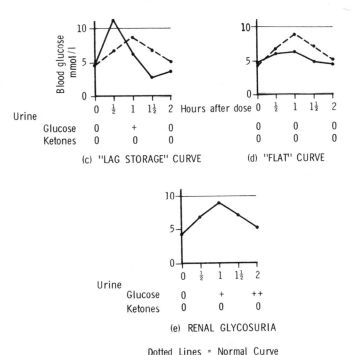

FIG. 22.—Glucose tolerance curves.

glycosuria may then not occur even with severe hyperglycaemia. Conversely, with proximal tubular abnormality, whether acquired or inherited, glycosuria may appear at blood glucose levels below 10 mmol/l (*renal glycosuria*).

It must be noted that tests such as Clinitest also detect reducing substances other than glucose. Those are considered in the Appendix and are of particular importance in the neonatal period.

The Oral Glucose Tolerance Test (see Appendix)

The measurement of blood glucose at intervals after a standard load of glucose, tests all the homeostatic mechanisms described earlier. Simultaneous testing for urinary glucose indicates the "renal threshold". The test is used, or misused, most often in the diagnosis of diabetes mellitus. Several patterns of response are described (Fig. 22).

(a) The "normal" GTT (Fig. 22a and Table XIX).—In Britain the usual criteria employed are those of the Medical and Scientific Section of the British Diabetic Association, for a 50 g glucose load and using whole blood. Table XIX compares these with the Fajans-Conn criteria

TABLE XIX

Upper Limits of Normal Glucose Tolerance (mmol/l with mg/dl in brackets)

	Fasting	1 hour	1½ hours	2 hours
1. British Diabetic Association:				
Whole blood (capillary)		10·0 (180*)		6·7 (120)
Whole blood (venous)		8·9 (160*)		6·1 (110)
Plasma or serum		10·3 (185)		7·8 (140)
2. Fajans-Conn:				
Whole blood (venous)	5·6 (100)	8·9 (160)	7·8 (140)	6·7 (120)
Plasma or serum	6·4 (116)	10·3 (185)	9·0 (162)	7·8 (140)

Note:
1. * This level is the upper limit at any time during the test, but the highest level is found usually at 1 hour.
2. It is *not* necessary that the level at 2 hours should return to that of the fasting patient.

commonly used in America which, despite a larger glucose load, are basically similar. It is important to note that these limits were established in young, active, healthy individuals.

In many cases, if the test is prolonged for 3 to 4 hours the level of glucose will be found to fall below the fasting level and return to normal at half to one hour later. This is considered further on p. 199.

(b) Impaired glucose tolerance (Fig. 22b).—Blood glucose levels exceeding those outlined in Table XIX are abnormal, that of the 2-hour specimen being the most important. Glucose intolerance occurs in diabetes mellitus *but may also be found* in people *confined to bed* and therefore inactive, in *obese* individuals and in *acute and chronic illness*. Glucose tolerance decreases with increasing age and is impaired by a carbohydrate-poor diet. A number of drugs such as thiazide diuretics, diphenylhydantoin, phenylephrine and oral contraceptives (p. 424) may occasionally mildly impair glucose tolerance. If the peak concentration exceeds the renal threshold for glucose, glycosuria will be present. Note that, because of the delay in urine reaching the bladder, maximum glycosuria is found in the next specimen *after* the highest blood glucose concentration is reached.

The limitations of the GTT in the diagnosis of diabetes are discussed on p. 191.

(*c*) **Lag storage curve** (Fig. 22*c*).—The peak blood glucose level may be higher than normal but the 2-hour value is within normal limits, or often low. This implies a delay in hepatic storage ("lag") without necessarily impairment of insulin response. The hypoglycaemia may be due to outpouring of insulin in response to the initial high glucose levels. Such a curve may be found:

in apparently normal individuals (reactive hypoglycaemia);

after gastrectomy or gastro-jejunostomy when rapid entry of glucose into the intestine leads to rapid absorption and sudden outpouring of insulin with "overswing";

in very severe liver disease, probably due to a decreased glycogenesis. This is rarely of diagnostic value;

rarely in thyrotoxicosis, probably due to rapid absorption of glucose.

If the highest glucose concentration exceeds the renal threshold, glycosuria will occur in the *next* specimen.

(*d*) **Flat glucose tolerance curve** (Fig. 22*d*).—Blood glucose levels fail to rise normally following a glucose load. This type of curve is sometimes considered to indicate malabsorption. It is, however, not uncommonly found in completely healthy individuals and interpretation is difficult. It can also occur due to lack of glucocorticoid in Addison's disease or due to lack of GH in hypopituitarism. It should never be used as the sole test of malabsorption.

(*e*) **Renal glycosuria** (Fig. 22*e*).—The blood glucose levels follow the normal pattern, but glucose is found in some or all of the urine specimens.

HYPERGLYCAEMIA AND DIABETES MELLITUS

Hyperglycaemia occurs:

in the syndrome of diabetes mellitus;

in patients receiving glucose-containing fluids intravenously;

temporarily in severe stress;

after cerebrovascular accidents.

DIABETES MELLITUS

The syndrome of diabetes mellitus is due to absolute or relative insulin deficiency. There are many causes and severity varies. The classification given below, based on the recommendations of the Medical and Scientific Section of the British Diabetic Association, combines aetiology, clinical features and the response to the oral glucose tolerance test.

Primary, Idiopathic or Essential Diabetes

This is the commonest form and has an hereditary basis.

Primary diabetes has been classified into several *stages* of severity based on the clinical features and the results of the glucose tolerance test. The terms used are those of the British Diabetic Association with the corresponding American Diabetes Association terminology in parentheses.

Potential diabetic (prediabetic).—A subject with a normal glucose tolerance test but a strong family history of the disease (e.g. the identical twin of a diabetic, or the offspring of two diabetic parents), or a woman who has given birth to a child of 4·5 kg or over.

Latent diabetic (suspected diabetic).—A subject with a normal glucose tolerance test, but who has had a diabetic type of glucose tolerance test during pregnancy, after cortisone administration, when obese, or at a time of stress such as during a severe infection.

Asymptomatic diabetic (latent or chemical diabetes).—A subject with a diabetic glucose tolerance test, but without symptoms of diabetes.

Clinical diabetic (overt diabetic).

In the British terminology *prediabetic* is reserved for the period in the life of a diabetic before the diagnosis is made and is thus applied retrospectively.

Secondary Diabetes

Pancreatic—absolute insulin deficiency due to pancreatic destruction in *chronic pancreatitis, pancreatic carcinoma* or *haemochromatosis* or following *total pancreatectomy.*

Presence of insulin antagonists, as in *acromegaly* (excess growth hormone), *Cushing's syndrome* or *steroid therapy* (excess glucocorticoids) and *pregnancy.*

Inhibition of insulin secretion due to excessive catecholamine activity (e.g. *phaeochromocytoma, thyrotoxicosis* or *severe stress* such as myocardial infarction) or associated with potassium depletion, and with diuretic therapy.

Aetiology of Primary Diabetes Mellitus

Diabetes mellitus can be classified on clinical grounds into two broad groups:

insulin-dependent diabetes is commoner in young people (juvenile-onset diabetes). The disease is severe, management is difficult and the patients are liable to develop ketoacidosis and coma. There is an absolute insulin deficiency.

insulin-independent diabetes is usually first diagnosed after the age of 40 (maturity-onset diabetes). Metabolic disturbances are less severe. Plasma insulin levels are lower than in corresponding weight-matched non-diabetic subjects and there is a delay in insulin secretion in response to glucose, maximum levels being reached later and persisting longer.

This classification is only a guide. Mild, obesity-associated diabetes may occur in children, and older subjects may present with, or progress to, a metabolically unstable type of disease. It is unlikely that there is a common aetiology.

In the insulin-independent diabetic obesity is usual and adequate control of the disease may be achieved by simple weight reduction. Obesity may be a precipitating factor, possibly in genetically pre-disposed individuals.

In the insulin-dependent diabetic genetic factors have been postulated but not defined. The histocompatibility antigens HL-A8 and W15 occur more frequently than in non-diabetics and may indicate susceptible individuals. Precipitating factors currently being evaluated include viral infections (particularly Coxsackie B_4) and immunological mechanisms which may damage the pancreatic islets. The role of the other islet hormones, glucagon and somatostatin, are also under investigation. For further information and other theories the interested student should consult the references at the end of the chapter.

The Clinical and Laboratory Features of Primary Diabetes Mellitus

Most of the metabolic changes of diabetes mellitus can be explained as a consequence of insulin deficiency and tend to parallel it in severity. In the *mildest* form insulin production in the fasting state may be adequate for normal metabolism and *fasting blood glucose levels are normal*. Following a meal, or administration of glucose, as in the glucose tolerance test, less glucose than normal is metabolised due to delayed insulin response. The abnormally high peripheral blood glucose levels therefore persist longer than normal, producing a *mild diabetic type of glucose tolerance curve*.

In more *severe* insulin deficiency, there is *fasting hyperglycaemia and glycosuria*. The osmotic diuresis leads to water and electrolyte depletion with the "classical" symptoms of diabetes, polyuria and polydipsia. There may be protein breakdown, causing muscle wasting. If the insulin deficiency is very severe, the patient, if untreated, progresses steadily to ketoacidosis, coma and death (see later).

Abnormalities in *lipid metabolism* secondary to insulin deficiency also occur. Lipolysis is stimulated and plasma FFA levels rise. In the liver

FFA are converted to acetyl CoA and ketones, or are re-esterified into endogenous triglycerides and incorporated into VLDL. Cholesterol synthesis is also increased with an increase in LDL and if insulin deficiency is very severe, such as in uncontrolled diabetes, chylomicrons may accumulate in the blood.

Vascular disease, due in part to the lipid abnormalities, is common, and may present as cerebrovascular disease or peripheral vascular insufficiency.

The biochemical basis of many of the other clinical features of diabetes mellitus is not clear. Abnormalities of small blood vessels particularly affect the retina and kidney. In the latter, nodular glomerulosclerosis (Kimmelstiel-Wilson kidney) may produce the nephrotic syndrome. Infections are common and may aggravate renal and peripheral vascular disease. Diabetic and potentially diabetic women tend to give birth to large babies.

Diagnosis of Diabetes Mellitus

The diagnosis of diabetes mellitus depends on the demonstration of carbohydrate intolerance, either as frank hyperglycaemia or as an impaired response to a glucose load.

Glycosuria.—Testing of the urine is the oldest method of diagnosing diabetes and certainly the simplest. It can, however, lead to false conclusions. Glucose normally appears in the urine at blood levels of about 10 mmol/l (180 mg/dl) and glycosuria is an indirect indicator of hyperglycaemia. While a positive result requires further investigation a negative test by no means excludes diabetes. There are two main reasons for *false negative* results:

the *timing of the urine specimen* tested is important. Early morning specimens of urine are used for many tests because the urine is usually most concentrated at that time of day. However, as it is collected after a period of fasting, the test will be positive for glucose only if the blood glucose level has exceeded 10 mmol/l (180 mg/dl) during this period— that is, in severe diabetes. A specimen of urine passed about one hour after a meal is the most valuable one for a screening test;

in the presence of a *reduced glomerular filtration rate* with normal tubular function (p. 185), ubular reabsorption may remove glucose at concentrations well above the usual "renal threshold" of 10 mmol/l. This is seen particularly in elderly people and in cases of undiagnosed diabetes with renal damage. As the "classical" symptoms of diabetes —polyuria and polydipsia—depend on glycosuria, these patients are "asymptomatic".

False positive tests may be found in *renal glycosuria* (p. 185)

Random blood glucose level.—This test is only of value if glucose levels are very high, when they are almost certainly due to diabetes. A normal result does not exclude the diagnosis. In either case a fasting blood glucose estimation should also be performed.

Fasting blood glucose level.—The blood glucose level after an overnight fast is commonly used to diagnose diabetes. While a glucose level of greater than 6·7 mmol/l (120 mg/dl) is strongly suggestive and one over 7·2 mmol/l (130 mg/dl) almost diagnostic, many diabetics have fasting glucose levels in the normal range. A high fasting blood glucose concentration indicates inadequate insulin output for even basal requirements and no further test is required. A normal fasting level may be present in mild diabetes (Fig. 22*b*) and under these circumstances further tests should be performed.

Two-hour post-glucose blood glucose level.—This is probably the simplest screening test for diabetes mellitus and is in fact a "modified" glucose tolerance test. By two hours after a meal the blood glucose has normally fallen below 6·7 mmol/l (120 mg/dl). Although a meal is often recommended as the load, a more standardised stimulus is that given by 50 g of glucose orally, as in the GTT. A normal result almost certainly excludes clinical diabetes.

Glucose tolerance test (GTT).—There are only two indications for performing a GTT in the diagnosis of diabetes:

Detection of mild diabetes when there are normal fasting and random glucose levels. The test is no more successful than a single estimation two hours after a glucose load.

To ascertain the renal threshold for glucose. Knowledge of this helps the patient to control insulin dosage by testing for urinary glucose and this is the most important indication for the test in a known diabetic *without constant glycosuria.*

There is little point in performing a GTT on a subject with markedly raised fasting or random glucose values and similarly in a patient with fasting hyperglycaemia and glycosuria, no further information on the renal threshold will reward the investigator or the patient for the discomfort of a GTT. Measurement of plasma insulin levels during a routine GTT is of *no* diagnostic value.

As stressed on p. 186 the criteria for "normal" glucose tolerance refer only to young, healthy active individuals. Uncritical application to ill, inactive or elderly people has resulted in overdiagnosis of diabetes with its medical and social implications. It has become clear that less than half the symptom-free subjects with impaired glucose tolerance progress to clinical diabetes on long-term follow up. In many, the glucose tolerance test reverts to normal. The difficulties of interpretation are so great that some workers will only diagnose diabetes mellitus in the presence of symptoms, or of frank fasting hyperglycaemia. It has not been proved that people in whom the only abnormality is a mildly impaired glucose tolerance benefit by drug therapy, although weight reduction is important in the obese. They should be followed at regular intervals in case they develop more severe diabetes.

Principles of Management of Diabetes Mellitus

The *management* of diabetes mellitus will be only briefly considered. Juvenile diabetics require *insulin* in amounts that vary according to circumstances. Maturity onset diabetics are frequently controlled by *diet* with *weight reduction*. In this group insulin secretion may be stimulated by the *sulphonylurea* drugs such as tolbutamide. *Biguanides*, such as phenformin, are also used to lower blood glucose; the mechanism of their action is not completely elucidated, but they inhibit gluconeogenesis either directly or indirectly. The danger of lactic acidosis has been mentioned on p. 182.

"Resistance" to insulin is found in *pregnancy* (possibly due to high levels of sex hormones) and in *infections* and *trauma* (possibly due to glucocorticoid excess). Should any of these conditions develop in a diabetic patient on insulin therapy, the dosage may have to be increased. A patient is usually termed "insulin resistant" if he needs more than 200 U a day, those needing 60 to 200 U being called "insulin insensitive".

Secondary Diabetes

Many cases of *Cushing's syndrome* or *acromegaly* show impaired glucose tolerance on testing, but in only about 20 per cent is the abnormality severe enough to require treatment. It is probable that these cases are latent diabetics. Generally the disease is mild and ketosis and vascular complications are uncommon. Treatment is directed at the primary disease.

After total *pancreatectomy*, or *extensive pancreatic destruction* by disease, insulin requirements are often less than those of a severe primary diabetic, possibly because glucagon-secreting α-cells are lost as well as insulin-secreting β-cells.

ACUTE METABOLIC COMPLICATIONS OF DIABETES

The diabetic patient may develop one of several complications requiring emergency therapy. The main ones are:
diabetic ketoacidosis;
hyperosmolal non-ketotic coma;
hypoglycaemia due to overtreatment (p. 198).

Diabetic Ketoacidosis

Diabetic ketoacidosis is an extension of the metabolic events outlined on p. 189. It may be precipitated by infections or gastro-intestinal upsets with vomiting. Insulin may mistakenly be withheld by the patient who reasons "no food, therefore no insulin". The consequences are due principally to two factors—*glycosuria* and *ketosis.*

Blood glucose levels are usually in the range of 20–40 mmol/l (400–700 mg/dl) but may be considerably higher. This results in marked *glycosuria* which produces an *osmotic diuresis* leading to loss of water and depletion of body electrolytes. Vomiting is frequently present in ketoacidosis and adds to the fluid and electrolyte depletion. The fluid loss involves both intra- and extracellular fluid with reduction of circulating blood volume, reduced renal blood flow and glomerular filtration rate, and severe cellular dehydration. As tubular reabsorption of glucose depends on volume of glomerular filtrate as well as concentration of glucose therein, reduced GFR allows greater proportional reabsorption of glucose. In some cases this may be complete and glucose is no longer present in the urine despite hyperglycaemia. In such cases the plasma urea level is usually raised and there is evidence of haemoconcentration such as elevated haematocrit and total protein levels. Both these findings are the result of *dehydration.*

As a result of intracellular glucose deficiency *lipolysis* is accelerated. More FFA are produced than can be metabolised by peripheral tissues and they are converted in the liver to ketones, or incorporated into endogenous triglycerides. Severe *hyperlipaemia* may be present. The hydrogen ions produced with ketones (other than acetone) are buffered by plasma bicarbonate. Urinary pH drops as the hydrogen ions are secreted by the mechanism described on p. 95: although H^+ secretion is, as always, linked with equimolar generation of bicarbonate, this cannot keep pace with utilisation in buffering and plasma bicarbonate levels fall. The developing metabolic acidosis stimulates the respiratory centre and breathing becomes deeper with a lowering of PCO_2. This respiratory compensation may maintain blood pH within the normal range for a while, but at the cost of further lowering of the plasma bicarbonate to very low levels. The deep sighing respiration (*Kussmaul respiration*) with the odour of acetone on the breath is a classical feature of diabetic ketoacidosis.

Plasma *potassium levels* are usually *raised* before treatment is started, due probably to the acidosis and to reduced entry of glucose into cells and to the low GFR. It must be realised, however, that urinary potassium loss has been high and that there is a *total body deficiency.* This will be revealed during treatment, as potassium re-enters the cells, with resultant, and possibly severe, hypokalaemia.

The losses in a typical case of diabetic ketoacidosis may amount to 6 or 7 litres of water and 300–500 mmol each of sodium and potassium.

The findings in diabetic ketoacidosis are therefore:

clinical — confusion and later coma (hyperosmolality)
   overbreathing (acidosis)
   dehydration (osmotic diuresis)
in blood—hyperglycaemia
   ketonaemia
   acidosis with low total CO_2 (bicarbonate)
   hyperkalaemia (occasionally normokalaemia)
   haemoconcentration (p. 42) and mild uraemia
in urine — glycosuria ⎫ while the urinary flow is adequate
            ketonuria ⎭
   low pH (unless there is renal failure)

Associated findings.—Changes in plasma phosphate levels parallel those of potassium. Low levels may be found for several days. Amylase concentration may be markedly elevated in both urine and plasma, and should not be interpreted as indicating acute pancreatitis. These may be found coincidently, but are of no importance in diagnosis, prognosis or treatment. They are mentioned to avoid the dangers of misinterpretation.

Hyperosmolal Non-ketotic Coma

In diabetic ketoacidosis there is always some hyperosmolality due to the hyperglycaemia, and many of the symptoms are probably related to it; the term "hyperosmolal" coma (or "precoma") is, however, usually confined to a condition in which there is marked hyperglycaemia, but no detectable ketonaemia or acidosis. (The reason for these different presentations is not clear.) It is more common in older subjects. Blood glucose levels are very high, often greater than 50 mmol/l (900 mg/dl). The resulting glycosuria produces an osmotic diuresis, with severe dehydration and electrolyte depletion. Hypernatraemia and uraemia are often present due to predominant water loss, and these aggravate extracellular hyperosmolality. Coma is thought to be the result of consequent cerebral cellular dehydration, which may also lead to hyperventilation: the consequent respiratory alkalosis and slight fall in plasma bicarbonate concentration should not be confused with that of metabolic acidosis.

Osmolarity can be calculated with sufficient accuracy for clinical purposes as described on p. 36.

Investigation of a Diabetic Presenting in Coma

A known diabetic may present in coma, or in a confused state ("precoma"), due to one of the metabolic complications of diabetes (ketoacidosis, hyperosmolal coma or hypoglycaemia), to a cerebrovascular accident, or to an unrelated cause.

The initial assessment is made (and therapy started) on *the clinical*

findings. Blood must be taken at once and sent for estimation of *glucose, potassium*, bicarbonate or arterial pH, sodium and urea. Urine may be tested for glucose and ketones.
The usual findings are shown in Table XX.

TABLE XX

Diagnosis	Clinical features	Blood glucose	Plasma [HCO$_3^-$]	Urine glucose	Urine ketones
Ketoacidosis	Dehydration Acidotic breathing	High	Low	+++	+++
Hyperosmolal coma	Dehydration May be hyper-ventilating	Very high	N or slightly reduced	+++	Neg.
Hypoglycaemia	Non-specific	Low	N	Neg.	Neg.
Cerebrovascular accident	Neurological May be hyper-ventilating	May be raised	May be low	May be +	Usually neg.
Other	Variable	Usually N	Variable	Usually neg.	Usually neg.

The following points should be noted:
(*a*) *It is unwise to rely solely on urine testing to diagnose hyperglycaemia.* The urine may have been in the bladder for some time and reflect earlier and very different blood glucose levels. Moreover, glycosuria (and ketonuria) may be absent despite high blood glucose levels if the GFR is much impaired by reduced vascular volume.

(*b*) Rapid assessment of blood glucose levels by Dextrostix (Ames) is often recommended to confirm the clinical diagnosis before blood glucose levels are available. *It is essential to remember that improperly stored reagent strips may give completely and dangerously fallacious results.* Again, more weight should be given to the clinical findings.

(*c*) Repeated arterial puncture is undesirable. If arterial pH is measured initially, plasma Tco$_2$ should also be estimated to provide a base-line for assessing progress.

(*d*) Mild uraemia and hypernatraemia (often severe) are usually present as a result of water depletion and contribute to the raised osmolality. These results influence therapy after the initial stage.

(*e*) Plasma ketone levels may be assessed semi-quantitatively (see Appendix) but are rarely required for diagnosis. However, in a severely acidotic patient without ketonuria they may give added information. Minimal ketonaemia in such a patient would indicate another cause for the acidosis, such as that associated with lactate production. As the initial therapy is similar, this investigation should not delay treatment.

(*f*) Cerebrovascular accidents may be associated with hyperglycaemia and glycosuria ("piqûre diabetes"). Stimulation of the respiratory centre may cause overbreathing and consequent respiratory alkalosis, with a low plasma bicarbonate level. This should not be confused with metabolic acidosis. Diagnosis depends on clinical findings.

In summary, institution of therapy in diabetic coma should be based on clinical diagnosis. Blood samples should be sent to the laboratory immediately, but treatment should never be delayed until the results are available. Side-room tests must be interpreted with caution.

Principles of Treatment of Diabetic Coma

Only the outline is discussed here. For details consult the references given at the end of the chapter. The treatment of hypoglycaemia is considered on p. 203.

Ketoacidosis.—*Repletion of fluid and electrolytes* should be vigorous. If the plasma sodium concentration is normal or low, isotonic saline is used initially and is continued if the sodium concentration remains normal. A careful watch should be kept for the development of hypernatraemia, and if it is found, hypotonic saline should be used. If acidosis is very severe, bicarbonate may be given, but it is unnecessary to correct the bicarbonate level entirely as it rapidly returns to normal following adequate fluid and insulin therapy. *Potassium* should be administered as soon as plasma levels start to fall. Urine output should be monitored: if it falls despite adequate rehydration, fluid and potassium should be given with care.

Insulin is given immediately by continuous infusion, or frequently repeated small doses. Alternatively it may be given half intravenously and half subcutaneously: the intravenous dose will act until restored circulation allows absorption from the subcutaneous site.

The factor which precipitated the coma (such as infection) should be sought and treated.

Frequent monitoring of blood glucose and plasma potassium and sodium levels is necessary to assess progress and to detect developing hypoglycaemia, hypokalaemia or hypernatraemia. The latter may delay the fall of plasma osmolality to normal, with persisting confusion or coma.

Hyperosmolal coma.—Treatment of hyperosmolal coma is similar to that of ketoacidosis, but should be more cautious as sudden reduction of extracellular osmolality may do more harm than good (p. 35). *Relatively small doses of insulin* should be given to reduce blood glucose levels slowly. These patients are often very sensitive to the action of insulin. Fluid should be replaced with hypotonic solutions but, if severe hypernatraemia is present, this, too, should be done cautiously.

HYPOGLYCAEMIA

Hypoglycaemia is defined as a true blood glucose level of below 2·2 mmol/l (40 mg/dl). The corresponding plasma glucose level is 2·5 mmol/l (45 mg/dl). Symptoms may develop at levels greater than this when there has been a rapid fall from a previously elevated, or even normal, value; by contrast, some people may show no symptoms at levels below 2·2 or even 1·7 mmol/l (30 mg/dl), especially if these have fallen gradually. As discussed earlier, cerebral metabolism is dependent on an adequate supply of glucose from the blood and the symptoms of hypoglycaemia resemble those of cerebral anoxia. Faintness, dizziness or lethargy may progress rapidly to coma and, if untreated, death or permanent cerebral damage may result. If the fall of blood glucose is rapid there may be a phase of sweating, tachycardia and agitation due to adrenaline secretion, but this phase may be absent if the fall is gradual. Existing cerebral or cerebrovascular disease may aggravate the condition. Rapid restoration of blood glucose concentration is essential.

CAUSES OF HYPOGLYCAEMIA

There is no completely satisfactory classification of the causes of hypoglycaemia as overlap between different groups occurs. A practical approach is based on the history. Particular attention is paid to drugs used and to the relation of symptoms to meals or a particular food. The patient can usually be provisionally allocated to one of two main groups.

Fasting hypoglycaemia.—Symptoms typically occur at night or in the early morning, or an episode may be precipitated by a prolonged fast or strenuous exercise. This implies excessive utilisation of glucose or an abnormality of the glucose-sparing or glucose-forming mechanisms (p. 175).

The main causes are:
  inappropriately high insulin levels due to tumour or hyperplasia of the pancreatic islet cells;
  deficiency of glucocorticoids;
  severe liver disease;
  non-pancreatic tumours.

Provoked hypoglycaemia.—Symptoms typically occur within 5 to 6 hours of a meal, related to a particular type of food, or are associated with medication. Substances that may provoke hypoglycaemia include:
  drugs (especially insulin);
  alcohol;

glucose (reactive hypoglycaemia);
galactose (in milk);
fructose (in sucrose-containing foods);
leucine (an amino acid in casein).

It is not always possible to distinguish between provoked and early fasting hypoglycaemia on the basis of history alone.

Because hypoglycaemia presents a particular problem in the early years of life this will be considered separately after a more general discussion.

HYPOGLYCAEMIA PARTICULARLY IN ADULTS

Symptoms can only be attributed to hypoglycaemia if hypoglycaemia is demonstrated. This statement may seem superfluous but it is not unknown for patients to be submitted to multiple tests for the differential diagnosis of hypoglycaemia that does not exist, or worse, to undergo treatment for it.

Once demonstrated the following main causes must be considered:

Insulin- or other drug-induced hypoglycaemia.—This is probably the commonest cause and unless deliberately concealed by the patient the offending drug should be easily identified. Hypoglycaemia in a diabetic may follow accidental overdosage, be due to changing *insulin* requirements, or to a failure to take food after insulin has been given. Self-administration for suicidal purposes is not unknown and homicidal use is a remote possibility. *Sulphonylureas* may also induce hypoglycaemia, especially in the elderly, and salicylate poisoning in children may be complicated by hypoglycaemia. Other drugs such as antihistamines have been suspected in some cases and a drug history is an important part of the investigation.

Insulinoma.—An insulinoma is a tumour of the islet cells of the pancreas. As with other endocrine tumours, hormone secretion is inappropriate and usually excessive. It may occur at any age and is usually single and benign but may rarely be malignant. Multiple tumours may be present and it may be part of the pluriglandular syndrome (p. 435). The tumour involves β-cells although other islet cells are usually present in it.

Symptoms may be sporadic, with trouble-free intervals. Hypoglycaemic attacks occur typically at night and before breakfast or may be precipitated by strenuous exercise. Personality or behavioural changes may be the first feature and many of these patients present to psychiatrists.

Alcohol-induced hypoglycaemia.—Hypoglycaemia may develop between 2 and 10 hours after ingestion of alcohol, usually in considerable amounts. It is described most frequently in chronic alcoholics when starvation or malnutrition is present, but may occur in young persons after first exposure. The probable mechanism is reduced hepatic output of glucose due to suppression of gluconeogenesis during metabolism of the alcohol. This is potentiated in the fasting or starved state when gluconeogenesis is the main source of blood glucose. Clinical differentiation from alcoholic stupor may be impossible without blood glucose estimation. Frequent infusions of glucose may be needed in treatment. It is difficult to reproduce the hypoglycaemia after infusion of alcohol unless the patient fasts for 12 to 72 hours.

Non-pancreatic tumours.—Although many types of malignant tumours—carcinoma (especially hepatocellular) as well as sarcoma—have been incriminated in the production of hypoglycaemia, it occurs most commonly in association with mesodermal tumours, resembling fibrosarcomata, that occur mainly retroperitoneally. They are slow-growing and may become very large. Hypoglycaemia may be the presenting feature. The mechanism is uncertain and may be secretion of an insulin-like substance or excessive utilisation of glucose (p. 440).

Functional (reactive) hypoglycaemia (sensitivity to glucose).—Some people develop symptoms of hypoglycaemia 2–4 hours after a meal or after a glucose load. Unconsciousness usually does not occur. The diagnosis is made by performing a prolonged (5–6 hour) glucose tolerance test (p. 201). A similar "reactive" hypoglycaemia may be seen after *gastrectomy*. These patients show a "lag storage" glucose tolerance curve and excessive insulin secretion (p. 187). A similar exaggerated insulin response is seen in some, but not all, cases of idiopathic functional hypoglycaemia.

Hypoglycaemia 3 to 5 hours after a carbohydrate meal may indicate early diabetes. It is possibly due to the delayed but exaggerated insulin response seen in the early stages of the disease.

True reactive hypoglycaemia, except after gastrectomy, is probably rare, and is too frequently diagnosed.

Endocrine causes.—Hypoglycaemia may occur in adrenal or pituitary insufficiency. Rarely it is the sole manifestation and these conditions are considered in Chapters V and VI.

Impaired liver function.—The functional reserve of the liver is so great that, despite its central role in the maintenance of blood glucose levels, hypoglycaemia is uncommon. It may occur in very severe *hepatitis* or liver necrosis—conditions in which the whole liver is affected. Its importance lies more in the recognition of hypoglycaemia as a cause of coma in such patients than as a problem of differential diagnosis of hypoglycaemia.

Diagnosis

A careful history and clinical examination should identify most cases of provoked hypoglycaemia and those due to underlying endocrine disease. Many non-pancreatic tumours are also clinically detectable. The appropriate tests for liver dysfunction and adrenocortical or pituitary hypofunction are described in Chapters V, VI and XIII. The remaining patients must be presumed to harbour an insulinoma and further tests aim at confirming or excluding this diagnosis.

The hallmark of an insulinoma is that insulin secretion continues when it should be cut off. As emphasised on p. 114, inappropriate hormone secretion is confirmed by measuring the hormone when levels should be low, in this case during hypoglycaemia.

One of the best diagnostic opportunities occurs if the patient is seen during an attack. It should be an absolute rule that *before glucose is administered* to a patient in a suspected hypoglycaemic attack *blood is taken for glucose and insulin*. This step may considerably shorten later investigations. *Hypoglycaemia with high levels of insulin* is seen only with insulinoma or islet cell hyperplasia, administration of insulin or, occasionally, of sulphonylureas, leucine sensitivity and post-gastrectomy hypoglycaemia.

Failing this opportunity, the following procedures are applied:

Fasting.—Blood is collected for glucose and insulin estimation after a 14 to 16-hour fast. In most patients with true fasting hypoglycaemia the blood glucose will fall below 2·2 mmol/l *with symptoms*. It may be necessary to repeat the fast on several occasions or, rarely, to prolong it for 72 hours followed by exercise, under careful supervision. If the hypoglycaemia is due to an insulinoma, insulin levels will be inappropriately high whereas in all other cases of fasting hypoglycaemia they will be low. Failure to demonstrate hypoglycaemia after this ordeal is strongly against the diagnosis of fasting hypoglycaemia.

This is usually the only test necessary.

Stimulation tests.—A number of substances that stimulate insulin secretion in normal persons do so to a greater extent in patients with insulinoma.

Intravenous glucagon test.—Following administration of intravenous glucagon there is a rise of plasma insulin levels due to direct stimulation of the islet cells. Blood glucose levels also rise due to stimulation of hepatic glycogenolysis, further stimulating insulin secretion. As they return to normal, insulin secretion ceases. In cases of insulinoma late hypoglycaemia may develop as insulin continues to be secreted. This test is not completely specific, but is positive in over 70 per cent of cases of insulinoma.

L-leucine test.—Oral administration of l-leucine produces excessive insulin secretion and hypoglycaemia in about 50 per cent of cases of insulinoma. It is the most specific provocative test available for this purpose.

A test using intravenous *tolbutamide* is potentially dangerous and rarely necessary.

Prolonged glucose tolerance test.—This test is restricted to the diagnosis of "functional" hypoglycaemia and has *no* place in the diagnosis of insulinoma. The glucose tolerance test (p. 209) is prolonged for 3 to 5 hours with blood samples taken every half hour. Interpretation is not easy. In many normal people the blood glucose level falls below fasting levels at some stage and even hypoglycaemic values may not cause symptoms. It has been suggested that the diagnosis be made only in patients in whom symptomatic hypoglycaemia occurs, and in whom a post-hypoglycaemic rise in plasma cortisol is demonstrated. This test should also detect diabetes mellitus presenting as post-prandial hypoglycaemia.

HYPOGLYCAEMIA IN CHILDREN

Hypoglycaemia in infancy is not uncommon and is important because permanent brain damage can result, especially in the first few months of life. Only the main causes will be outlined.

(*a*) **Neonatal period.**—Blood glucose levels in the neonatal period as low as 2·2 mmol/l (about 40 mg/dl) or, in the first 72 hours, 1·7 mmol/l (about 30 mg/dl) may be considered "normal" and are often asymptomatic. Such hypoglycaemia often lasts only 24 hours. In premature (less than 2·5 kg) infants, levels below 1·1 mmol/l (about 20 mg/dl) may occur without clinical evidence of hypoglycaemia. Signs of hypoglycaemia at this age are convulsions, tremors and attacks of apnoea with cyanosis; treatment in such cases is urgent, and such symptomatic hypoglycaemia may last for up to a week. Neonatal hypoglycaemia occurs particularly:

in babies of diabetic mothers. If the fetus is exposed to hyperglycaemia during pregnancy, islet-cell hyperplasia occurs and the consequent hyperinsulinism may produce hypoglycaemia when the supply of excess glucose from the mother is removed after parturition. Severe hypoglycaemia has been noted in babies whose mothers have been treated with *sulphonylurea drugs* during pregnancy. In some series the incidence of hypoglycaemia in babies of diabetic mothers is 50 per cent or higher. Most of these infants show no clinical signs and the significance of these low levels is controversial. Similar islet-cell hyperplasia and neonatal hypoglycaemia may be found with severe erythroblastosis fetalis:

in babies suffering from intra-uterine malnutrition (for instance, infants of toxaemic mothers, or the smaller of twins). These babies are usually "small for dates" and may show a tendency to hypoglycaemia during the first week, possibly due to poor liver glycogen stores. Prematurity is an aggravating factor as most of the liver glycogen is laid down after 36 weeks. They usually show clinical signs of hypoglycaemia.

(*b*) **Early infancy.**—Soon after birth, or after the introduction of milk or sucrose, hypoglycaemia may be due to:

Glycogen storage disease.—A deficiency of one of the several enzymes involved in glycogenesis or glycogenolysis results in the accumulation of normal or abnormal glycogen. In the commonest of these diseases (von Gierke's) *glucose-6-phosphatase* is deficient. As this enzyme is essential for the conversion of glucose-6-phosphate to glucose, *fasting hypoglycaemia* occurs. The further metabolic consequences include *ketosis* and *endogenous hyperlipoproteinaemia* (excessive lipolysis due to intracellular glucose deficiency), *lactic acidosis* (excessive glycolysis) and *hyperuricaemia* (p. 375).

Hepatomegaly occurs due to the accumulated glycogen.

The diagnosis is made directly by demonstrating the absence of the enzyme in a liver biopsy specimen, or indirectly by the failure of blood glucose levels to rise after glucagon administration (which stimulates glycogenolysis). Infusion of galactose or fructose, normally converted to glucose, also fails to raise blood glucose levels in these patients as the G-6-P formed cannot be converted to glucose.

Treatment is to take frequent meals to maintain blood glucose levels and prevent cerebral damage.

Galactosaemia.—Galactose is necessary for the formation of cerebrosides, some glycoproteins and, during lactation, of milk. Any excess is rapidly converted to glucose.

The commonest form of galactosaemia is due to deficiency of *hexose-1-phosphate uridylyltransferase* (previously known as *galactose-1-phosphate uridyltransferase*).

The condition becomes apparent only after milk has been added to the infant's diet. The main features are:

　　vomiting and diarrhoea with failure to thrive;
　　hepatomegaly leading to jaundice and cirrhosis;
　　cataract formation;
　　mental retardation;
　　renal tubular damage (Fanconi syndrome); hypoglycaemia.

Galactose in the urine is a cause of a positive reaction with Benedict's solution or Clinitest tablets. This feature is dependent on ingestion of galactose and may be absent if the subject is not receiving milk. Tubular damage may result in generalised aminoaciduria.

The *diagnosis* is made by identifying the urinary sugar as galactose by chromatography and by demonstrating a deficiency of the relevant enzyme in the erythrocytes. The latter test should be done on cord blood in all newborn infants with affected siblings.

Treatment.—Galactose (in milk and milk products) should be eliminated from the diet. Sufficient galactose (as UDP-galactose) is synthesised endogenously for the body's needs.

Hereditary fructose intolerance.—This is a *rare* cause of hypogly-caemia. Symptoms start after the introduction of sucrose or fruit juice to the diet. Fructose administration leads to symptoms of hypoglycaemia in half to one hour, accompanied by nausea or vomiting and abdominal pain. There is failure to thrive and progressive liver damage with hepatomegaly, jaundice and ascites. Cirrhosis may develop. Diagnosis is based on the demonstration of fructosuria and hypoglycaemia after oral or intravenous administration of fructose (the basic defect is a deficiency of the enzyme *fructose-1-phosphate aldolase* and accumulation of *fructose-1-phosphate*). The hypoglycaemia is probably due to in-hibition of glycogenolysis and gluconeogenesis by fructose-1-phosphate.

(c) **Infancy.**—**Idiopathic hypoglycaemia of infancy** consists of a group of cases, probably of multiple aetiology. Symptoms usually develop after fasting or after a febrile illness. The diagnosis is made by excluding other causes. There is a high incidence of brain damage.

In some cases there is excessive insulin secretion and differentiation from insulinoma is not possible.

Leucine sensitivity.—In the first six months of life *casein* may pre-cipitate severe hypoglycaemia. This is due to its content of the amino acid, leucine. Leucine sensitivity is probably due to excessive stimulation of insulin secretion and there is often a familial incidence. The condition appears to be self-limiting and does not usually persist above the age of 6 years. Diagnosis is confirmed by the demonstration of hypoglycaemia within half-an-hour after an oral dose of leucine or casein. Treatment is a low leucine diet.

Normal individuals do not show a significant drop in blood glucose after leucine, but a number of patients with insulinoma do demonstrate leucine sensitivity.

Ketotic hypoglycaemia is the commonest cause in the second year of life and develops after fasting or a febrile illness. These children were usually "small-for-dates" babies. Ketonuria precedes the hypoglycaemia (as it does in starvation) and the diagnosis can be established by feeding a ketogenic diet (high fat, low calorie) for 48 hours, during which clinical hypoglycaemia occurs.

"Adult" causes, including insulinoma, must be considered at all times.

Treatment of hypoglycaemia is by the urgent intravenous administra-tion of 50 per cent glucose solution *after withdrawal of a blood sample for diagnosis* (see above). Some cases may need to be maintained on a glucose infusion until the cause has been established and treated.

The treatment of *insulinoma* is by surgical removal if possible. If not a combination of diazoxide and chlorothiazide may maintain normo-glycaemia.

SUMMARY

1. Glucose is the main product of dietary carbohydrate.

2. The brain is very dependent on extracellular glucose as an energy source, and maintenance of blood glucose levels is important for normal cerebral function.

3. After a carbohydrate-containing meal, excess glucose is:
    stored as glycogen in liver and muscle;
    converted to fat and stored in adipose tissue.
    Insulin stimulates these processes.

4. During fasting:
    glycogen breakdown in the liver (and kidney) releases the glucose into the blood;
    triglyceride breakdown in adipose tissue releases glycerol, which can be converted to glucose by the liver;
    triglyceride breakdown also releases fatty acids, which can be metabolised by most tissues *other than the brain*.

5. The liver converts excess fatty acids to ketoacids. Ketones can be used as an energy source by the brain and other tissues.

6. If ketoacid formation exceeds the capacity of homeostatic mechanisms, ketoacidosis may develop.

7. Anaerobic glycolysis produces lactic acid:
    lactic acid production occurs temporarily in contracting muscles;
    lactic acid is produced in hypoxic tissues. The hypoxic liver becomes a major lactic acid producing rather than a lactic acid consuming organ, and lactic acidosis results.
    Other factors increasing glycolysis or reducing utilisation of lactic acid may also cause lactic acidosis. Some drugs, such as biguanides (e.g. phenformin) reduce lactic acid utilisation.

8. Diabetes mellitus is the result of relative or absolute insulin deficiency. It is characterised by hyperglycaemia (either fasting, or only after a glucose load). In severe diabetes mellitus excessive lipolysis may cause ketosis and later acidosis. Coma may be due to hyperosmolality, and in both ketoacidosis and hyperosmolal coma there is severe water and electrolyte depletion.

9. Hypoglycaemia may occur during fasting, or be provoked by drugs or by some foods.

10. In adults *insulinoma* is the *most important* cause of fasting hypoglycaemia; it is diagnosed by demonstrating high plasma insulin levels despite hypoglycaemia. *Insulin administration* is the *commonest* cause of hypoglycaemia.

11. Factors causing hypoglycaemia in childhood vary with age. Many inborn errors of carbohydrate metabolism may cause hypoglycaemia.

12. It may be necessary to supply energy substrate intravenously over

long periods (Appendix, p. 212). Glucose minimises lipolysis and ketosis, but usually needs to be supplemented by fat solutions to provide enough energy in a manageable volume of fluid.

FURTHER READING

NEWSHOLME, E. A. (1976). Carbohydrate metabolism *in vivo*: regulation of the blood glucose level. *Clinics in Endocr. & Metab.*, **5**, 543,

SIPERSTEIN, M. D. (1975). The glucose tolerance test: a pitfall in the diagnosis of diabetes mellitus. *Advanc. intern. Med.*, **20**, 297.

FELIG, P., Ed. (1977). Symposium on Diabetes Mellitus. *Arch. intern. Med.*, **137**, 461.
This symposium includes articles on aetiology and biochemical abnormalities.

KREBS, H. A., WOODS, H. F., und ALBERTI, K. G. M. M. (1975). Hyperlactataemia and lactic acidosis. In *Essays in Medical Biochemistry*, **1**, 81 (Eds. Marks, V., and Hales, C. N.) London: The Biochemical Society and the Association of Clinical Biochemists.

FAJANS, S. S., and FLOYD, J. C. (1976). Fasting hypoglycaemia in adults. *New Engl. J. Med.*, **294**, 766.

HOFELDT, F. D. (1975). Reactive hypoglycaemia. *Metabolism*, **24**, 1193.

PAGLIARA, A. S., KARL, I. E., HAYMOND, M., and KIPNIS, D. M. (1973). Hypoglycaemia in infancy and childhood. *J. Pediat.*, **82**, 365 and 558.

APPENDIX

Estimation of Blood Glucose

Glucose levels may be assessed by enzymatic methods, which are specific for this sugar, or by measurement of total reducing substances. Because of the presence of non-glucose reducing substances, the latter methods may over-estimate glucose by $0.3-1.1$ mmol/l or more. The percentage error is greatest at low levels of glucose and *may mask hypoglycaemia*. It is important to know which method your laboratory uses.

The supply of glucose to cells depends on extracellular concentrations and these are reflected in *plasma* levels. However, largely for historical reasons, whole blood is often used for estimation. Since intracellular levels are kept low by metabolism, inclusion of erythrocytes in the sample "dilutes" the plasma, giving results 10 per cent to 15 per cent lower than those of plasma or serum (the actual figure depending on the haematocrit).

Either whole blood or plasma must be assayed immediately, or mixed with an inhibitor of glycolysis. *In vitro* metabolism of glucose by the cellular elements of the blood will produce falsely low levels; tubes containing fluoride or iodoacetate (both of which inhibit glycolysis) with an anticoagulant are used.

Unless otherwise indicated the glucose values given in this book are those of "true glucose" measured on whole blood.

Glycosuria

Glycosuria is best detected by enzyme reagent strips such as Clinistix (Ames). Directions for use are supplied with the strips. Clinistix detects about 5.6 mmol/l (100 mg/dl) of glucose in urine, but is not suitable for the relatively accurate quantitation required for control of diabetic therapy. The concentration of glucose in normal urine varies from 0.1 to 0.8 mmol/l (1–15 mg/dl).

False negative results may occur if the urine contains large amounts of ascorbic acid, as may occur after therapeutic doses, or after injection of tetracyclines which contain ascorbic acid as a preservative.

False positive results may occur if the urine container is contaminated with detergent.

Reducing Substances

Although Benedict's test for urinary reducing substances is still used in many laboratories, it has been replaced in most ward side-rooms by Clinitest tablets (Ames) or Clinistix (Ames). Clinistix contains the enzyme glucose oxidase, and is specific for glucose; it is used for detection of glycosuria. Clinitest, like Benedict's test, on which it is based, gives a positive result with any reducing substance. It is semiquantitative and is valuable in the control of diabetic therapy, whereas Clinistix is unsatisfactory for this purpose. Urine screening in the newborn and infants should be done by Benedict's or Clinitest because of the diagnostic importance of non-glucose reducing substances, such as galactose in this age group.

The tests are carried out as follows:

(a) *Benedict's qualitative solution.* 0·5 ml (about 8 drops) of urine is boiled with 5 ml of reagent for a few minutes. A positive result is the formation of a yellow-red precipitate that colours the urine greenish (+) to orange-red with heavy precipitate (+ + + +).

(b) *Clinitest tablets* (Ames)—used as directed. These tablets are more convenient than Benedict's solution for patients controlling their own insulin dosage, but are caustic and require careful handling.

These two tests are less sensitive than Clinistix but are more accurate for quantitation. Benedict's test is slightly more sensitive than Clinitest and therefore has a greater incidence of "false" positive results, e.g. due to drug metabolites.

+represents approximately 10–30 mmol/l (0·2–0·5 g/dl)
+ +represents approximately 30–60 mmol/l (0·5–1 g/dl)
+ + +represents approximately 60–120 mmol/l (1–2 g/dl)
+ + + +represents over 120 mmol/l (2 g/dl)
expressed as glucose

Causes of a Positive Benedict's test or Clinitest

The most important causes of a positive Benedict's test or Clinitest are:

glucose	common
glucuronate	
lactose	common in pregnancy
fructose	
galactose	rare
pentoses	
homogentisic acid	

Slight reduction may occur if urate or creatinine is present in high concentration.

Most causes of a positive Clinitest in the adult are due to glucose. Clinistix will confirm the presence or absence of glucose. Non-glucose reducing substances are further identified by chromatography and specific tests.

The significance varies with the substance.

Glucose.—Glycosuria may occur in:

diabetes mellitus;
glucose infusion;
renal glycosuria.

Glucose may be present in the urine when blood levels are normal. This is due to a low renal threshold and may be found in *pregnancy* or as an *inherited* (autosomal dominant) characteristic. The condition is usually benign. It is also found in generalised proximal tubular defects (Fanconi syndrome). Differentiation from diabetes mellitus is made by simultaneous blood and urine glucose estimation or, if necessary, by a glucose tolerance test.

Glucuronates.—A large number of drugs, such as salicylates, and their metabolites are excreted in the urine after conjugation with glucuronic acid in the liver (p. 285). These glucuronates are relatively common reducing substances in the urine.

Galactose.—Galactose is found in the urine in galactosaemia (p. 202).

Fructose.—Fructose may appear in the urine after very large oral doses of the sugar, or excessive fruit ingestion, but in general fructosuria occurs in two rare inborn errors of metabolism, both transmitted by autosomal recessive genes.

Essential fructosuria is seen almost exclusively in Jews and is a harmless condition.

Hereditary fructose intolerance (p. 203) is a serious disease characterised by hypoglycaemia that may lead to death in infancy.

Lactose.—Lactosuria may occur in:
late pregnancy and during lactation;
congenital lactase deficiency (p. 276).

Pentoses.—Pentosuria is very rare. It may occur in:
alimentary pentosuria after excessive ingestion of fruits such as cherries and grapes. The pentoses excreted are arabinose and xylose;
essential pentosuria, a rare recessive disorder characterised by the excretion of xylulose due to a block in the metabolism of glucuronic acid. It is seen usually in Jews and is harmless.

Homogentisic acid.—Homogentisic acid appears in the urine in the rare inborn error alkaptonuria (p. 360). It is usually recognisable by the blackish precipitate formed.

Ketones

(*a*) **Ketonuria.**—Most simple urine tests for ketones detect acetoacetate, because of the greater sensitivity of reagents to this than to acetone, and because acetoacetic acid is present in greater concentration. β-Hydroxybutyrate is not measured.

Rothera's test.—5 ml of urine is saturated with a mixture of ammonium sulphate and sodium nitroprusside (as a powder). 1–2 ml concentrated ammonia is added and mixed. A purple-red colour indicates the presence of acetoacetate (or acetone).

Ketostix and *Acetest* (Ames) are strips and tablets respectively, impregnated with the reagents of Rothera's test. Instructions are supplied with the reagents. They are more convenient to use.

Rothera's test detects about 0·5 mmol/l (5 mg/dl) acetoacetate while the tablet tests are slightly less sensitive (10–20 mg/dl). After a fast of several hours ketonuria may be detected by these methods. Occasional colour reactions resembling, but not identical to, that of acetoacetate, may be given by phthalein compounds such as phenolphthalein, bromsulphthalein (BSP) or phenolsulphthalein (PSP).

Gerhardt's test.—3 per cent ferric chloride is added drop by drop to 5 ml of urine. A red–purple colour develops if acetoacetate is present in concentrations of 2·5–5·0 mmol/l (25–50 mg/dl) or greater.

This test is less sensitive than those described above and, if the latter are negative, need not be done. If, however, ketone bodies are found, Gerhardt's test gives additional information about the severity of the ketosis.

A large number of compounds that may be found in urine give a colour with ferric chloride. The commonest are salicylates which give an almost identical colour. Boiling the urine for several minutes will destroy ketones and subsequent testing will give a negative result. Salicylates are unchanged by this treatment.

(b) **Ketonaemia.**—The severity of ketonaemia may be rapidly assessed using the reagents for Rothera's test or, more conveniently, using Ketostix (Ames).

1. Serum or plasma is obtained by centrifuging a specimen of blood.

2. Ketostix is dipped into the serum, immediately removed, and excess serum drained against the side of the tube.

3. Exactly 15 seconds later compare any purple colour developing with the colour chart on the reagent bottle. The colour developed is semiquantitative.

$$+1\text{--}2 \text{ mmol/l} (10\text{--}20 \text{ mg/dl}) \text{ acetoacetate}$$
$$++3\text{--}5 \text{ mmol/l} (30\text{--}50 \text{ mg/dl}) \qquad ,,$$
$$+++8\text{--}12 \text{ mmol/l} (80\text{--}120 \text{ mg/dl}) \qquad ,,$$

Note that β-hydroxybutyrate does *not* react.

$+++$results may be further assessed by diluting the serum in water, e.g. 1 drop serum and 3 drops water gives a final dilution of $\frac{1}{4}$. The test result can then be recorded as $+++(\frac{1}{4})$.

Oral Glucose Tolerance Test

The indications for and interpretation of the glucose tolerance test are discussed on p. 185; only the procedure is given here.

1. The patient fasts overnight (at least 10 hours). Water, but no other beverage, is permitted.

2. A fasting blood sample for glucose is withdrawn and a specimen of urine obtained.

3. 50 g glucose (children 1·75 g/kg body weight to a maximum of 50 g) is dissolved in water. The patient drinks this within a period of five minutes.

4. Further samples for glucose are taken 30, 60, 90, 120 minutes after glucose.

5. Urine samples are obtained at 60 and 120 minutes.

During the test the patient should be at rest and may not smoke.

Factors influencing the result of the GTT.—*Dose of glucose.* The use of a dose of 50 g as recommended by the British Diabetic Association is probably adequate. In America a dose of 1·75 g/kg ideal body weight or a standard dose of 100 g is usually given, and is claimed to be a more sensitive test. The larger dose may cause nausea and variable gastric emptying (compare xylose absorption test, p. 282) and does not improve diagnostic accuracy.

Capillary or venous blood. Although there is very little difference in glucose levels between these two types of sample during fasting, at the high levels found during the GTT capillary tends to give higher results than venous blood and to return to fasting levels more slowly. These differences should be allowed for in the interpretation of the test.

Age. There is deterioration of glucose tolerance with increasing age and mild impairment is less significant in older than in younger subjects (compare urea, p. 467). This may, in part, be due to a poor diet.

Previous diet. No special restrictions are necessary if the patient has been on a normal diet for 3–4 days. If, however, the test is performed after a period of carbohydrate restriction, such as a reducing diet, abnormal glucose tolerance may be shown. The probable reason for this is that metabolism is set in the "fasted" state favouring gluconeogenesis. Mildly abnormal tests should be repeated after appropriate dietary preparation.

Time of day. Most glucose tolerance tests are performed in the morning and the values quoted are for this time. There is evidence that tests performed in the afternoon show higher blood glucose values and the accepted normals may not be applicable. This phenomenon may be due to a circadian variation in islet cell activity.

Tests for Insulinoma

Several points apply to all the tests:

1. As many blood samples are to be taken, an indwelling needle, kept open by a saline infusion, is required (see p. 127).

2. Samples are taken for both *glucose* and *insulin*. Glucose is taken into a fluoride tube and insulin into a plain tube or one with an anticoagulant. (N.B. Heparin interferes with some insulin estimations—check with your laboratory.) The insulin tube is kept at 4°C until it is delivered to the laboratory.

3. These tests may be dangerous. 50 per cent glucose for intravenous administration *must be immediately available* and given if dangerous hypoglycaemia develops. Should it be necessary to terminate the test, a blood specimen should be taken, if possible, *before glucose administration*.

4. The tests are done with the patient lying down. At the conclusion of the test the patient should be given food.

5. These tests should be preceded by a normal diet for at least three days.

Intravenous glucagon test.—1. The patient fasts overnight.

2. A *basal blood sample* for glucose and insulin is drawn.

3. One mg of glucagon (children 30 μg/kg body weight to a maximum of 1 mg) is injected intravenously.

4. Further samples for glucose and insulin are taken 5, 10, 15, 20, 30, 60, 120 and 180 minutes after glucagon.

Response:

1. *Blood glucose concentration.* The blood glucose rises by about 40 per cent and is followed in some patients with insulinoma, but not in normal subjects, by a fall to below the basal value.

2. *Plasma insulin concentration.* Plasma insulin rises in 5–20 minutes to 100 mU/l. The response in patients with insulinoma is much greater.

A similar excessive rise may be seen in states of insulin resistance, e.g. obesity. The test is positive in about 70 per cent of cases of insulinoma.

L-leucine test.—1. The patient fasts overnight.

2. A *basal blood sample* for glucose and insulin is drawn.

3. L-leucine (200 mg/kg body weight) is taken orally mixed with cold water.

4. Further blood samples for glucose and insulin are taken 30, 60, 90 and 120 minutes after l-leucine.

Response:

1. *Blood glucose concentration.* Within 30 minutes the blood glucose falls by up to 1·4 mmol/l (about 25 mg/dl) in normal subjects, or may fall to hypoglycaemic levels in cases of insulinoma.

2. *Plasma insulin concentration.* The normal insulin response is an increase of up to 30 mU/l within 30 minutes. An excessive response is seen in about 50 per cent of patients with insulinoma. This response is very specific and is seen only in patients with insulinoma, in subjects who are taking sulphonylureas and in leucine-sensitivity hypoglycaemia in children (p. 200).

PRINCIPLES OF INTRAVENOUS FEEDING

The principles discussed at the beginning of this chapter have an important application in the management of parenteral feeding. More details will be found in the references quoted at the end of this section.

Degradation of carbohydrate, fat and, to a lesser extent, the carbon chains of some amino acids, supplies the energy needs of the body by coupling the breaking of their energy-rich bonds with ATP synthesis. Some of the constituent elements, carbon, hydrogen and oxygen, leave the body as carbon dioxide and water: a normal male loses energy at the rate of about 8400 kJ (2000 kcal) a day as heat. These losses are usually balanced by dietary intake of equivalent amounts of carbohydrate, fat and protein. If output exceeds intake, energy stores are drawn upon: once these are depleted structural compounds, such as the protein of cells and muscle, may be used to supply energy.

The energy needs of the ill patient may be much higher than normal, and yet he may not be able to take any or all his nutritional requirements by mouth.

Short-term Intravenous Therapy (a period of days)

A well-nourished individual has enough energy stored as hepatic glycogen to last for at least a day. Stores will be replenished when he can eat again. During such a short fast energy is derived from carbohydrate as well as fat, and ketosis is usually mild. Maintenance of electrolyte and fluid balance is all that is necessary, for instance, in the immediate post-operative period, or during an acute gastro-intestinal disturbance. Although a litre of 5 per cent glucose (dextrose) contains 850 kJ (205 kcal) it is probably unnecessary to give even this, and certainly not all the administered fluid should be in this form (p. 50).

If this so-called *catabolic phase* is likely to be followed by rapid recovery, amino acid infusion is unnecessary. Much protein breakdown is due to stress-induced cortisol secretion, sometimes contributed to by tissue damage, and nitrogen is lost in the urine, mostly as urea; the body is unable to use administered nitrogen efficiently until the factors causing negative balance are removed. Moreover the short-term nitrogen loss will be replenished when the patient starts eating. For similar reasons supplementation of other cellular constituents is ineffective in the catabolic phase.

Medium-term Intravenous Feeding (a period of weeks)

If prolongation of fasting is anticipated intravenous feeding should be used from the outset.

Hepatic glycogen stores are relatively small, and once they are used up energy is derived almost entirely from triglyceride. At this stage ketoacidosis may become a problem unless carbohydrate or substrate for gluconeogenesis is provided: in addition, at least enough energy substrate should be given to maintain balance and to replete hepatic glycogen. If it is not possible to do this by nasogastric tube, "food" must be given intravenously (*total parenteral nutrition*). 10 litres of 5 per cent glucose would be needed merely to provide normal daily energy requirements, and an even larger volume would be necessary to keep an ill patient in balance. It is impossible to infuse such volumes and a more concentrated energy source is therefore needed.

40 per cent glucose contains 6800 kJ/l, but is very hyperosmolar, and is therefore irritant to veins; this leads to local thrombosis. Insertion of a central venous catheter so that the fluid is delivered to a site of more rapid flow may overcome this problem, but is technically more difficult and carries a greater risk of infection. Severe illness may cause some insulin resistance and utilisation of such large amounts of glucose may be inefficient unless exogenous insulin is given; the consequent high extracellular levels result in urinary wastage, and may cause dangerous hyperosmolality (p. 35). Blood glucose levels must be monitored frequently to ensure that control is adequate. Fructose, sorbitol and xylitol are not insulin dependent, but may cause lactic acidosis in such ill patients (p. 182).

Fat may therefore be used to supplement glucose infusions. 10 per cent "Intralipid" (soybean oil emulsion) contains 4620 kJ/l, and even a 20 per cent solution is so little hyperosmolar that it can be infused into a peripheral vein. The fat particles are similar in size to, and are metabolised in the same way as chylomicrons. Provided that some carbohydrate is given at a constant rate throughout the day there is no risk of dangerous ketoacidosis. The grossly *lipaemic plasma may interfere with laboratory tests* (p. 37), but this is easily avoided if *non-lipid containing solutions are infused for the few hours before sampling.*

Administration of carbohydrate and fat, by minimising the utilisation of amino acids for gluconeogenesis, reduces nitrogen loss in the urine. However, there is always some daily nitrogen loss and if this is not replaced healing may be delayed, there may be increased liability to infection and, eventually, there may even be profound muscle wasting and osteoporosis. Mixtures containing essential *amino acids* should be infused into any patient unable to eat for long. Because most nitrogen is lost as urinary urea, estimation of 24-hour urea nitrogen output can, if glomerular function is normal, be used as a guide to replacement. However, once tissue repair predominates, amino acid utilisation increases (the *anabolic phase*): urinary nitrogen loss may fall abruptly and much more than that retained may be needed for maximum regeneration. The onset of this phase is heralded by a sense of relative well-being and by increasing weight.

The composition of some of the available solutions is given on p. 63.

Long-term Intravenous Feeding (a period of months)

In addition to the above, vitamin supplementation is advisable in long-term parenteral feeding. Calcium rarely needs to be given if vitamin D intake is adequate: low plasma calcium levels are usually due to reduction of the protein-bound rather than of the ionised fraction. Magnesium supplementation may be necessary and is included in some amino acid solutions.

During a long illness tissue destruction leads to loss of intracellular constituents, such as phosphate and some trace metals (for example zinc and copper). Phosphate may need to be added to the infusion. Urinary trace metal loss is high, with quantitative excretion of supplements during the catabolic phase, but falls as anabolism becomes predominant: if parenteral feeding has been very prolonged deficiencies may become apparent at this time and weight gain may be impaired if supplements are not given. A more detailed discussion of this controversial topic will be found in the references.

Fluid and electrolyte balance should be controlled throughout as described in Chapters II and III. Potassium supplementation is almost always necessary especially during the anabolic phase.

FURTHER READING

FARQUHARSON, S. (1977). Parenteral Nutrition. In: *Tracheostomy and Artificial Ventilation in the Treatment of Respiratory Failure*, p. 129. Eds. Feldman, S. A. and Crawley, B. E. London: Edward Arnold.

FISCHER, J. E., Ed. (1976). *Total Parenteral Nutrition*. Boston: Little, Brown and Co.

Chapter X

PLASMA LIPIDS AND LIPOPROTEINS

THE association of abnormalities of plasma lipids with atheromatosis and ischaemic heart disease has stimulated much research on the subject. Current understanding of the physiology and pathology of the plasma lipids is based mainly on the concept of lipoproteins, the form in which lipids circulate in the blood. The first part of this chapter will describe and correlate the terms in current usage.

TERMINOLOGY AND CLASSIFICATION

CHEMICAL CLASSIFICATION OF PLASMA LIPIDS

The chemical structures of the four forms of lipid present in plasma are illustrated in Fig. 23.

Fatty acids are straight-chain compounds, the length of the chain depending on the number of carbon atoms in the molecule. Fatty acids may be *saturated* (containing no double bonds), or *unsaturated* (with one or more double bonds). The main saturated fatty acids in plasma are *palmitic* (16 carbon atoms) and *stearic* (18 carbon atoms). Fatty acids may be *esterified* with glycerol to form glycerides, or they may be free, when they are called *free fatty acids* (FFA) or *non-esterified fatty acids* (NEFA). In the blood FFA are carried mainly on albumin. Free fatty acids are an immediately available energy source and provide much of the energy requirements of the body. This aspect of fat metabolism is further considered on p. 178.

Triglycerides consist of glycerol, each molecule of which is esterified with three fatty acids. They are an important energy store, and make up about 95 per cent of the lipids of adipose tissue.

Phospholipids are complex lipids, resembling triglycerides, but containing phosphate and a nitrogenous base. The major phospholipids in plasma are *lecithin* and *sphingomyelin*. The phosphate and nitrogenous bases are water soluble, a fact that is important in lipid transport.

Cholesterol has a steroid structure and other steroids are synthesised from it. In the plasma, cholesterol occurs in two forms. About two-thirds is esterified with fatty acid (Fig. 23) to form *cholesterol ester*, and the remainder is free. For routine clinical purposes the estimation of total cholesterol is adequate.

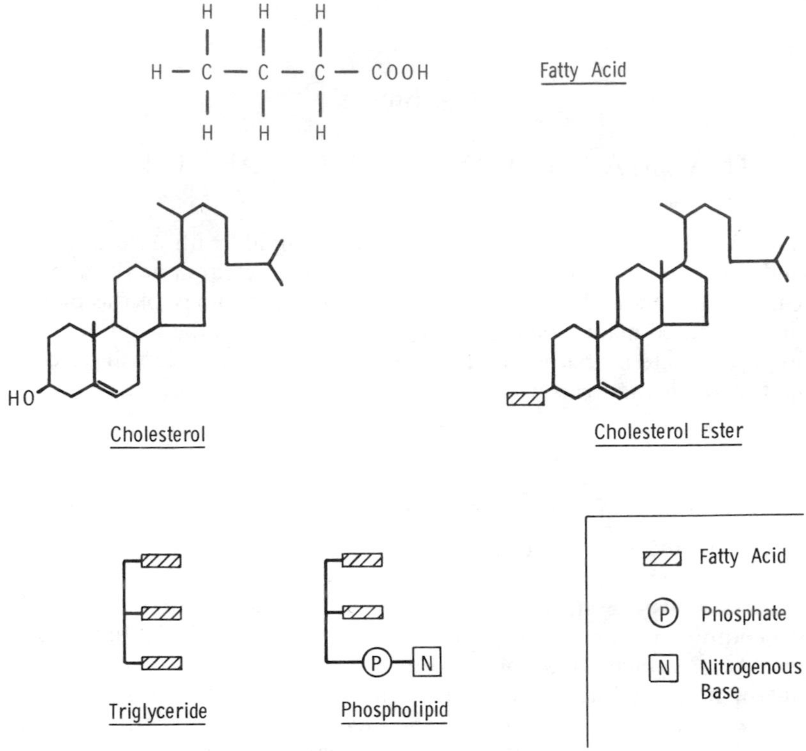

H — C — C — C — COOH   Fatty Acid

Cholesterol

Cholesterol Ester

Triglyceride

Phospholipid

▨ Fatty Acid

Ⓟ Phosphate

N Nitrogenous Base

FIG. 23.—Plasma lipids.

LIPOPROTEINS

Lipids are carried in plasma combined with protein. Although many lipids are insoluble in water, the *lipoprotein* complex is so arranged that the water soluble (polar) groupings of protein and phospholipid enclose a core of insoluble (non-polar) groups. All lipoproteins contain cholesterol, phospholipids, triglycerides and one or more specific proteins (apoproteins). Differences in the proportions of lipid and protein, however, produce different *classes* of lipoproteins which may be separated by physical techniques. *In vivo* there is interchange between different lipids and apoproteins, and there is, therefore a state of flux between classes of lipoproteins; it is probable that the apoprotein regulates the metabolism of the lipoprotein. The three main apoproteins are named apo-A, apo-B and apo-C.

Classification of Lipoproteins

The major lipoprotein classes may be separated and classified in a number of different ways.

Ultracentrifugation separates the molecules on the basis of their density and thus, indirectly, their composition. After centrifugation at high speeds some of the lipoproteins sediment with the other plasma proteins and are called *high density lipoproteins* (*HDL*). The remainder, of lower density, tend to float, the rate of flotation being expressed in S_f (Svedberg flotation) units. The greater the content of triglycerides, the lower the density of the lipoprotein class and the higher its S_f number.

The major lipoprotein classes so determined (Fig. 24) are:

high density lipoprotein (HDL) which has a high protein content (mainly apo-A) and contains mostly cholesterol and phospholipids. Apo-C is associated with the complex;

low density lipoprotein (LDL) contains mostly cholesterol and apo-B;

very low density lipoprotein (VLDL) contains large amounts of triglycerides, some cholesterol and phospholipids and both apo-B and apo-C;

chylomicrons are very large particles, consisting almost entirely of triglycerides, with small amounts of other lipids, apo-B and apo-C.

Ultracentrifugation is the definitive method, but is not generally available. *When it is necessary* to establish lipoprotein patterns for diagnosis, other techniques are used in most laboratories.

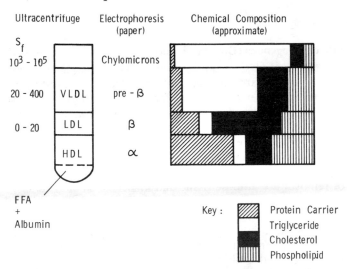

FIG. 24.—Correlation of plasma lipid and lipoprotein nomenclature.

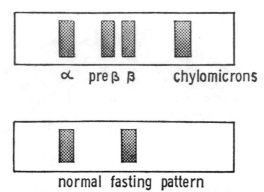

α preβ β   chylomicrons

normal fasting pattern

FIG. 25.—Lipoprotein electrophoretic strips (diagrammatic).

Electrophoresis separates the complexes by means of the electrical charge on the protein carrier. The technique is similar to that of protein electrophoresis (p. 307) except that a different staining method is used to render the lipid visible. Four fractions may be recognised, α, β and pre-β (corresponding to HDL, LDL and VLDL respectively), and chylomicrons (Fig. 25).

These nomenclatures are correlated in Fig. 24.

METABOLISM OF LIPOPROTEINS (Fig. 26)

Plasma lipids are derived from food (*exogenous*) or are synthesised in the body (*endogenous*).

The Fate of Exogenous (Dietary) Lipids

Triglycerides, cholesterol and phospholipids released by digestion of dietary fat (p. 263) are absorbed into the cells of the intestinal wall, where they combine with apo-B to form *chylomicrons*. These particles enter the lymphatics and pass, via the thoracic duct, into the blood stream. Because of their large size, they scatter light and account for the turbidity of plasma seen in blood samples taken after a fatty meal (postprandial hyperlipaemia).

Many tissues are capable of metabolising chylomicrons but the main sites of removal are adipose tissue and muscle. The enzyme *lipoprotein lipase* hydrolyses the triglycerides in the chylomicrons and the released fatty acids are taken up by the cells. This reduces the size of the lipo-protein complex, and the plasma clears. The *remnant particles*, now much smaller and containing mostly cholesterol, pass to the liver where they are metabolised further.

At the end of this process dietary triglycerides have been delivered

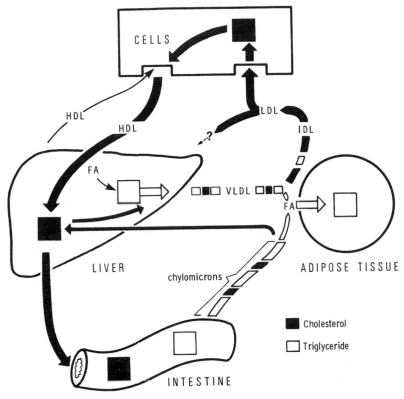

Fig. 26.—Metabolism of lipoproteins. (FA = fatty acid.)

to adipose tissue and muscle, and cholesterol and phospholipids to the liver.

The Fate of Endogenous Lipids

Many tissues can synthesise cholesterol but most endogenous cholesterol is formed in the liver. Triglycerides, also, are synthesised in the liver from FFA derived from adipose tissue; in subjects with a high carbohydrate intake, FFA may be synthesised directly from glucose within the liver.

The endogenous cholesterol and triglyceride combine with apo-B and are transported from the liver as VLDL. In peripheral tissues triglycerides are removed in the same way as from chylomicrons. The smaller complex is further broken down, via complexes of intermediate density (IDL), to form LDL. Most of the plasma cholesterol is found in LDL.

Many, possibly most, cells have specific surface receptors for LDL. The molecule is bound and enters the cell where the cholesterol is released and becomes available to the cell for membrane synthesis. This binding and uptake of LDL *suppresses* endogenous synthesis of cholesterol.

At the conclusion of this process, endogenous triglycerides and cholesterol have been delivered to peripheral cells for local energy and synthetic requirements.

The Role of HDL

Despite the suppressive effect of LDL uptake on cell cholesterol synthesis, some cholesterol is formed peripherally and accumulates, together with cholesterol derived from cell membrane breakdown. Cholesterol can leave the body only by excretion in the bile and the only known carrier of cholesterol from peripheral tissues to the liver is HDL.

HDL, probably synthesised in the liver, is rich in phospholipids. It passes into the blood stream and takes up cholesterol from cells. An enzyme, *lecithin-cholesterol acyltransferase* (*LCAT*), is associated with HDL and esterifies this cholesterol with a fatty acid from the phospholipid lecithin (in HDL). This esterified cholesterol may be transferred to other lipoproteins.

HDL has another important function. In the fasting state it carries most of the apo-C found in plasma. When VLDL or chylomicron levels rise, apo-C is transferred to these particles where it activates lipoprotein lipase. When triglyceride hydrolysis has been accomplished, the apo-C returns to HDL.

INHERITED DISORDERS OF LIPID METABOLISM

Several rare inherited disorders illustrate the metabolic pathways outlined above.

Lipoprotein lipase deficiency leads to accumulation of chylomicrons. This condition usually presents in childhood with eruptive xanthomata, lipaemia retinalis, hepatomegaly and abdominal pain. As chylomicrons consist mainly of *exogenous* triglycerides, symptoms are successfully controlled by restriction of dietary fat.

In **LCAT deficiency** free cholesterol accumulates in the tissues and almost all the plasma cholesterol is *unesterified.*

In **HDL deficiency** (Tangier disease) cholesterol *esters* accumulate in tissues, particularly of the reticulo-endothelial system. This results in characteristic large, orange-yellow tonsils, hepatomegaly and lymphadenopathy.

In **LDL deficiency** (abetalipoproteinaemia) apo-B is absent. The formation of LDL, VLDL and chylomicrons, all of which contain apo-B, is impaired and lipids cannot be transported from the intestine or liver. The condition is characterised clinically by intestinal malabsorption with steatorrhoea, progressive ataxia, retinitis pigmentosa and acanthocytosis.

LDL receptor deficiency causes accumulation of LDL in the blood and excessive synthesis of cholesterol by cells (see familial hypercholesterolaemia, p. 225).

HYPERLIPIDAEMIA

(Syn. Hyperlipoproteinaemia)

Hyperlipidaemia should be suspected in patients with *premature cardiovascular disease* or *xanthomatosis*. It is often first detected by finding a raised cholesterol level or a turbid plasma in a patient without specific symptoms.

Before we discuss the diagnostic approach to hyperlipidaemia we will consider these two "classical" manifestations in greater detail.

Xanthomatosis

Patients with markedly raised plasma lipid levels may develop yellowish deposits of fat in their tissues. Several different types of these deposits or *xanthomata* exist and usually seem to be correlated with elevation of different lipid fractions in the plasma, although this correlation is not a constant one.

Eruptive xanthomata are crops of small, itchy, yellow nodules that occur in association with raised triglycerides, either VLDL or chylomicrons. They disappear rapidly if plasma lipid levels are restored to normal.

Tuberous xanthomata are yellow plaques, in some cases large and disfiguring, found mainly over the elbows and knees. They occur with elevated LDL, including the abnormal lipoprotein of Type III (see below) and, less commonly, VLDL.

Deposits in tendons (*xanthoma tendinosum*), eyelids (*xanthelasma*) and cornea (*corneal arcus*) at an early age occur typically with elevated LDL.

Planar xanthomata, raised yellow linear deposits in the creases of the palm, are seen typically with the abnormal lipoprotein of Type III hyperlipoproteinaemia ("broad beta disease").

Atheromatosis

One of the most controversial topics in medicine is the relationship of plasma lipids to the development of atheromatosis. Very high plasma levels of LDL and VLDL are associated with early and marked atheromatosis; the effect of moderate elevation is less obvious. It is usually assumed that it is beneficial to reduce levels to "normal"; it is very difficult to prove that this is so because of the multiple factors involved. Nevertheless it has been accepted that a raised plasma cholesterol value (which indicates an increase in LDL) is a risk factor in the development of cardiovascular disease. Because management of patients with cardiovascular disease often includes reducing plasma cholesterol, it is important to understand the factors affecting these levels.

As mentioned above, cholesterol is both absorbed from the diet and synthesised in the body. Absorption is largely proportional to intake in the range commonly found in the usual diet. In affluent societies this is about 1·5 to 2·0 mmol (600 to 800 mg) per day, with meat, dairy products and particularly egg yolk as the main sources. Most of the *endogenous* cholesterol is synthesised in the liver and the rate of synthesis is controlled by the amount of cholesterol reaching it, much of which is from the diet (*exogenous*). Synthesis in peripheral tissues is regulated by plasma LDL levels (p. 220). Thus endogenous synthesis is reduced if intake and absorption of cholesterol is high. This suppression, however, is not complete and plasma cholesterol values are usually higher with a high cholesterol intake than with a low one. The *nature of fat* in the diet influences plasma cholesterol: *saturated fatty acids*, which are found chiefly in animal fats, as is cholesterol, increase plasma levels while *polyunsaturated fatty acids*, in vegetable oils, lower them. A diet substituting vegetable for animal fat is used to treat raised LDL levels.

The amount of cholesterol added daily to the body pool is balanced by excretion in the bile. Part is degraded by the liver to bile acids and bile salts (p. 299), while the remainder is excreted as cholesterol and, together with dietary cholesterol, is available for reabsorption. *Cholestyramine*, which binds bile salts and interferes with their reabsorption, can be used to reduce plasma cholesterol and LDL levels.

While there is considerable evidence of a *positive* correlation between LDL and cardiovascular disease, there is an even stronger *negative* correlation with HDL. The higher the plasma HDL level, the lower the risk. HDL seems to have a protective function which, because of the physiological role of HDL as a carrier of cholesterol from peripheral tissues for excretion, is not surprising. Numerous factors known to diminish the risk of cardiovascular disease are in fact associated with high HDL levels. Such factors may be hormonal (levels are higher in women during the reproductive years than in men); physical exercise tends to increase HDL levels and carbohydrate-rich diets to lower them.

DIAGNOSIS OF HYPERLIPIDAEMIA

There should be a logical progression in the evaluation of a patient suspected of having an abnormality of plasma lipids.

Is there hyperlipidaemia?

If so, is it primary or secondary?

If primary, what is the nature of the abnormality?

Is there Hyperlipidaemia?

Hyperlipidaemia is diagnosed by measuring cholesterol and triglyceride levels in plasma. While cholesterol levels do not vary greatly after fat intake, triglyceride levels do, so that specimens for analysis of both should be taken after the patient has fasted (see Appendix).

Plasma cholesterol at birth (cord blood) is usually below 2·6 mmol/l (100 mg/dl). Levels increase slowly throughout childhood, mostly during the first year, but do not usually exceed 4·1 mmol/l (160 mg/dl). In most affluent populations a further progressive increase occurs after the second decade. This increase is greater in men than in women during the reproductive years. The "normal" (95 per cent) upper limit in many societies is as high as 8·4 mmol/l (330 mg/dl) in the fifth and sixth decades. This rise is not seen in less affluent societies where the incidence of ischaemic heart disease is much lower. There is much controversy about the lipid level above which treatment should be instituted. A plasma cholesterol value of greater than 7·0 mmol/l (about 275 mg/dl) is clearly undesirable; some authorities put the limit at 6·5 mmol/l (250 mg/dl), particularly in a young patient. Similarly, fasting triglyceride values of greater than 2·0 mmol/l (about 175 mg/dl) should be considered abnormal.

Primary or Secondary?

Many cases of hyperlipidaemia are *secondary* to other disease and may be corrected by treatment of that disease. The main secondary causes of abnormal plasma lipids are shown in Table XXI. In general *diabetes mellitus, hypothyroidism* and the *nephrotic syndrome* should be sought in all instances of hyperlipidaemia. *Dysglobulinaemia*, either paraproteinaemia or a marked increase in γ-globulin as found in systemic lupus erythematosus, should also be considered. Abuse of *alcohol* is a common cause of secondary hyperlipidaemia.

The Nature of the Abnormality

If no secondary cause is found, the hyperlipidaemia must be considered a *primary* abnormality and requires treatment as such. Selection of appropriate therapy requires a more accurate definition of the nature

TABLE XXI

THE CAUSES OF SECONDARILY RAISED PLASMA CHOLESTEROL AND TRIGLYCERIDE
LEVELS (Where both are indicated it does not imply that they always
occur together)

	Cholesterol	Triglycerides
Diabetes mellitus	+	+
Nephrotic syndrome	+	+
Hypothyroidism	+	+
Dysglobulinaemia	+	+
Pancreatitis		+
Renal failure		+
Excessive alcohol intake		+
Pregnancy	+	
Oral contraceptives		+
Biliary obstruction	+	
Glycogen storage disease		+
Acute porphyria	+	

of the basic lipoprotein disturbance. In most cases this poses no difficulty.

On the basis of the initial chemical estimation we can distinguish three main groups:

predominant cholesterol elevation;

predominant triglyceride elevation;

both cholesterol and triglyceride proportionately increased.

A predominant increase in plasma cholesterol is almost always due to increased LDL levels. Such sera are usually *clear* as the LDL particle is small and does not scatter light. Turbidity in a fresh sample implies an increase in triglycerides, either as VLDL or chylomicrons, or both. It is important to distinguish which of these two large particles is involved as the treatment differs. To do so the turbid plasma sample is left standing for 18 hours at 4°C. During this time the large, low density chylomicrons form a creamy layer on the surface. The smaller, denser VLDL particles do not rise and the sample remains diffusely turbid. With gross hyperlipidaemia the distribution may be less clear.

In cases with mixed lipoprotein abnormalities in particular, it may be necessary to proceed to lipoprotein electrophoresis or ultracentrifugation.

The full investigation of primary hyperlipidaemia does not end with definition of the lipoprotein abnormality. Family studies are necessary both to identify the disorder (see below) and to detect other affected, possibly asymptomatic, individuals.

The Primary Hyperlipidaemias

Numerous classifications of hyperlipidaemia have been proposed. One such, that of Fredrickson, (see Appendix) was adopted as a basis for the World Health Organisation classification. Experience has shown however, that while the Fredrickson classification describes *lipoprotein patterns*, it does not define inherited disease entities and contributes little to routine diagnosis and treatment.

Current methods of classifying hyperlipidaemia include the following entities.

Hypercholesterolaemia

Primary hypercholesterolaemia is probably always genetic in origin. It includes familial hypercholesterolaemia, polygenic hypercholesterolaemia and familial combined hyperlipidaemia (see later).

Familial (monogenic) hypercholesterolaemia is the most lethal of the inherited hyperlipidaemias because of the increased incidence of cardiovascular disease. It is transmitted as an autosomal dominant; homozygotes rarely survive beyond the age of 20 and heterozygotes have a 10–50-fold risk. The basic lesion is a deficiency of LDL receptors on peripheral cell membranes, with resultant failure of metabolism of LDL and uncontrolled intracellular cholesterol synthesis. Heterozygotes have about half the normal number of receptors. Plasma cholesterol levels (in LDL) are extremely high (often greater than 15 mmol/l) in homozygotes and in the region of 8–13 mmol/l in heterozygotes. Triglyceride levels are usually normal, but may be slightly increased. Tendon xanthomata, and xanthelasma, which are typical of hypercholesterolaemia, develop in early childhood in homozygotes but only in the second or third decade, or later, in heterozygotes.

The defect is present at birth and raised cholesterol levels may be found in cord blood.

When large numbers of family members are studied, the plasma cholesterol levels can be separated into three groups—normal (unaffected), elevated (heterozygotes) and very high (homozygotes). Because of this clear demonstration of single gene influence, familial hypercholesterolaemia is called *monogenic*.

Polygenic hypercholesterolaemia.—Familial hypercholesterolaemia accounts for less than 5 per cent of cases of primary hypercholesterolaemia in most countries. In the majority of hypercholesterolaemic families, plasma cholesterol values show a continuous distribution with a higher than normal mean value. Because this differs from the clear single gene pattern described above, it is presumed to be due to a number of genes affecting LDL or cholesterol synthesis or disposal. It is therefore called *polygenic*. Environmental and dietary factors probably have an effect too. Cell membrane LDL receptors are normal. These people have an increased risk of cardiovascular disease but xanthomata are less common.

In both these groups the lipoprotein pattern (see Appendix) is of Type IIa or, less often, IIb.

Hypertriglyceridaemia

Hypertriglyceridaemia is more often secondary than primary, and may be familial or sporadic.

Familial hypertriglyceridaemia, due to an increase in VLDL (Type IV pattern) appears to be transmitted as an autosomal dominant trait, usually appearing only in the fourth or later decades of life. There is probably an increased incidence of cardiovascular disease and it is associated with obesity, hyperuricaemia and glucose intolerance. Superimposed secondary factors such as diabetes mellitus or excessive alcohol intake may produce very marked elevation of VLDL, often with hyperchylomicronaemia (Type V pattern). The basic lesion in these families is not known.

Familial hyperchylomicronaemia, (Type I pattern) due to lipoprotein lipase deficiency, occurs in children and is described on p. 220. It does not carry an increased risk of cardiovascular disease.

Familial Combined Hyperlipidaemia

This is probably the commonest inherited hyperlipidaemia and is probably transmitted as a single gene. LDL, VLDL, or both, are involved and about a third of *affected* family members have hypercholesterolaemia, a third hypertriglyceridaemia, and a third both. The lipoprotein patterns include Types IIa, IIb, IV and, rarely, V. As with familial hypertriglyceridaemia the lipid abnormalities only become apparent in the third or fourth decade or later. There is an increased risk of cardiovascular disease in these families, irrespective of the lipoprotein pattern. While the high triglyceride levels may cause eruptive xanthomata, tuberous and tendinous xanthomata are uncommon.

Broad Beta Disease

A rare but probably distinct familial hyperlipidaemia is the accumulation in the plasma of a lipoprotein rich in both cholesterol and triglycerides. It may represent abnormal formation or impaired catabolism of IDL (p. 219). The electrophoretic pattern (Type III) shows a single broad band in the β- pre-β-region, hence the name. Confirmation of the diagnosis, however, requires more sophisticated studies. There is a high incidence of cardiovascular, particularly peripheral vascular, disease. Tuberous xanthomata are common and there is a characteristic deposition of lipid in the palmar creases.

Principles of Treatment

The treatment of hyperlipidaemia is a highly controversial subject, covered more fully in the references at the end of the chapter. It should be clear from the discussion in this chapter that a number of different disease entities produce similar plasma lipid and lipoprotein changes, probably by different mechanisms. Appropriate therapy depends on knowledge of the basic defect and its consequences: this knowledge is largely unavailable. As an example let us consider the patient heterozygous for familial hypercholesterolaemia. It has been shown that intracellular cholesterol synthesis is normally suppressed *when the plasma LDL levels are elevated*. It now becomes important to know which is the dangerous factor— the intracellular synthesis of cholesterol or the raised plasma LDL levels. If the former, lowering plasma LDL would, by no longer suppressing intracellular synthesis, aggravate the condition. Because this and similar information is not available, current therapeutic regimes are directed at lowering elevated plasma lipid levels. The principles may be summarized briefly.

1. An attempt should be made to correct such aggravating causes as excessive alcohol intake.

2. **Diet.**—Obesity should be treated. Carbohydrate restriction is particularly important in patients with *endogenous hypertriglyceridaemia*;

exogenous hypertriglyceridaemia (hyperchylomicronaemia) requires restriction of dietary fat;

hypercholesterolaemia should be treated by restricting the intake of cholesterol and saturated fat. As these are both found in animal fat, this should be largely replaced by vegetable fats which contain unsaturated fatty acids.

3. **Drugs.**—Dietary measures are often adequate for lowering triglyceride levels but do not usually suffice in patients with hypercholesterolaemia. In such cases, drugs are used. Only two will be mentioned. *Cholestyramine* is a resin that binds bile salts (formed from cholesterol) and prevents reabsorption. Although there is a compensatory increase in synthesis, this drug often reduces plasma cholesterol levels. It is, however, not pleasant to take.

Clofibrate reduces endogenous triglyceride levels, and, less successfully, the raised cholesterol levels in some cases of polygenic hypercholesterolaemia. It is highly successful in the management of broad beta disease. The mechanism of its action is not clear.

It is probable that future developments will be directed at the basic defects in the hyperlipidaemias and at increasing plasma HDL levels.

SUMMARY

1. Lipids in plasma are carried in the form of lipoproteins.

2. Plasma lipids may be classified according to their chemical structure, or as lipoproteins by electrophoresis or ultracentrifugation. Chemical estimations are most frequently available but lipoprotein analysis offers the best understanding of abnormalities.

3. The chemical lipid fractions are cholesterol, triglycerides, phospholipids and free fatty acids.

4. The main lipoproteins are high-density lipoprotein (HDL), low-density lipoprotein (LDL), very low-density lipoproteins (VLDL) and chylomicrons. The corresponding electrophoretic nomenclature is α, β, pre-β and chylomicrons.

5. Cholesterol occurs mainly in HDL and LDL and triglycerides in VLDL and chylomicrons.

6. Hyperlipidaemia may be primary or secondary to other disease. Only in primary hyperlipidaemia is it necessary to define the lipoprotein abnormality more fully. This can often be done by simple techniques.

7. Different genetic defects may produce similar lipoprotein abnormalities. Extensive family studies are required to differentiate them. This is rarely practicable.

8. The treatment of primary hyperlipidaemia is primarily regulation of the diet. If necessary drugs may be added. The treatment is largely empirical.

FURTHER READING

LEWIS, B. (1976). *The Hyperlipidaemias*. Oxford: Blackwell Scientific Publications.
LEVY, R. I. (1975). Pathophysiology of lipid transport disorders. *Postgrad. med. J.*, **51**, (Suppl. 8), 16.
This issue contains a number of relevant articles on hyperlipidaemia.
MOTULSKY, A. G. (1976). The genetic hyperlipidaemias. *New Engl. J. Med.*, **294**, 823.
Lancet (1976). Editorial: HDL and CHD, **2**, 131.

APPENDIX

Correct Blood Sample

Plasma lipid levels and lipoprotein patterns are labile and affected by eating, changes in diet, posture and stress. For the initial evaluation and follow up studies it is *essential* that the sample be taken under *standard conditions*. It may, in some cases, be necessary to repeat the analysis to establish the most consistent pattern. The following points are important.

1. The patient must have fasted for 14 to 16 hours.

2. For the two weeks preceding the test the patient should have been on his "normal" diet and his weight should have remained constant.

3. The patient must not be on any treatment that affects his plasma lipids.

4. As in the case of all large particles, lipoprotein concentrations are affected by venous stasis (p. 455) and posture. Plasma cholesterol concentration may be up to 10 per cent higher in the upright than in the recumbent position; triglyceride concentration may change slightly more than this. A standardised collection procedure, e.g. sitting for 30 minutes before blood is taken, is important in serial estimations to assess the effect of treatment.

5. Diagnosis and typing of hyperlipidaemia is preferably deferred for three months after a myocardial infarction or major operation.

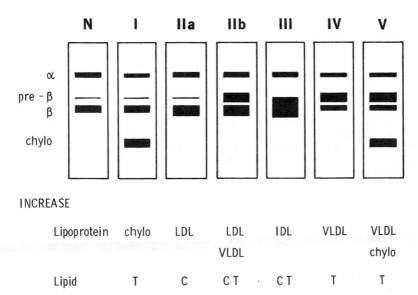

FIG. 27.—Fredrickson classification of lipoprotein patterns. (N = normal; Roman numerals = Fredrickson types; T = triglyceride; C = cholesterol; chylo = chylomicrons.)

6. The blood sample should not be heparinised and must be separated as soon as possible.

The Fredrickson (WHO) Classification

This classification of lipoprotein patterns remains a useful way of describing the abnormality found, but does not define distinct disease entities (Fig. 27).

Chapter XI

CALCIUM, PHOSPHATE AND MAGNESIUM METABOLISM

MOST of the body calcium is in bone, and a significant deficiency causes bone disease. However, the extraosseous fraction, although amounting to only 1 per cent of the whole, is of great importance because of its effect on neuromuscular excitability and on cardiac muscle. Most calcium salts in the body contain phosphate. This radical is very important in its own right (for instance in the formation of "high energy" phosphate bonds), but clinically, apart from its buffering power, it is mainly of interest because of its association with calcium.

More than half the body magnesium is in bone: most of the remainder is intracellular but, as in the case of calcium, the small extracellular fraction (about 1 per cent) is important because of its neuromuscular action. Because of this similarity in distribution and physiological action, calcium and magnesium metabolism will be discussed in one chapter.

CALCIUM METABOLISM

TOTAL BODY CALCIUM

The total amount of calcium in the body depends on the balance between intake and loss.

Factors Affecting Intake

The amount of calcium entering the body depends on the amount in the *diet*, which normally amounts to about 25 mmol (1 g) a day. Between 25 and 50 per cent of this is absorbed. *Vitamin D*, in the form of 1, 25-dihydroxycholecalciferol (p. 235), is required for adequate calcium absorption.

Factors Affecting Loss

Calcium is lost in urine and faeces. *Faecal calcium* consists of that which has not been absorbed, together with that which has been secreted into the intestine. Calcium in the intestine, whether endogenous or exogenous in origin, may be rendered insoluble by large amounts of *phosphate*, *fatty acids* or *phytate* (in cereals). Oral phosphate may be used therapeutically to control calcium absorption and reabsorption

　　　　CLINICAL CHEMISTRY

(p. 252): an excess of fatty acids in the intestinal lumen in steatorrhoea contributes to calcium malabsorption.

Urinary calcium depends on the amount of calcium circulating through the glomerulus, on renal function, parathyroid hormone levels and to a lesser extent on urinary phosphate excretion. The amount of calcium circulating through the glomerulus is increased after a calcium load or during decalcification of bone not due to calcium deficiency (e.g. osteoporosis or acidosis). Such decalcification rarely causes hypercalcaemia because of the rapid renal clearance of calcium. Conversely, hypercalcaemia, whatever the aetiology, causes hypercalcuria if renal function is normal. Glomerular insufficiency reduces calcium loss in the urine, even in the presence of hypercalcaemia.

PLASMA CALCIUM

Estimation of plasma calcium should be accurate to about 0·05 mmol/l (0·2 mg/dl) and changes of less than this are probably insignificant. Calcium circulates in the plasma in two main states. The albumin-bound fraction accounts for a little less than half the total calcium as measured by routine analytical methods: it is physiologically inactive and is probably purely a transport form comparable to iron bound to transferrin (p. 382). Most of the remaining plasma calcium is ionised (Ca^{++}), and this is the physiologically important fraction (compare with thyroxine, p. 162, and cortisol, p. 135). Ideally, ionised rather than total calcium should be measured, but techniques for this are not available in most routine laboratories. Whenever a plasma calcium value is assessed an effort should be made to decide whether the ionised fraction is normal by comparing the value for total calcium with that for protein (preferably albumin) on the same specimen. The total (but not ionised) calcium concentration is lower in the supine than in the erect position, because of the effect of posture on plasma protein concentration.

The ionisation of calcium salts is greater in acid than in alkaline solution. In alkalosis tetany may occur in the presence of normal levels of total calcium because of the reduction in the ionised fraction. In clinical states associated with prolonged acidosis (for instance after transplantation of the ureters into the colon or in renal tubular acidosis, p. 97), the increased solubility of bone salts removes calcium from bone and leads to osteomalacia, even in the presence of an adequate supply of the ion to bone and of normal parathyroid function.

Control of Plasma Calcium

The level of circulating ionised calcium is controlled within narrow limits by parathyroid hormone (PTH), secreted by the parathyroid glands. Calcitonin, produced in the thyroid, has an opposite action to that of PTH on calcium levels: its importance in calcium homeostasis is less certain than that of PTH. Both these hormones control ionised calcium by using bone as a reservoir from which plasma levels can be controlled. Vitamin D increases calcium absorption from the intestine, and in physiological amounts appears to be necessary for the action of PTH.

Action and control of parathyroid hormone.—Parathyroid hormone raises the plasma ionised calcium concentration. It has two direct actions on plasma calcium and phosphate levels:

by acting directly on osteoclasts it releases bone salts into the extracellular fluid. This action tends to increase both calcium *and* phosphate in the plasma;

by acting on the renal tubular cells it decreases the reabsorption of phosphate from the glomerular filtrate, causing phosphaturia. This action tends to *decrease* the plasma phosphate: this in turn, by a mass action effect, increases release of phosphate salts (and therefore calcium) from bone. PTH also increases renal tubular reabsorption of calcium.

In addition to these direct effects of PTH on plasma calcium and phosphate levels it may, either directly or through its action on phosphate (p. 235), stimulate the renal production of active vitamin D.

The secretion of PTH, like that of insulin from the pancreas, is not controlled by any other endocrine gland. It is influenced only by the concentration of calcium ions circulating through it. Reduction of this concentration increases the rate of release of hormone, and this increase is maintained until the ionised calcium level returns to normal, when hormone production stops.

Bearing in mind the actions of PTH on bone and on the renal tubules, and the fact that its secretion is controlled by the level of ionised calcium, the consequences of most disturbances in calcium metabolism on plasma calcium and phosphate levels can be predicted. A low ionised calcium from any cause (other than hypoparathyroidism), by stimulating the parathyroid, causes phosphaturia: this loss of phosphate overrides the tendency to hyperphosphataemia due to the direct action of PTH on bone, and plasma phosphate levels are normal, or even low. Conversely, a high calcium concentration due to any cause (except excess of PTH) cuts off production of the hormone and causes a high phosphate level. As a general rule, therefore, calcium and phosphate levels tend to vary in the same direction unless there is an inappropriate excess or deficiency of PTH (as in primary hyperparathyroidism or hypoparathyroidism), or in the presence of renal failure when the phosphaturic effect is absent (see, however, "Action of vitamin D").

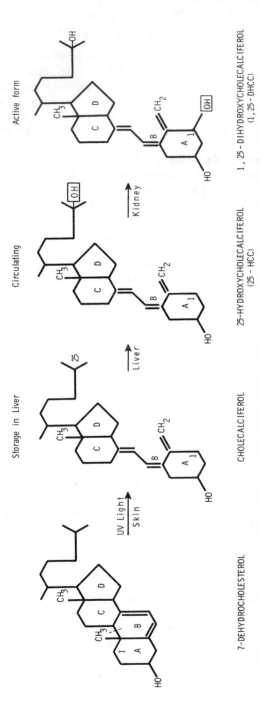

FIG. 28.—Formation of active "vitamin D" in the body from vitamin D_3.

Action and metabolism of vitamin D (Fig. 28).—Vitamin D (D_3: cholecalciferol) may be formed in the skin from 7-dehydrocholesterol by the action of ultraviolet light; this is the form in animal tissues, particularly liver, and is therefore the most important dietary "vitamin D". This term is also applied to ergocalciferol (D_2), which can be obtained from plants after artificial ultraviolet irradiation. These vitamins D are transported in the blood bound to specific carrier proteins: they are inactive until metabolised.

In the *liver* the molecule is hydroxylated to *25-hydroxycholecalciferol* (*25-HCC*), the main circulating form of the vitamin. Any excess of vitamin D is stored in the organ, or is converted to inactive forms and excreted.

Further hydroxylation of 25-HCC by the *kidney* to *1, 25-dihydroxycholecalciferol* (*1, 25-DHCC*) is necessary to confer biological activity on the vitamin. 1, 25-DHCC acts on the intestine, increasing calcium absorption; in conjunction with PTH it releases calcium from bone. The kidney is thus an endocrine organ, manufacturing and releasing the hormone 1, 25-DHCC; failure of the final hydroxylation probably explains the hypocalcaemia of renal disease.

The lag between administration of therapeutic vitamin D and its action (p. 258) may be due to the time taken for the two-phase hydroxylation.

Vitamin D has a phosphaturic effect. Usually in the hypercalcaemia of vitamin D excess this effect is over-ridden by the reduction in phosphate excretion following reduction of PTH secretion. Occasionally, however, phosphate levels are low or low normal when hypercalcaemia is due to this cause.

Control of renal vitamin D metabolism.—The enzyme 1-α-*hydroxylase* catalyses the hydroxylation of 25-HCC in the kidney. Its activity, and hence the production of 1, 25-DHCC, is *stimulated* by:

a *low* circulating *phosphate* concentration;

an *increase* in circulating *PTH* (perhaps because of its phosphate lowering effect);

oestrogens;

prolactin;

growth hormone.

(The last three hormones mediate increased 1, 25-DHCC production, and therefore increased calcium absorption, during pregnancy, lactation and growth).

1, 25-DHCC production is *reduced* by *high* circulating *phosphate* levels. This effect may contribute to deficiency in renal glomerular failure.

Interrelationship of parathyroid hormone and vitamin D. The action of PTH on bone is impaired in the absence of 1, 25-DHCC.

A reduction of extracellular ionised calcium levels stimulates PTH production by the parathyroid glands. PTH stimulates 1, 25-DHCC synthesis, and the two hormones act synergistically on the bone "reservoir" releasing calcium into the circulation; 1, 25-DHCC also increases calcium absorption. In short-term homeostasis the bone effect is more important: after prolonged hypocalcaemia more efficient absorption becomes important. As soon as the ionised calcium concentration is corrected PTH production ceases, and that of 1, 25-DHCC is reduced.

Action and control of calcitonin.—Calcitonin reduces plasma calcium by decreasing osteoclastic activity, and therefore bone resorption. It also has a phosphaturic effect. It is produced in the C cells of the thyroid and its secretion is stimulated by high ionised calcium levels. Its clinical importance is doubtful. Circulating concentrations may be very high in patients with medullary carcinoma of the thyroid, but abnormalities of plasma calcium concentration have very rarely been reported associated with these high levels.

Action of thyroid hormone on bone.—Thyroid hormone excess may be associated with the histological lesion of osteoporosis, and with increased faecal and urinary excretion of calcium, probably due to release from bone. Unless there is very gross excess of thyroid hormone, effects on plasma calcium levels are overridden by the homeostatic reduction of PTH secretion and hypercalcaemia rarely occurs. Some cases of post-thyroidectomy hypocalcaemia may be due, not to parathyroid damage, but to re-entry of calcium into depleted bone when the circulating thyroid hormone levels fall.

Calcium homeostasis follows the general rule that body content is regulated by control of plasma levels. In the presence of normal parathyroid, kidney and intestinal function, and of sufficient supply of calcium and vitamin D, this is usually adequate. In the absence of one or more of these, control of plasma levels may be effected at the expense of the total amount of calcium in the body (compare sodium and water, p. 41).

CLINICAL EFFECTS OF DISORDERS OF CALCIUM METABOLISM

Clinical Effects of Excess Ionised Calcium

On the kidney.—The most dangerous consequence of prolonged, mild hypercalcaemia is *renal damage*. The high ionised calcium causes the solubility product of calcium phosphate in the plasma to be exceeded, and this salt is precipitated in extra-osseous sites. The most important of these sites is the kidney. Calcification of the renal tubules may reduce their ability to reabsorb water, so that the presenting symptom of hypercalcaemia may be *polyuria*. It is because of this danger of renal damage that every attempt should be made to diagnose the cause of even mild hypercalcaemia and treat it at an early stage. The high ionised calcium concentration in the glomerular filtrate may, if circumstances favour it, result in precipitation of calcium salts in the urine: the patient will present with *renal calculi*, and these may occur without significant renal parenchymal damage. All patients with renal stones should have plasma calcium estimated on more than one occasion.

On neuromuscular excitability.—High ionised calcium levels depress neuromuscular excitability in both voluntary and involuntary muscle. The patient may complain of *constipation* and *abdominal pain*. Muscular hypotonia may also be present. The *anorexia, nausea* and *vomiting* of hypercalcaemia are more probably due to a central effect.

On the heart.—Hypercalcaemia causes changes in the electrocardiogram. If severe (as a rough guide, more than 3.75 mmol/l (15 mg/dl)) it carries the risk of sudden *cardiac arrest*, and for this reason should be treated as a matter of urgency.

Clinical Effects of Reduced Ionised Calcium Levels

Low ionised calcium levels (including those associated with a normal total calcium in alkalosis, p. 232) cause increased neuromuscular activity leading to *tetany*.

Prolonged hypocalcaemia, even when mild, interferes with the metabolism of the lens and causes *cataracts*: it may also cause mental depression and other *psychiatric symptoms*. Because of the danger of cataracts, asymptomatic hypocalcaemia should be sought when there has been a known risk of damage to the parathyroid glands (for example, after partial or total thyroidectomy).

Effects of High Parathyroid Hormone Levels on Bone

High PTH levels with a normal supply of vitamin D and calcium (for example, *primary hyperparathyroidism*).—Parathyroid hormone acts on the osteoclasts of bone, increasing their activity. In cases in which the supply of vitamin D and calcium is normal and in which PTH has been circulating in excess over a long period of time the effects are as follows:

clinical:
  bone pains;
  bony swellings.
radiological:
  generalised decalcification;
  subperiosteal erosions;
  cysts in the bone.
histological:
  increased number of osteoclasts.

A secondary osteoblastic reaction may occur: osteoblasts manufacture the enzyme alkaline phosphatase which, when these cells are increased in number, is released into the extracellular fluid in abnormal amounts and leads to a *rise in plasma alkaline phosphatase activity*.

The effects of excess parathyroid hormone on bone are only evident in long-standing cases. In disease of short duration (for instance, when the excess of PTH is due to malignant disease, or to early primary hyperparathyroidism) they are absent. Osteoblastic reaction is a late manifestation and plasma *alkaline phosphatase levels are usually normal or only moderately raised* in primary hyperparathyroidism.

High PTH levels in vitamin D and calcium deficiency (causing *rickets and osteomalacia*).—Vitamin D must be present in adequate amounts for the complete action of PTH on bone. In cases of vitamin D deficiency PTH levels may be very high, but the effects described above are seen to only a minor degree. Calcium deficiency means that osteoblastic reaction is ineffective, and uncalcified osteoid is a characteristic histological finding: this osteoblastic activity is marked and increases plasma alkaline phosphatase levels.

The effects are therefore as follows:
clinical:
bone pains;
rarely bony swelling.
radiological:
generalised decalcification;
pseudofractures;
rarely subperiosteal erosions or cysts in bone.
histological:
uncalcified osteoid: wide osteoid seams;
occasionally increased numbers of osteoclasts.
in the plasma:
raised alkaline phosphatase activity.

These changes in the bone are also seen in certain errors of renal tubular reabsorption of phosphate (p. 245).

DISORDERS OF CALCIUM METABOLISM

In this section disorders of calcium metabolism will be considered in relation to circulating parathyroid hormone levels. They can be summarised as follows:
diseases associated with high PTH levels.
inappropriate secretion of PTH (*with raised plasma calcium concentration*);
appropriate secretion of PTH (*with normal or low plasma calcium concentration*).

diseases associated with low PTH levels.

disease of the parathyroid gland (*with low plasma calcium concentration*);

appropriate suppression of PTH secretion by *high plasma calcium levels* due to causes other than inappropriate PTH levels.

Diseases Associated with High Circulating Parathyroid Hormone Levels

Diseases due to inappropriate secretion of parathyroid hormone.—The term "inappropriate secretion of parathyroid hormone" is used here to mean the production of the hormone under circumstances in which it should normally be absent (i.e. in the presence of normal or high ionised calcium concentrations). This syndrome should be compared with that of "inappropriate" ADH secretion (p. 52).

Such inappropriate secretion occurs in three circumstances:

primary hyperparathyroidism;

tertiary hyperparathyroidism;

(In these two cases the hormone is produced by the parathyroid gland.)

ectopic production of PTH ("pseudohyperparathyroidism"; "quaternary" hyperparathyroidism).

The findings, if renal function is normal, are those of excess PTH production, that is a raised plasma calcium concentration, with a normal or low plasma phosphate level. With the onset of renal damage, usually due to hypercalcaemia, both of these tend to return to normal, the phosphate because of the inability of the kidney to respond to the phosphaturic effect of PTH and the calcium because of the lowering of plasma calcium in renal failure (p. 242). Diagnosis at this stage may be very difficult.

The *clinical features* of these cases are due to:

excess of circulating ionised calcium (p. 236);

the effect of PTH on bone (p. 237).

Differences between these syndromes depend on the duration of the disease more than any other factor.

Primary hyperparathyroidism.—Primary overaction of the parathyroid glands is usually due to an adenoma, which may occasionally be present in an ectopic gland. Parathyroid tumours are almost always benign, but very rarely primary hyperparathyroidism is due to parathyroid carcinoma. Occasionally it results from diffuse hyperplasia of the glands.

Primary hyperparathyroidism presents most commonly with signs and symptoms associated with high ionised calcium levels, such as nausea, anorexia and abdominal pain, or with renal changes (calculi, polyuria or even renal failure). *In many of these cases bone changes are absent and the plasma alkaline phosphatase concentration is normal or only slightly elevated.* Patients presenting with overt bony lesions, on the other hand, often show no renal changes. It may be that when conditions in the urine are favourable for calcium precipitation (for instance an alkaline pH), or when a patient complains of symptoms early, the disease is usually diagnosed before bony changes occur: if these warning signs are absent the patient presents with the overt picture of bone disease. However, it should be pointed out that some of the renal cases have not developed bone disease after remaining untreated for many years: the reason for the two types of presentation is not fully understood.

Occasionally the patient is admitted as an emergency with abdominal pain, vomiting and constipation due to severe hypercalcaemia.

Tertiary hyperparathyroidism.—Tertiary hyperparathyroidism is the name given to the condition in which the parathyroid has been under prolonged feedback stimulation by low ionised calcium levels, and the resulting hypersecretion has become partially autonomous. Except for the history of previous hypocalcaemia, the disease is identical with primary hyperparathyroidism. It is usually diagnosed when the cause of the original hypocalcaemia is removed (e.g. by renal transplantation or correction of calcium or vitamin D deficiency). The typical biochemical findings of excess PTH are superimposed on those of osteomalacia, and because the condition is of long duration the plasma alkaline phosphatase level is usually very high. In some cases calcium levels later return to normal.

Ectopic production of parathyroid hormone.—Sometimes malignant tumours of non-endocrine tissues manufacture peptides foreign to them: PTH can be one of these. The production of hormones at these sites is not subject to feedback control, and the initial findings are those of inappropriate PTH excess. Because of the nature of the underlying disease the condition is rarely of long standing, and the bony lesions due to excess circulating PTH are not evident: the alkaline phosphatase, however, may be raised because of bony or hepatic metastases, or both.

The subject of ectopic hormone production is discussed more fully in Chapter XXII.

Secondary hyperparathyroidism ("appropriate" secretion of parathyroid hormone).—In secondary hyperparathyroidism, the parathyroids are stimulated to produce hormone in response to low plasma ionised calcium levels. This secretion is therefore "appropriate" in that it is necessary to restore ionised calcium levels to normal. If it is effective in doing so the stimulus to production is removed, and the glands are "switched off". In secondary hyperparathyroidism, therefore, plasma calcium concentration is low (if the increased production of hormone is inadequate to correct the hypocalcaemia) or normal: it is *never* high. Hypercalcaemia, or a plasma calcium level in the upper normal range, suggests either a diagnosis of primary hyperparathyroidism (and this may be a *cause* of renal failure), or that prolonged calcium deficiency has led to tertiary hyperparathyroidism.

If renal function is normal the high PTH levels cause phosphaturia with hypophosphataemia. In all cases of secondary hyperparathyroidism, except that due to renal failure, both the plasma calcium and phosphate levels tend to be low.

If the PTH feedback mechanism is normal any factor tending to lower the ionised calcium concentration causes secondary hyperparathyroidism. In long-standing cases, if the bone cells can respond to PTH, decalcification of bone occurs, leading to *osteomalacia* in adults, or *rickets* in children (before fusion of the epiphyses). The findings are described on p. 238. In very prolonged or severe depletion the high concentration of PTH may fail to maintain plasma calcium levels and the effects of reduced ionised calcium concentration may be seen (p.237).

Secondary hyperparathyroidism with osteomalacia or rickets occurs in associated chronic calcium and vitamin D deficiency. This may be due to:

reduced dietary intake of vitamin D and calcium in *malnutrition*;

reduced absorption of vitamin D in *steatorrhoea*;

reduced metabolism of vitamin D to 1,25-DHCC due to *renal disease*;

possible impairment of intestinal response to 1,25-DHCC due to *anticonvulsant therapy.*

Secondary hyperparathyroidism without osteomalacia or rickets occurs in:

early calcium and vitamin D depletion;

most cases of pseudohypoparathyroidism (very rare).

Osteomalacia and rickets without secondary hyperparathyroidism may rarely occur in phosphate depletion due to:

renal tubular disorders of phosphate reabsorption (p. 245).

Calcium and vitamin D deficiency.—1,25-DHCC is essential for normal absorption of calcium from the intestinal tract and calcium depletion is more commonly the result of deficiency of intake of vitamin D than of calcium itself. The commonest cause of calcium and vitamin D deficiency in Britain is the malabsorption syndrome: dietary deficiency is more important in the world as a whole. In the presence of steatorrhoea, vitamin D, which is fat soluble, is poorly absorbed. In addition calcium combines with the excess of fatty acids in the intestine to form insoluble soaps. Patients on prolonged *anticonvulsant therapy* (barbiturates and hydantoin) sometimes develop hypocalcaemia, and even osteomalacia; circulating 1, 25-DHCC levels are normal in such cases, and it has been suggested that the drugs may block the effect of the vitamin on the intestinal mucosa.

Malnutrition is rarely selective, and protein deficiency may lead to a reduction of the protein-bound fraction of the calcium. The ionised calcium level may not, therefore, be as low as that of total calcium would indicate.

Patients in renal failure are relatively resistant to the action of vitamin D and the resultant malabsorption of calcium is undoubtedly an important factor in the development of hypocalcaemia. The discovery that vitamin D (cholecalciferol) is inactive until finally hydroxylated in the kidney may explain this "resistance". Both renal impairment and high circulating phosphate levels may contribute to reduced 1,25-DHCC production (p. 235). The previously puzzling fact that the onset of hypocalcaemia is very rapid in renal failure in spite of adequate hepatic stores of vitamin D might now be explained: if active vitamin is formed in the kidney as required, failure of this mechanism would produce rapid effects. By contrast, dietary deficiency of the vitamin, with normally functioning kidneys, would not be evident until all these stores had been utilised. Low plasma protein levels sometimes contribute to the reduction of total calcium. Tetany rarely occurs, possibly because the accompanying acidosis keeps the ionised calcium above tetanic levels.

The resulting low plasma ionised calcium concentration stimulates PTH production, with the consequences described above. However, in renal glomerular failure, there is hyperphosphataemia. In chronic cases a rising alkaline phosphatase level heralds the onset of the bone changes of osteomalacia or rickets.

Osteomalacia, usually mild, is common in chronic liver disease, especially when there is cholestasis. Although 25-HCC levels are usually low, they can be corrected by vitamin D supplements. It is therefore unlikely that there is sufficient impairment of 25-hydroxylation to account for the osteomalacia. Factors such as reduced vitamin D intake and impaired absorption due to bile salt deficiency are probably more important.

Pseudohypoparathyroidism.—In this rare inborn error circulating levels of parathyroid hormone are high. However, because in most cases neither the kidney nor bone can respond to the hormone, the biochemical changes of hypoparathyroidism, rather than of hyperparathyroidism, are present.

Acute pancreatitis.—*Temporary* hypocalcaemia frequently follows an attack of pancreatitis. In most of these cases there is a parallel fall in protein, and ionised levels are probably normal. It has been suggested that precipitation of calcium soaps (calcium salts of fatty acids) following the local release of excess fatty acids from fats by the liberated lipase, contributes to the hypocalcaemia. It does not usually cause symptoms or require treatment, and is not usually a problem of differential diagnosis of hypocalcaemia.

Diseases Associated with Low Circulating Parathyroid Hormone Levels

A low level, or absence, of circulating PTH leads to the changes described on p. 233.

Primary parathyroid hormone deficiency (hypoparathyroidism).— Hypoparathyroidism may be due to primary atrophy of the parathyroid glands (probably of autoimmune origin—compare Hashimoto's disease, p. 168), or more commonly to surgical damage, either directly to the glands or to their blood supply during partial thyroidectomy (during total thyroidectomy or laryngectomy removal of the glands is almost inevitable). Evidence for asymptomatic hypocalcaemia should always be sought after partial thyroidectomy because of the danger of cataracts (p. 237): early post-operative parathyroid insufficiency may recover, but a low calcium level persisting for more than a few weeks should be treated. Even with highly skilled surgery there is a danger of parathyroid damage during this operation. Post-thyroidectomy hypocalcaemia may not always be due to parathyroid damage, but may be due to the rapid entry of calcium into depleted bone when the thyroid hormone levels fall to normal (see p. 236): it may be these cases that appear to recover from hypoparathyroidism.

The clinical symptoms are those due to a low ionised calcium concentration (p. 237).

Secondary suppression of parathyroid hormone secretion.—Just as hypocalcaemia from any cause (except hypoparathyroidism) leads to secondary hyperparathyroidism, so hypercalcaemia (unless due to inappropriate PTH secretion) suppresses PTH secretion, with the consequences described on p. 233.

The causes of hypercalcaemia with suppression of PTH secretion are:
vitamin D excess;
idiopathic hypercalcaemia of infancy;
sarcoidosis;
thyrotoxicosis;
malignant disease of bone and myeloma (possibly);
milk-alkali syndrome.

The *clinical picture* in these cases is due to the high ionised calcium level (p. 236). Decalcification of bone only occurs in this group in the hypercalcaemia of thyrotoxicosis and in sarcoidosis. The alkaline phosphatase level is usually normal in conditions in which bone is not involved.

*Hypercalcaemia due to vitamin D excess.—*Overdosage with vitamin D increases calcium absorption and may cause dangerous hypercalcaemia. PTH secretion is suppressed, and phosphate levels are usually normal. However, as vitamin D has a phosphaturic effect (p. 235) plasma phosphate levels may occasionally be in the low normal range.

Such overdosage may be caused by over-vigorous treatment of hypo-calcaemia, and this therapy should always be controlled by frequent plasma calcium and alkaline phosphatase estimations. Occasionally patients overdose themselves with vitamin D. In any obscure case of hypercalcaemia a drug history should be taken.

The hypercalcaemia associated with two diseases is thought to be due to oversensitivity to the action of vitamin D. These diseases are:
idiopathic hypercalcaemia of infancy;
hypercalcaemia of sarcoidosis.

Idiopathic hypercalcaemia of infancy.—In these days of milk and vita-min supplements rickets is a rare disease in Great Britain. By contrast, the number of infants presenting with hypercalcaemia of obscure origin increased during the 1950s. The incidence of this syndrome has declined since the recommended dosage of vitamin D for infants was reduced in 1957, although the correlation between these two variables is uncertain. This is still a rare disease. An even rarer and more severe form—an inborn error—is associated with, amongst other stigmata, mental deficiency.

Hypercalcaemia of sarcoidosis.—Hypercalcaemia as a complication of sarcoidosis may also be due to vitamin D sensitivity. *Chronic beryl-lium poisoning* produces a granulomatous reaction very similar to that of sarcoidosis, and may also be associated with hypercalcaemia: beryl-lium is used in the manufacture of fluorescent lamps and in several other industrial processes.

Other diseases associated with hypercalcaemia are:
Hypercalcaemia of thyrotoxicosis.—Thyroid hormone probably has a direct action on bone (p. 236). In rare cases of very severe thyrotoxi-cosis significant hypercalcaemia may occur. If the hypercalcaemia fails to respond when hyperthyroidism is controlled, the possibility of co-existent hyperparathyroidism should be borne in mind, since some patients have multiple endocrine abnormalities.

Malignant disease of bone.—The hypercalcaemia due to ectopic PTH production by malignant tumours has been described on p. 240. A different syndrome has been reported in cases with multiple bony metastases, or with myeloma: in these the parallel rise of phosphate is said to indicate that the hypercalcaemia is caused by direct breakdown of bone by the local action of malignant deposits. Critical examination of the findings in most of these cases shows a rise of plasma urea accompanying that of phosphate, and such findings are compatible with the action of excess PTH with renal glomerular failure. Moreover, there is little correlation between the incidence of hypercalcaemia and the rate of extension of the bony lesions. Probably most, if not all, such cases are due to ectopic PTH production.

The raised alkaline phosphatase level associated with bony secondary deposits is due to the resultant stimulation of local osteoblasts (p. 238). This osteoblastic reaction does not occur in myelomatosis, in which the bone alkaline phosphatase level is therefore normal. This fact may be a useful pointer to the diagnosis of myeloma in the presence of extensive bony deposits of unknown origin.

The paraproteins of myeloma do not bind calcium to any significant extent, and are unlikely to account for hypercalcaemia in these cases.

Milk-alkali syndrome.—Hypercalcaemia has been reported due to massive intake of calcium in milk, when alkalis are also being taken, during treatment of peptic ulcer. The hypercalcaemia is probably the consequence of the very high calcium intake, and the alkalis, by causing glomerular damage, may temporarily aggravate the hypercalcaemia by delaying excretion of the excess ion.

This is an exceedingly rare cause of hypercalcaemia nowadays. Most modern antacids are acid adsorbents rather than alkalis, and do not cause alkalosis. The syndrome should not be diagnosed in the absence of alkalosis, nor until other causes have been excluded. In true milk-alkali syndrome the calcium level falls rapidly when therapy is stopped.

Calcium stimulates gastrin secretion (p. 434) and there is a well-documented association between peptic ulceration and parathyroid adenomata: the association of hypercalcaemia with medication for dyspepsia should be assumed to be due to primary hyperparathyroidism until proved otherwise.

Diseases of Bone Not Affecting Plasma Calcium Levels

Diseases producing biochemical abnormalities, or important in the differential diagnosis from those already mentioned, are:

osteoporosis.—In this disease the primary lesion is a reduction in the mass of bone matrix with a secondary loss of calcium. Clinically it may resemble osteomalacia, but normal plasma calcium, phosphate and alkaline phosphatase concentrations (especially the latter) are more in favour of osteoporosis;

Paget's disease of bone.—Calcium and phosphate levels are rarely affected in Paget's disease. The alkaline phosphatase level is typically very high;

renal tubular disorders of phosphate reabsorption.—This group of diseases comprises a number of inborn errors of renal tubular function in which *phosphate is not reabsorbed normally* from the glomerular filtrate. They usually present as cases of *rickets* or *osteomalacia*, but, unlike the usual form of this syndrome, response to vitamin D therapy is poor: large doses, together with oral phosphate, may be needed to produce an effect. The syndrome has, therefore, been called "*resistant rickets*". The defect in phosphate reabsorption may be an isolated one, as in *familial hypophosphataemia*, or part of a more general reabsorption defect as seen in the *Fanconi syndrome* (p. 14). In these cases phosphate deficiency is the primary lesion, and failure to calcify bone may be due to this. Plasma levels of phosphate are usually very low and fail to rise when vitamin D alone is given: there is a relative phosphaturia. The osteomalacia is reflected in the high plasma alkaline phosphatase levels. Plasma calcium concentration is usually normal, and in this respect the findings differ from those of classical osteomalacia; because of the normocalcaemia there is rarely evidence in the bone of hyperparathyroidism.

DIFFERENTIAL DIAGNOSIS

In the preceding sections diseases of calcium metabolism have been discussed according to their aetiology, in an attempt to explain biochemical and clinical findings. Clinically they more often present as a problem of differential diagnosis of the causes of an abnormal plasma calcium concentration.

DIFFERENTIAL DIAGNOSIS OF HYPERCALCAEMIA

Primary hyperparathyroidism must, in its early stages, exist with plasma calcium levels within the normal range, but diagnosis at this time is almost impossible. A plasma calcium concentration in the upper normal range, with that of phosphate in the low normal range, in a patient with recurrent calcium-containing renal calculi, is suggestive of primary hyperparathyroidism: in such a case plasma calcium levels should be estimated at three-monthly intervals and if primary hyperparathyroidism is present, these will eventually become unequivocally raised.

In the presence of hypercalcaemia the causes must be differentiated. These fall into three groups:
raised protein-bound, with normal ionised, calcium;
raised ionised calcium due to inappropriate PTH excess;
raised ionised calcium due to other causes, and associated with appropriately low PTH levels.

Plasma Levels

Raised protein-bound calcium.—Mild hypercalcaemia may be due to a raised albumin concentration, with no change in that of ionised calcium (p. 232). The only *in vivo* cause of this is dehydration with haemoconcentration (p. 42), and in such cases the clinical state of the patient, and the raised plasma *protein* and *haemoglobin* levels, should provide the clue. Such hypercalcaemia will disappear when the patient is rehydrated.

A more common cause of a raised protein bound calcium is the arte-factual one caused by prolonged stasis during venepuncture (p. 455). When hypercalcaemia is found, especially if mild, the analysis should be repeated on a specimen taken without stasis.

In both the *in vivo* and artefactual causes of a raised protein-bound calcium, plasma phosphate tends to rise because of leakage from cells (compare potassium).

Calcium levels, especially serial ones, should never be interpreted without an accompanying protein level. If the cause of a change in protein concentration is uncertain, the serum albumin should be esti-mated, as this is the calcium-binding fraction. In most cases, however, total protein estimation is adequate.

Raised ionised calcium.—Hypercalcaemia found in a specimen taken without stasis from a well-hydrated patient can be assumed to be due to a rise in the physiologically active ionised calcium. It should be noted that a normal total plasma calcium in the presence of significant hypo-proteinaemia also suggests an increase in the ionised fraction.

An accompanying low plasma *phosphate* is suggestive of inappropriate PTH secretion due to primary or tertiary hyperparathyroidism, or to ectopic hormone production. In the presence of a high plasma urea level the presence of hyperphosphataemia does not exclude these causes, and because renal disease also tends to lower calcium levels the plasma urea, as well as phosphate and protein, should always be estimated when calcium metabolism is being investigated. If this estimation is normal an increase of both calcium and phosphate suggests that these three causes are unlikely.

In primary hyperparathyroidism the plasma *alkaline phosphatase* con-centrations rise significantly only at a relatively late stage of the disease. A finding of a very high level with no clinical evidence of severe bone disease makes the diagnosis improbable. In the hypercalcaemia of malig-nant disease the level of this enzyme may rise because of bony or hepatic metastases, and other signs of these should be sought. The alkaline phosphatase concentration is normal in myelomatosis (p. 324) unless there is hepatic infiltration by the malignant plasma cells: in such a rare case the enzyme can be shown to be of hepatic origin. In sarcoidosis, as in malignant disease, the plasma alkaline phosphatase level may in-crease because of bony or hepatic infiltration. In the rare hypercalcaemia of severe thyrotoxicosis there may be some elevation of plasma concentra-tions of the enzyme, but in all other uncomplicated causes of hyper-calcaemia it is normal.

A marked increase in plasma *protein* level, in the absence of clinical dehydration or of stasis in taking the blood, suggests myelomatosis as a cause, and this will be evident in the electrophoretic pattern. If this cause is considered bone marrow aspiration must be performed, and the presence of Bence Jones protein sought (p. 322), even in the presence of normal serum protein fractions.

Steroid Suppression Test

This test is by far the most useful one in the differential diagnosis of significant hypercalcaemia (plasma calcium above about 3 mmol/l) of obscure origin: at lower levels of plasma calcium it is difficult to interpret. In most cases it differentiates primary or tertiary hyperparathyroidism from any other cause. Large doses of hydrocortisone (or cortisone) result in a fall of high plasma calcium levels to within the normal range in almost all cases *except primary* (or tertiary) *hyperparathyroidism*. Exceptions do occasionally occur, but in most of them the clinical diagnosis is clear and the test unnecessary (for instance in very advanced malignant disease of bone or very severe hyperthyroidism). Details of the test are given in the Appendix (p. 257).

The reason for this difference in reaction to steroids is not clear. It is particularly interesting that the calcium level in many cases of ectopic PTH production is suppressed, whereas it fails to be so in those in which the hormone is produced in the parathyroid glands.

Urinary Calcium

This frequently requested estimation is of little diagnostic value in the differential diagnosis of hypercalcaemia. In the presence of normal renal function, hypercalcaemia from any cause, by increasing the load on the glomerulus, causes hypercalcuria: although excess PTH, by increasing renal tubular reabsorption of calcium, might be expected to reduce urinary excretion below that appropriate to the plasma level, the "normal" range is too wide for the test to be helpful. Causes of a high urinary calcium excretion without hypercalcaemia are the so-called "*idiopathic hypercalcuria*", some cases of *osteoporosis* in which calcium cannot be deposited in normal amounts in bone because of a deficient matrix, and acidosis, in which ionisation of calcium is increased.

Calcium excretion may be normal or low if renal glomerular function is impaired, even in the presence of hypercalcaemia (whether due to hyperparathyroidism or to other causes).

Plasma Parathyroid Hormone Levels

Parathyroid hormone levels may be measured by radioimmunoassay. The specificity of the assay is still uncertain, and not all proven cases of primary hyperparathyroidism can be shown to have raised PTH levels. In our experience careful attention to the plasma calcium levels, together with the clinical history and the other findings discussed in this chapter, still provide the most useful information.

TABLE XXII
Differential Diagnosis of Hypercalcaemia

Diagnosis	Plasma Phosphate	Plasma Proteins	Plasma Alk. phos.	Steroid suppression	Comments
Due to Raised Ionised Calcium					
Group 1. Inappropriate Parathyroid Hormone					
Primary or tertiary hyperparathyroidism	N or ↓	N	N or ↑	No	Calcium and phosphate changes masked in uraemia.
Malignancy with ectopic hormone production	N or ↓	N (↑ myeloma)	N or ↑ (N in myeloma)	Yes (usually)	
Group 2. Appropriately low PTH					
Vitamin D					
Vitamin D overdosage	Variable	N	N	Yes	Rare
Sarcoidosis	Variable	γ glob. ↑	N or ↑	Yes	
Idiopathic hypercalcaemia of infancy	Variable	N	N	Yes	Very rare.[HCO$_3$⁻]↑
Milk-Alkali Syndrome	N or ↑	N	N	Yes	
Thyrotoxicosis	N or ↑	N	N or ↑	Yes (usually)	Severe thyrotoxicosis. Thyroid function tests abnormal.
Due to Raised Protein Bound Calcium					
Dehydration	N or ↑	↑	N		Clinical signs of dehydration. Urea ↑ Corrected by rehydration.
Artefactual (excess stasis)	N or ↑	↑	N		Specimen taken without stasis gives normal values.

Hypercalcaemia due to raised protein-bound calcium should *NOT* be treated.

TESTS USED IN THE DIFFERENTIAL DIAGNOSIS OF HYPOCALCAEMIA

As in the case of hypercalcaemia, the causes of hypocalcaemia fall into three groups:

reduced protein-bound, with normal ionised, calcium;
reduced ionised calcium due to primary PTH deficiency;
reduced ionised calcium due to other causes and associated with appropriately high PTH levels.

Plasma Levels

Reduced protein-bound calcium.—As in the case of hypercalcaemia, hypocalcaemia due to a low protein-bound fraction must be distinguished from that due to a low ionised fraction.

Just as protein-bound calcium concentrations may be raised in haemoconcentration due to dehydration, so they may be reduced in overhydration. In such cases the protein and calcium levels will return to normal as hydration is corrected.

In any condition associated with significant hypoalbuminaemia (p. 309) the protein-bound calcium may be reduced. In malnutrition, whether due to malabsorption or to a deficient diet, there may be an accompanying reduction of ionised calcium due to associated vitamin D and calcium deficiency. In such cases treatment should not aim to restore the total calcium level to normal, but to maintain that of alkaline phosphatase within the normal range (i.e. to prevent osteomalacia).

Reduced ionised calcium levels.—If plasma albumin levels are normal, or only slightly reduced, a significant reduction in total calcium concentration can be assumed to be due to that of ionised calcium.

Accompanying high *phosphate* levels suggest either reduced circulating PTH, or the failure of bone and kidney to respond to it in the very rare pseudohypoparathyroidism. Hypophosphataemia with hypocalcaemia is associated with calcium and vitamin D deficiency.

High plasma *alkaline phosphatase* levels also suggest that calcium deficiency is the primary abnormality, which has led to increased PTH secretion and resultant decalcification of bone (secondary hyperparathyroidism). This is most commonly due to malabsorption or to dietary deficiency. A dietary history should be taken and tests for malabsorption (p. 281) performed if indicated by the bowel history.

A high plasma *urea* suggests that the hypocalcaemia is due to renal failure, and caution should be exercised in basing treatment on plasma calcium levels alone.

Plasma Parathyroid Hormone Levels

In most cases estimation of PTH levels is unnecessary: the present sensitivity of the method is such that low levels are less significant than

TABLE XXIII

DIFFERENTIAL DIAGNOSIS OF HYPOCALCAEMIA

Diagnosis	Plasma				Comments
	Phosphate	Proteins	Urea	Alk. phos.	
Due to Low Ionised Calcium					
Hypoparathyroidism	↑	N	N	N	
Calcium and vitamin D deficiency	↓	N or ↓	N	↑	May be accompanied by low protein-bound calcium.
Renal failure	↑	N or ↓	↑	N or ↑	Should be treated with caution in absence of bone disease.
Pseudohypoparathyroidism	↑	N	N	N	Very rare.
Due to Low Protein-Bound Calcium					
Overhydration	N	↓	N or ↓	N	Responds to fluid restriction.
Hypoalbuminaemia	N	↓	N or ↓	N	

Hypocalcaemia due to lowered protein-bound calcium should *NOT* be treated.

high ones. In the very rare cases in which pseudohypoparathyroidism is suspected this estimation may sometimes help. Very high levels of PTH together with hypocalcaemia are incompatible with the diagnosis of primary hypoparathyroidism. If secondary hyperparathyroidism has been excluded and if hyperphosphataemia is present, they are suggestive of end organ unresponsiveness (p. 242).

BIOCHEMICAL BASIS OF TREATMENT

Hypercalcaemia

Mild hypercalcaemia.—If raised calcium levels of less than about 3·75 mmol/l (15 mg/dl) are present in a patient without serious symptoms of hypercalcaemia, there is no *immediate* need for urgent therapy. However, treatment should be instituted as soon as a diagnosis is made because of the danger of renal damage.

The patient should be rehydrated and, if possible, the cause should be treated (e.g. a parathyroid adenoma, a primary malignant lesion, hyperthyroidism). If this is not possible hydrocortisone will reduce the levels within a day or two in most cases not due to primary (or tertiary) hyperparathyroidism (see "Steroid Suppression Test", p. 248). It may sometimes be relatively ineffective in malignant hypercalcaemia.

Oral sodium phosphate will tend to precipitate calcium phosphate in the intestine, preventing reabsorption of secreted calcium and thus removing calcium from the body. This treatment may be preferable to steroid therapy in many cases. It may be used, in conjunction with steroids, to treat intractable hypercalcaemia of late malignancy. The effect of oral phosphate, like that of steroids, is only apparent after 24 hours. The osmotic effect of large doses sometimes causes diarrhoea.

Severe hypercalcaemia.—If plasma calcium levels exceed about 3·75 mmol/l (15 mg/dl) treatment is indicated as a matter of urgency, because of the danger of cardiac arrest. The exact level varies in different subjects. If there is any doubt about the degree of urgency the electrocardiogram should be inspected for abnormalities associated with hypercalcaemia.

Most measures available for lowering plasma calcium acutely (i.e. within a few hours) depend, partially at least, on precipitation of insoluble calcium salts. Since some of this precipitation may occur in the kidney, the treatment carries a slight risk of initiating or aggravating renal failure, and this risk should always be weighed against the danger of cardiac arrest. It is small in subjects with normal or only slightly raised plasma urea levels: although many cases show a transient rise of plasma urea at the start of treatment, this usually subsides rapidly.

Solutions for intravenous administration contain a mixture of sodium and potassium phosphates (see Appendix, p. 257).

As with most other extracellular constituents, rapid changes in calcium level may be dangerous because time is not allowed for equilibration across cell membranes (e.g. see urea, p. 23 and sodium, p. 59). The aim of emergency treatment should be to lower calcium temporarily to safe levels, while initiating treatment for mild hypercalcaemia. A too rapid reduction of calcium concentration may induce tetany, or, more seriously, hypotension, even though the calcium is within or above normal levels. (Tetany in the presence of normocalcaemia may also occur during the rapid fall of serum calcium concentration after removal of a parathyroid adenoma.) A slow reduction of calcium level also reduces the risk of renal calcification.

Hypocalcaemia

Asymptomatic hypocalcaemia.—Hypocalcaemia, whatever the cause, if asymptomatic or accompanied by only mild clinical symptoms, is usually treated with large doses of vitamin D by mouth. It is difficult to give enough oral calcium, by itself, to make a significant lasting difference to plasma calcium levels, and vitamin D, by increasing absorption of calcium, is usually adequate without calcium supplementation.

The hypocalcaemia of renal failure should be treated with caution in the absence of signs of osteomalacia (raised alkaline phosphatase, etc.), because of the danger of ectopic calcification in the presence of hyperphosphataemia. Hypocalcaemia associated with low protein levels should not be treated.

Hypocalcaemia with severe tetany.—In the presence of severe tetany hypocalcaemia should be treated, as an emergency, with intravenous calcium (usually as the gluconate).

MAGNESIUM METABOLISM

Magnesium is present with calcium in bone salts, and tends to move in and out of bone with calcium. It is also present in all cells of the body in much higher concentrations than in the extracellular fluid, and therefore tends to enter and leave cells under the same conditions as do potassium and phosphate.

Magnesium can be lost in large quantities in the faeces in diarrhoea.

PLASMA MAGNESIUM AND ITS CONTROL

About 35 per cent of the plasma magnesium, like calcium, is protein bound. However, less is known about the importance of this than in the case of calcium.

The mechanism of control of magnesium levels is poorly understood, but may involve the action of PTH and perhaps aldosterone.

CLINICAL EFFECT OF ABNORMAL PLASMA MAGNESIUM LEVELS

Hypomagnesaemia

This causes symptoms very similar to those of hypocalcaemia. If a patient with tetany has normal calcium, protein and bicarbonate (or, more accurately, blood pH) levels, blood should be taken for magnesium estimation. If there is good reason to suspect magnesium deficiency (e.g. in the presence of severe diarrhoea), and if the estimation cannot be performed reasonably quickly, intravenous magnesium should be administered as a therapeutic test (see Appendix, p. 259). Less severe magnesium deficiency should be treated orally (Appendix, p. 259).

Hypermagnesaemia

This causes muscular hypotonia, but as this condition is rarely seen in isolation, symptoms are difficult to distinguish from those of co-existent abnormalities (e.g. hypercalcaemia).

CAUSES OF ABNORMAL PLASMA MAGNESIUM LEVELS

Hypomagnesaemia

Excessive loss of magnesium.—Excessive loss of magnesium occurs in *severe*, prolonged diarrhoea, and this is by far the most important cause of a clinical disturbance of magnesium metabolism requiring treatment.

Hypomagnesaemia accompanied by hypocalcaemia.—Magnesium tends to move in and out of bones in association with calcium, and hypocalcaemia is often accompanied by hypomagnesaemia. This may occur in hypoparathyroidism, or during the fall of calcium after the removal of a parathyroid adenoma, especially if severe bone disease is present.

Hypomagnesaemia accompanied by hypokalaemia.—Since magnesium moves in and out of cells with potassium, hypomagnesaemia tends to occur in association with hypokalaemia. These conditions include diuretic therapy and primary aldosteronism (p. 53). Such hypomagnesaemia is rarely of clinical importance.

Hypermagnesaemia

The commonest cause of hypermagnesaemia is probably renal glomerular failure (when plasma potassium is also high). It rarely, if ever, requires treatment on its own, and it responds to measures to treat the underlying condition (e.g. haemodialysis). Magnesium salts should never be administered in renal failure.

SUMMARY

Calcium Metabolism

1. About half the calcium in plasma is bound to protein, and half is in the ionised form.

2. The ionised calcium is the physiologically important fraction and calcium levels should be interpreted together with protein levels.

3. Plasma calcium levels are controlled by parathyroid hormone. Parathyroid hormone secretion is increased if ionised calcium concentrations are reduced.

4. Parathyroid hormone acts:
   on bone, releasing calcium and phosphate into the plasma: if prolonged, this increases osteoblastic activity and increases plasma alkaline phosphatase levels;
   on kidneys, causing phosphaturia, which lowers the plasma phosphate.

5. Symptoms and findings in diseases of calcium metabolism can be related to circulating levels of ionised calcium, to levels of parathyroid hormone, and to renal function.

6. Excessive "inappropriate" levels of parathyroid hormone are present in primary parathyroid disease, or if the hormone is produced at ectopic sites. In these circumstances plasma calcium is high.

7. Increased "appropriate" levels of parathyroid hormone are present whenever plasma calcium levels fall.

8. Parathyroid hormone levels are low in hypoparathyroidism (associated with a low serum calcium), or in the presence of hypercalcaemia other than that of inappropriate parathyroid hormone production. The causes and differential diagnosis of hypercalcaemia and hypocalcaemia are summarised in Tables XXII and XXIII.

Magnesium Metabolism

1. Hypomagnesaemia may cause tetany in the absence of hypocalcaemia.

2. The commonest cause of significant hypomagnesaemia is severe diarrhoea.

3. Magnesium levels tend to follow those of calcium (in and out of bone) and potassium (in and out of cells).

FURTHER READING

TOMLINSON, S., and O'RIORDAN, J. L. H. (1978). The parathyroids. *Brit. J. hosp. Med.*, **19**, 40.

HAUSSLER, M. R., and McCAIN, T. A. (1977). Basic and clinical concepts related to vitamin D metabolism and action. *New Engl. J. Med.*, **297**, 974 and 1041.

WILLS, M. R. (1974). Hypercalcaemia. *Brit. J. hosp. Med.*, **11**, 279.

APPENDIX

DIAGNOSIS

STEROID SUPPRESSION TEST IN DIFFERENTIAL DIAGNOSIS OF HYPERCALCAEMIA
(Dent, C. E., and Watson, L. *Lancet*, 1968 **2**, 662)

All specimens for calcium estimations should be taken without venous stasis (p. 455).

Procedure

1. At least two specimens are taken on two different days before the test starts for estimation of plasma calcium, protein and urea concentrations.
2. The patient takes 120 mg of oral hydrocortisone a day in divided doses (40 mg, 8-hourly), for 10 days.
3. Blood is taken for estimation of plasma calcium, protein and urea concentration on the 5th, 8th and 10th day.
4. After the 10th day the dose of hydrocortisone is gradually reduced, as usual after steroid therapy.

Calculations

During administration of steroids there may be significant fluid retention. This will dilute the protein-bound calcium and may lead to an apparent fall of calcium. To correct this it is desirable to "correct" total calcium results for changes in protein concentration, and a "standard" total protein of 72 g/l is used in the calculation.

For every 3·7 g/l that the protein concentration is *below* 72 g/l, 0·06 mmol/l (0·25 mg/dl) is *added* to the calcium level.

For every 3·7 g/l that the protein concentration is *above* 72 g/l, 0·06 mmol/l (0·25 mg/dl) is *subtracted* from the calcium level.

This correction is a rough one, and is inaccurate in the presence of very abnormal plasma protein concentrations.

Interpretation

Failure of plasma calcium concentration to fall to within the normal range by the end of the test is strongly suggestive of *primary hyperparathyroidism*. For exceptions see p. 248.

TREATMENT

HYPERCALCAEMIA

Emergency Treatment of Hypercalcaemia

The solution for intravenous infusion contains a mixture of mono- and dihydrogen phosphate such that the pH is 7·4.

Na_2HPO_4 (anhydrous)—11·50 g⎫
KH_2PO_4 (anhydrous)— 2·58 g⎭ made up to 1 litre with water.

500 ml of this should be infused over 4–6 hours. This 500 ml will contain a total of 81 mmol of sodium, 9·5 mmol of potassium and 50 mmol of phosphorus.

Long-Term Oral Phosphate Treatment of Hypercalcaemia

Oral phosphate is given as the disodium or dipotassium salt. The choice depends on the serum potassium level.

1. The *solution* should contain 32 mmol of phosphorus in 100 ml.

This is　　　46 g/litre of Na_2HPO_4 (anhydrous)
or　　　　　56 g/litre of K_2HPO_4 (anhydrous)

The dose is 100–300 ml per day in divided doses. 100 ml contains approximately 65 mmol of sodium or potassium respectively.

2. *Phosphate Sandoz Effervescent tablets* contain:

Phosphorus　　　16 mmol (500 mg)⎫
Sodium　　　　　21 mmol ⎬per tablet
Potassium　　　　3 mmol⎭

The dose is 1–6 tablets daily.

3. Alternatively the phosphate can be given as *sodium cellulose phosphate* (Whatman) 5 g (the contents of one sachet), three times a day.

Note that hypokalaemia is a common accompaniment of hypercalcaemia. When this is present the phosphate preparation of choice is K_2HPO_4.

HYPOCALCAEMIA

Emergency Treatment of Hypocalcaemia

Calcium Gluconate Injection (B.P.)—0·23 mmol (9 mg) calcium/ml.
Dose—10 ml intravenously in the first instance.

Long-Term Treatment of Hypocalcaemia

Vitamin D Therapy

Warning.—There may be several days' lag in response to either starting or stopping therapy with vitamin D: plasma calcium levels may continue to rise for weeks after stopping therapy. Treatment should be carried out with caution, using intelligent anticipation based on laboratory assessment of plasma calcium and, in osteomalacia, alkaline phosphatase levels. *Patients on maintenance doses should be seen at regular intervals, because requirements may change, with the danger of the development of hypercalcaemia.* Calcium levels must be assessed with those of plasma protein.

Osteomalacia with raised alkaline phosphatase levels.—1. *Initially*—500 000 IU (12·50 mg) daily orally or IM. Rarely 750 000 to 1 000 000 IU (18·75–25·00 mg) may be needed.

2. When the plasma calcium level reaches about 1·75 mmol/l (7·0 mg/dl), and the alkaline phosphatase is near normal, reduce to 100 000 to 250 000 (2·50–6·25 mg) daily. At this stage bone calcification is nearing normal, and hypercalcaemia may develop rapidly if high dosage is continued.

3. *Maintenance doses* vary between 25 000 and 100 000 IU (0·63–2·50 mg),

and are determined by trial and error. The aim should be to keep plasma calcium, phosphate, and alkaline phosphatase levels normal.

Hypocalcaemia of hypoparathyroidism with normal alkaline phosphatase levels.—The lower dosage of 100 000 IU (2·50 mg) daily should be used from the outset, monitoring being based on plasma calcium levels; again, the exact dose is found by trial and error.

Oral Calcium Tablets

Calcium Gluconate (B.P.C.) (600 mg) = 1·35 mmol (54 mg) of calcium per tablet.

Calcium Gluconate Effervescent (B.P.C.) (1000 mg) = 2·25 mmol (90 mg) of calcium per tablet.

"Sandocal" Effervescent (4·5 g calcium gluconate) = 10 mmol (400 mg) of calcium per tablet.

Calcium Sandoz Chocolate (1·5 g calcium gluconate) = 3·38 mmol (135 mg) of calcium per tablet.

Calcium Lactate (B.P.) (300 mg) = 1·4 mmol (55 mg) of calcium per tablet.

Calcium Lactate (B.P.) (600 mg) = 2·8 mmol (110 mg) of calcium per tablet.

MAGNESIUM

Emergency Treatment of Magnesium Deficiency

Magnesium Chloride ($MgCl_2.6H_2O$)—20 g/100 ml.

This contains 1 mmol of magnesium per ml and may be added to other intravenous fluid. If renal function is normal, up to 40 mmol (40 ml) may be infused in 24 hours.

Oral Magnesium Therapy

Magnesium Chloride ($MgCl_2.6H_2O$)—20 g/100 ml.

The dose is 5–10 ml, q.d.s. Each 5 ml contains 5 mmol of magnesium. Note, however, that magnesium is poorly absorbed.

Chapter XII

INTESTINAL ABSORPTION:
PANCREATIC AND GASTRIC FUNCTION

THE most important function of the gastro-intestinal tract is the digestion and absorption of nutrients. Many substances which are absorbed from the food also enter the intestinal tract, so that net absorption may be less than true absorption (insorption): for instance, dietary fat is almost completely absorbed under physiological circumstances, and almost all that found in normal faeces is derived from intestinal cells—a fact of importance in interpretation of faecal fat values (p. 281).

Digestion of larger molecules involves the action of intestinal enzymes: if these are to act under optimal conditions the nutrient molecules must be dispersed as much as possible, firstly by mechanical action such as chewing, and secondly by mixture of the food with fluid. Enzymes act best in the presence of certain electrolytes and at certain pH values: the fluid secreted contains these ions and varying amounts of hydrogen ion. Before the smaller water-soluble molecules can pass through the intestinal cells they must be in solution and some of these, too, require fluid at the correct pH. The very large amounts of electrolyte and water secreted into the gastro-intestinal tract have already been mentioned, and it has been pointed out that normally about 99 per cent of the sodium and water are reabsorbed (p. 32). In this context the gastro-intestinal tract acts in much the same way as the kidney (if we consider that gastro-intestinal secretions are in some ways analogous to the glomerular filtrate): water, sodium and potassium are reabsorbed throughout the small intestine, as they are in the proximal renal tubule. Final adjustment is made in the colon where, for instance, aldosterone stimulates sodium reabsorption and potassium secretion as it does in the distal renal tubule, and where final water reabsorption occurs. Disturbances of water and electrolyte metabolism and of hydrogen ion homeostasis are therefore common in diarrhoea due to extensive small intestinal or colonic disease, or when there is direct loss of fluid and electrolyte from the upper intestinal tract through vomiting or fistulae. Such disturbances also occur if absorption from the upper intestinal cells is so grossly impaired that the amounts of fluid and electrolyte entering distal parts exceed the reabsorptive capacity: this effect is similar to that of an increased glomerular filtration rate and of osmotic diuretics on the volume and composition of urine (p. 39 and p. 9). These disturbances of electrolyte, water and hydrogen ion homeostasis have been discussed in Chapters II, III, and IV.

In this chapter we are concerned with disturbances of the absorptive mechanisms, either due to abnormality of the absorptive cells of the small intestine, or to failure of normal digestion of food. Unless malabsorption is gross, causing severe intestinal hurry, water, electrolyte and hydrogen ion disturbances are relatively unimportant in malabsorption syndromes. The effects are largely those of disturbed nutrition, since nutrient cannot pass through the intestinal cells.

NORMAL DIGESTION AND CONVERSION OF NUTRIENT TO AN ABSORBABLE FORM

Complex molecules such as those of protein, polysaccharide and fat are usually broken down by digestive enzymes. This process starts in the mouth, where food is mechanically broken down by chewing and is mixed with saliva containing *amylase*. In the stomach further fluid is added and the low pH initiates protein digestion by *pepsin*. The stomach also secretes *intrinsic factor* necessary for absorption of vitamin B_{12}. However, quantitatively by far the most important digestion takes place in the duodenum and upper jejunum, where large volumes of alkaline fluid are added to the already liquid food. Pancreatic enzymes in this fluid convert protein to amino acids and small peptides, polysaccharides to mono- and disaccharides, and oligosaccharides (consisting of a small number of monosaccharide units), and fat to monoglycerides and fatty acids.

Severe generalised malabsorption due to failure of *digestion* is most commonly due to pancreatic disease.

NORMAL ABSORPTION

Normal absorption depends on:

the integrity and normal surface area of absorptive cells;

the presence of the substance to be absorbed in an absorbable form (and therefore on normal digestion);

a normal ratio of speed of absorption to speed of passage of contents through the intestinal tract.

The *absorptive area* of the intestine is normally very large. Macroscopically the mucosa forms *folds*, increasing the area considerably. Microscopically, these folds are covered with *villi*, lined with absorptive cells: this further increases the area about eight-fold.

If these villi are flattened, as they are in, amongst other conditions, gluten-sensitive enteropathy, the absorptive area is much reduced.

Each intestinal absorptive cell has on its surface a large number of

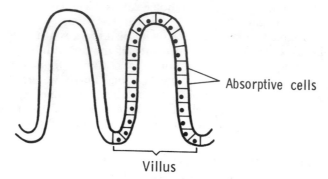

Absorptive cells

Villus

minute projections (*microvilli*) detectable with the electron microscope, further increasing absorptive area about 20-fold.

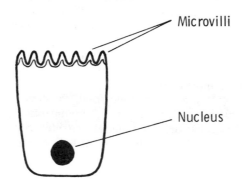

Microvilli

Nucleus

Minute spaces exist between the microvilli (*microvillous spaces*).

To be absorbable, the substances must be in the form of relatively *small molecules*, such as result from normal digestion. Vitamin B_{12} absorption requires the formation of a complex with intrinsic factor, which can attach to the intestinal cells and bring the substance in a spatially advantageous position for absorption. The method of absorption depends on whether the molecule is *water soluble* or *lipid soluble*. Lipid-soluble nutrients frequently share the mechanisms for fat absorption. Absorption may be *active*, in which case it can occur against a physicochemical gradient, or *passive* when it passes along physicochemical gradients.

LIPID ABSORPTION

Digestion of Neutral Fats

Neutral fats (triglycerides) contain glycerol esterified with three fatty acids. These fatty acids are usually three different ones (R_1, R_2, R_3).

$$CH_2OR_1$$
$$CHOR_2$$
$$CH_2OR_3$$

In shorthand notation this can be written

Bile salts emulsify fat entering the duodenum. They inhibit pancreatic *lipase*, but this inhibition is counteracted by a peptide coenzyme, *colipase*. Colipase also reduces the pH optimum of lipase from about 8·5 to about 6·5. A lipase-colipase-bile salt complex forms at the fat/water interface, and the lipase hydrolyses the fatty acid-glycerol bonds, especially in positions 1 and 3. The end result is production of some diglycerides, but mainly the 2-monoglycerides together with free fatty acids.

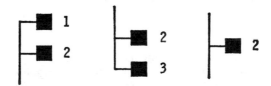

Micelle Formation

The resultant monoglyceride and fatty acids aggregate with bile salts to form a *micelle*: the micelle also contains free *cholesterol* (liberated by hydrolysis in the lumen from cholesterol esters) and *phospholipids*, as well as *fat-soluble vitamins* (A, D and K). The diameter of the micelle is between 100 and 1000 times as small as that of the emulsion particle and is small enough to pass through the microvillous spaces: it is also negatively charged, which is necessary if it is to pass through these spaces.

Lipids in the Intestinal Cell

In the intestinal cell triglycerides are resynthesised from monoglycerides and fatty acids. Cholesterol is re-esterified. The triglycerides, cholesterol esters and phospholipids, together with the fat soluble vitamins, are coated with a layer of protein (manufactured in the cell) to form *chylomicrons* (p. 218). These are readily suspended in water and pass into the lymphatic circulation, probably through the cell wall.

Absorption of Free Fatty Acids

Some short- and medium-chain free fatty acids pass through the intestinal cell into the portal blood stream.

From this account it will be seen that absorption of the following constituents of food,

| neutral fat | phospholipids |
| cholesterol | fat-soluble vitamins, |

depends on:

normal emulsification of fats, and therefore the presence of *bile salts*;

normal digestion of neutral fat by *lipase* and therefore normal pancreatic function;

normal *intestinal mucosa* for formation of chylomicrons and a normal absorptive area.

Bile salts are synthesised in the liver from cholesterol. They enter the intestinal tract in the bile and are actively reabsorbed in the distal ileum. They are recirculated to the liver and resecreted into the bile (the "enterohepatic circulation"). Resection of the distal ileum may prevent reabsorption and re-use of bile salts, and stimulate their resynthesis from cholesterol. Although the plasma cholesterol concentration may fall as a result, this is probably of little clinical importance.

Some intestinal bacteria convert bile salts to bile acids and thus inactivate them: some of the bile acids formed may damage the cells of the intestinal wall.

CARBOHYDRATE ABSORPTION

Polysaccharides such as starch and glycogen are hydrolysed by amylase (salivary and pancreatic, the latter being of the greater importance) mainly to 1:4 disaccharides such as maltose (glucose+glucose): a few larger branch-chain saccharides remain.

Disaccharides (maltose, sucrose (glucose+fructose) and lactose (glucose+galactose)) are hydrolysed to their constituent monosaccharides by the appropriate disaccharidase (maltase, sucrase or lactase): isomaltase hydrolyses isomaltose (two glucose molecules joined by 1:6 linkages). These enzymes are not present in significant amount in intestinal secretions, but are on the surface ("brush border") of the intestinal cell where the hydrolysis takes place.

Monosaccharides are absorbed in the duodenum. Glucose and galactose are probably absorbed by a common active process, while fructose is absorbed by a different mechanism.

Thus carbohydrate absorption depends on:

the presence of *amylase*, and therefore on normal pancreatic function (polysaccharides only);

the presence of *disaccharidases* on the intestinal cell (disaccharides);

normal *intestinal mucosa* with normal active transport mechanisms (monosaccharides).

Note that absorption of polysaccharides requires normal function of all three mechanisms.

PROTEIN ABSORPTION

The diet is not the only source of protein in the intestinal lumen: a significant proportion of that absorbed originates from intestinal secretions and desquamated mucosal cells.

Ingested protein is broken down by pepsin, followed by trypsin and the other proteolytic enzymes of pancreatic juice. The products are amino acids and small peptides. Many peptides are further hydrolysed by peptidases on the brush border.

Amino acids are actively absorbed in the small intestine.

Small peptides are actively absorbed into the cell intact: their absorption is independent of that of amino acids and, with a few exceptions, they are hydrolysed intracellularly.

Protein absorption therefore depends on:

the presence of *pancreatic proteolytic enzymes* and therefore on normal pancreatic function;

normal *intestinal mucosa* with normal active transport mechanisms.

VITAMIN B_{12} ABSORPTION

Vitamin B_{12} can be absorbed only when it has formed a complex with *intrinsic factor*, a glycoprotein secreted by the stomach: in this form it can bind to the intestinal cell where it is absorbed, mainly in the lower ileum. Its absorption therefore depends on normal:

gastric secretion;

intestinal mucosa in the lower ileum.

Some intestinal bacteria require vitamin B_{12} for growth and prevent its absorption by mechanisms probably involving competition with the intestinal cells for it. Normal absorption therefore also depends on normal *intestinal flora*.

ABSORPTION OF OTHER WATER-SOLUBLE VITAMINS

Most of the water-soluble vitamins (C and all those of the B group except B_{12}) are absorbed, probably by special mechanisms, mainly in the upper small intestine. Clinical deficiencies of all but folate are relatively uncommon in malabsorption syndromes, probably because absorption of these vitamins is independent of that of fat.

ELECTROLYTE AND WATER ABSORPTION

Much electrolyte and water absorption is really reabsorption of the contents of intestinal secretions.

Sodium and potassium are probably absorbed by an active process throughout the small intestine, much as they are in the renal tubule: many other active transport mechanisms depend on metabolic energy provided by the sodium pump. In the colon sodium-potassium exchange is stimulated by aldosterone.

Chloride absorption in the small intestine probably follows the electrochemical gradient created by absorption of sodium and other cations. In the colon chloride is absorbed in exchange for bicarbonate (see p. 97).

Water is probably absorbed passively, as in the proximal renal tubule, along an osmotic gradient created by absorption of sodium, sugars, amino acids and other solutes. If these solutes cannot be absorbed normally water is "held" in the intestinal lumen (see "dumping syndrome" p. 272 and disaccharidase deficiency p. 275). Final water absorption occurs in the colon.

CALCIUM AND MAGNESIUM ABSORPTION

Calcium is actively absorbed in the upper small intestine, particularly the duodenum and upper jejunum. Normal calcium absorption depends on its presence in an ionised form (it is inhibited by the formation of insoluble salts with phosphate, fatty acids and phytate), and on the presence of vitamin D in the form of 1, 25-dihydroxycholecalciferol (p. 235). Much of the faecal calcium is endogenous and is derived from calcium in intestinal secretions.

Magnesium is also absorbed by an active process and may share in the calcium pathway.

Normal calcium and magnesium absorption depends on:

a *low concentration of fatty acids*, phosphate and phytate in the intestine;

the absorption and metabolism of *vitamin D* and therefore on normal fat absorption;

normal *intestinal mucosa*.

Iron Absorption

Iron is absorbed by an active process in the duodenum and upper jejunum, and absorption is stimulated by anaemia (p. 382). Some of the iron absorbed into the intestinal cell enters the blood but some stays in the cell and is lost into the intestine when this is desquamated (p. 380).

We are now in a position to understand the consequences of disorders of digestion and absorption. From the above account it should be noted that:

pancreatic failure affects absorption of large molecules only (fats, polysaccharides and protein);

disease of the intestinal mucosa affects absorption of small molecules and products of digestion of large molecules;

absence of *bile salts* causes malabsorption of fats and of those substances sharing mechanisms for fat absorption.

MALABSORPTION SYNDROMES

From the clinical point of view the malabsorption syndromes can be divided into those associated with generalised malabsorption of fat (*steatorrhoea*), protein, carbohydrate and other nutrients, and those associated with failure to absorb one or more specific substances.

Generalised Malabsorption (Associated with Steatorrhoea)

Intestinal Disease

Reduction of absorptive surface or *general impairment of transport mechanisms*:

gluten sensitivity causing coeliac disease:

tropical sprue;

idiopathic steatorrhoea;

extensive surgical resection of small intestine.

Extensive infiltration or inflammation of the small intestinal wall (for example, Crohn's disease, amyloidosis, scleroderma).

Increased rate of passage through the small intestine:

post-gastrectomy;

carcinoid syndrome.

Pancreatic Dysfunction (Failure of Digestion)

Chronic pancreatitis.

Fibrocystic disease of the pancreas.

Inactivation of lipase by low pH in the Zollinger-Ellison syndrome (p. 434).

FAILURE OF ABSORPTION OF SPECIFIC SUBSTANCES

Altered bacterial flora (mainly fat, p. 264; and vitamin B_{12}, p. 265):
  blind-loop syndrome
  diverticula
  surgery
  neomycin therapy.
Biliary obstruction (fat and substances dependent on fat absorption, p. 264).
Local disease or surgery (for instance, disease of the terminal ileum affects vitamin B_{12} absorption).
Gastric atrophy causing malabsorption of vitamin B_{12} (pernicious anaemia).
Disaccharidase deficiency:
  congenital
  acquired.
Protein-losing enteropathy.
Isolated transport defects resulting from inborn metabolic errors (p. 357).

GENERALISED MALABSORPTION

Gluten-sensitive enteropathy (coeliac disease) occurs at any age. Sensitivity to gluten (in wheat germ) causes flattening of intestinal villi and a considerable reduction in absorptive area. The flattening of the villi may be demonstrated on intestinal biopsy specimens. These cases respond to treatment with a gluten-free diet, but the improvement may not be evident for some months.

In **tropical sprue** there is also flattening of the villi, but these cases do not respond to a gluten-free diet. They do respond to broad-spectrum antibiotics and folate, suggesting that a bacterial factor is important in the aetiology of the disease.

The term **idiopathic steatorrhoea** should be reserved for those cases of steatorrhoea which do not respond either to a gluten-free diet or to broad-spectrum antibiotics, and for which no cause can be found.

Extensive surgical resection of the small intestine may so reduce the absorptive area as to cause malabsorption.

Extensive infiltration and inflammation of the small intestinal mucosa may damage its ability to carry out absorptive processes. This may be aggravated by the presence of altered bacterial flora in these conditions.

After gastrectomy normal mixing of food with fluid, acid and pepsin does not occur in the stomach, so that the activity of enzymes in the small intestine is less efficient than usual. Passage of intestinal contents through the duodenum is also more rapid than usual. In spite of this, post-gastrectomy malabsorption is rarely severe, and clinically "dumping" and hypoglycaemic attacks are more troublesome (p. 272).

The carcinoid syndrome (p. 431) is due to the excessive production of 5-hydroxytryptamine (5-HT) by tumours of argentaffin cells usually arising in the small intestine, and by their metastases, usually in the liver. This is a rare disease which even more rarely presents a problem of differential diagnosis of malabsorption. The malabsorption is probably the result of increased intestinal motility induced by 5-HT. Diagnosis depends on the presence of the other clinical features of the disease, and on an increased excretion of 5-hydroxyindole acetic acid in the urine.

In pancreatic disease (due most commonly to *chronic pancreatitis*) failure of absorption is due to failure of digestion and predominantly affects large molecules. *Fibrocystic disease of the pancreas (cystic fibrosis; mucoviscidosis)* is an inherited disease, usually presenting in early childhood but, more rarely, first recognised in the adult patient. Pancreatic and bronchial secretions are viscid and, by blocking pancreatic ducts and bronchi, cause obstructive disease of these organs. Sweat glands are also affected and the diagnosis depends on the demonstration of an *increase in concentration of sweat sodium*, often to about twice normal (p. 46). The patient may present with pulmonary disease, or with malabsorption.

Results of Generalised Malabsorption

The most obvious finding common to generalised intestinal and pancreatic malabsorption is *steatorrhoea* (more than 18 mmol (5 g) of fat in the stools per day—see Appendix, p. 281). In gross steatorrhoea the stools are usually pale, greasy and bulky and, if the condition is severe, there may be diarrhoea. In intestinal malabsorption fat can be acted on by lipase, but the products cannot be absorbed normally. In pancreatic steatorrhoea the intestinal wall is normal but fat cannot be digested. The collections of specimens for faecal fat estimation must be carefully controlled if a reliable answer is to be obtained (Appendix, p. 281).

Malabsorption of fat is always accompanied by malabsorption of the substances listed on p. 264. *Vitamin D deficiency* impairs calcium absorption and this is one of the causes of a low level of circulating ionised calcium and of osteomalacia in malabsorption syndromes. *Vitamin K* is required for hepatic synthesis of prothrombin and other clotting factors, and severe cases of malabsorption may develop a haemorrhagic diathesis associated with a prolonged prothrombin time: this prothrombin deficiency, unlike that of liver disease, can be reversed by parenteral administration of vitamin K. *Vitamin A* deficiency is rarely clinically evident, although malabsorption of an oral dose of vitamin A can be demonstrated. Malabsorption of lipids causes low plasma *cholesterol* levels—an incidental finding of no diagnostic importance.

In intestinal malabsorption the *low ionised calcium level and osteomalacia* due to vitamin D malabsorption is aggravated by the formation of insoluble calcium soaps with unabsorbed fatty acids. If ionised calcium levels have been low for a sufficiently long time to cause osteomalacia the *plasma alkaline phosphatase* concentration will rise. Secondary hyperparathyroidism due to the low ionised calcium level causes phosphaturia with a low plasma phosphate (p. 241). Rarely the ionised calcium levels fall low enough to cause *tetany*.

Protein malabsorption occurs in intestinal disease because intestinal transport mechanisms are impaired: in pancreatic disease the digestion of protein is severely impaired. In either case prolonged disease causes generalised *muscle and tissue wasting* and *osteoporosis* (p. 245), and a lowering of all protein fractions in the blood. The low albumin may cause oedema (p. 55) and results in a reduction of *protein-bound calcium* (p. 232). The total calcium level may therefore give a false idea of the severity of the hypocalcaemia. Reduced *antibody formation* (immunoglobulins) predisposes to infection.

Thus in long-standing generalised malabsorption, whether pancreatic or intestinal, the *clinical picture* may be:

bulky, fatty stools, with or without diarrhoea;
general wasting and malnutrition (protein deficiency);
osteoporosis and osteomalacia (protein and calcium deficiency);
oedema (albumin deficiency);
haemorrhages (vitamin K deficiency);
tetany (calcium deficiency);
recurrent infections (immunoglobulin deficiency).

The *laboratory findings* may be:

increased excretion of fat in the stools;
hypocalcaemia (ionised and protein-bound)—with hypophosphataemia;
raised alkaline phosphatase levels in cases with osteomalacia;
generalised lowering of the concentration of all protein fractions;
sometimes a low urea (due to decreased production from amino acids);
hypocholesterolaemia;
prolonged prothrombin time.

The complete picture is only seen in advanced cases of the syndrome. It is important to realise that generalised malabsorption may present as osteomalacia, with bone pains, or with anaemia.

Differential Diagnosis of Generalised Intestinal and Pancreatic Malabsorption

Tests of pancreatic function are generally unsatisfactory (see p. 272), and the differential diagnosis of steatorrhoea is usually made on indirect evidence. Remember that pancreatic malabsorption affects large molecules predominantly.

Malabsorption of fat occurs in both types of disease. Theoretically, in pancreatic malabsorption fat should be predominantly present as triglyceride (neutral fat), while in intestinal disease free fatty acids should be present. However, the estimation of so-called "split" and "unsplit" fat in the stools provides no useful information because of the action of fat-splitting bacteria in the lower intestinal tract.

Differences in carbohydrate metabolism.—Polysaccharide absorption is impaired in both conditions. However, some of the carbohydrate in the diet is in the form of mono- and disaccharides and these can be absorbed in pancreatic disease when neither intestinal disaccharidase activity nor active monosaccharide absorption is affected. Hypoglycaemia is rare in either condition, but is probably more common in intestinal malabsorption: although in this type of malabsorption impaired absorption of a glucose load may cause a "flat" *glucose tolerance curve* this finding is so non-specific that it is of little diagnostic use. In pancreatic disease this curve may be normal, but, as insulin is of pancreatic origin, the curve may even be diabetic in type.

Because glucose is rapidly metabolised in the body, interpretation of the glucose tolerance curve as an index of the type of malabsorption may be difficult. *Xylose* is a pentose, not rapidly metabolised in mammalian tissues and filtered by the renal glomerulus. There is an active transport mechanism for xylose in the upper small intestinal mucosa, although this is relatively inefficient (p. 282). If an oral dose of xylose is given, and if the intestinal mucosa is normal, it will be absorbed and appear in the urine: if, however, there is disease involving the upper small intestinal mucosa it will not be absorbed normally and less will appear in the blood and urine. This is the basis of the *xylose absorption test* which is usually normal when malabsorption is due to disease of the pancreas and abnormal in intestinal disease: however, disease involving only the ileum (such as Crohn's disease) may be associated with normal xylose absorption.

Anaemia is more common in intestinal than in pancreatic malabsorption since iron, vitamin B_{12} and folate are not digested by pancreatic enzymes before absorption. In intestinal malabsorption the blood and bone marrow films typically show a mixed iron deficiency and megaloblastic picture. Malabsorption of iron, vitamin B_{12} and folate may be demonstrable and the absorption of vitamin B_{12} is not increased by the simultaneous administration of intrinsic factor, as it is in pernicious anaemia (see p. 275). Anaemia may be aggravated by protein deficiency.

The differential diagnosis is summarised in Table XXIV (p. 274).

The Post-Gastrectomy Syndrome

Malabsorption after gastrectomy is usually mild. However, rapid passage of the contents of the small gastric remnant into the duodenum may have two clinical consequences:

the "dumping syndrome".—Soon after a meal the patient may experience abdominal discomfort and feel faint and sick. The syndrome is thought to be due to the sudden passage of fluid of high osmotic content into the duodenum. Before this abnormally large load can be absorbed, water passes along the osmotic gradient from the extracellular fluid into the lumen of the intestine. The reduction in plasma volume causes faintness and the large volume of duodenal fluid causes abdominal discomfort;

post-gastrectomy hypoglycaemia.—If a meal containing much glucose passes more rapidly than normal into the duodenum glucose absorption is very rapid. The blood glucose level rises suddenly and causes an outpouring of insulin. The resultant "over-swing" of blood glucose concentration may cause hypoglycaemic symptoms which typically occur at about two hours after a meal. The glucose tolerance curve in this type of case is "lag storage" in type.

Both these disabilities can be mitigated if meals low in carbohydrate content are taken "little and often".

Tests of Exocrine Pancreatic Function

It is convenient to digress here to discuss tests of the ability of the pancreas to secrete digestive juices. Unfortunately such tests of pancreatic function are very unsatisfactory. In chronic pancreatic hypofunction the diagnosis is usually made by the indirect methods described above. Carcinoma of the pancreas is very difficult to diagnose unless the lesion is in the head of the organ and causes obstructive jaundice: extensive gland destruction may cause late onset diabetes. Acute pancreatitis, however, may usually be diagnosed by estimation of plasma amylase levels.

Plasma enzymes.—Plasma *amylase* levels are *not affected in chronic pancreatic disease*, possibly because of the presence of amylase of salivary gland origin, and the estimation is useless: in acute pancreatitis, however, levels are usually raised (see below). Plasma *lipase* estimation has been used to detect chronic pancreatic hypofunction.

Faecal trypsin levels are extremely variable, probably because of bacterial action. The estimation is of no value in diagnosis of pancreatic hypofunction in adults. In infants with diarrhoea its absence is suggestive of fibrocystic disease of the pancreas.

Duodenal enzymes.—Measurement of pancreatic enzymes and bicarbonate in duodenal aspirate before and after stimulation of the pancreas with secretin or pancreozymin (hormones stimulating the pancreas which are normally secreted during digestion) is not very suitable for routine use because of difficulty in positioning the duodenal tube correctly and in quantitative sampling of the secretions. In the *Lundh test* a mixture of corn oil, milk powder and glucose is given, and samples of duodenal secretion analysed for trypsin: this test, too, is only useful if performed by an expert.

Acute Pancreatitis

In acute pancreatitis necrosis of the cells of the organ results in release of their enzymes into the peritoneal cavity and blood stream. The presence of pancreatic juice in the peritoneal cavity causes *severe abdominal pain* and *shock*: this picture is common to many acute abdominal emergencies. A vicious circle is set up as more pancreatic cells are digested by the released enzymes.

Acute pancreatitis is most commonly the result of obstruction of the pancreatic duct, or of regurgitation of bile along this duct. The most important predisposing factors are *alcoholism* and *biliary tract disease*. *Trauma* to the pancreas by damaging the cells may also initiate the vicious circle. There is an association between acute pancreatitis and *hypercalcaemia*: evidence for the latter should be sought.

Typically plasma amylase values increase five-fold or more. However, it is important to realise that concentrations of up to, and even above, this value may be reached in any acute abdominal emergency, but especially after gastric perforation into the lesser sac: very high levels may occur in renal glomerular failure when the enzyme cannot be excreted normally, and these are usually asymptomatic. Conversely, levels in acute pancreatitis may not reach very high levels, and usually fall very rapidly as the enzyme is lost in the urine. High plasma amylase levels are therefore only a rough guide to the presence of acute pancreatitis and normal or only slightly raised values do not exclude the diagnosis.

Malabsorption probably does occur during acute pancreatitis but is of little importance in the acute phase. Chronic pancreatic failure may follow a severe attack and is especially probable following repeated attacks.

FAILURE OF ABSORPTION OF SPECIFIC SUBSTANCES

Altered Bacterial Flora (Malabsorption of Vitamin B_{12} and Fat)

The "blind-loop syndrome".—This syndrome is associated with stagnation of intestinal contents with a consequent alteration of bacterial flora, and occurs when the loops are the result of surgery, or in the presence of *diverticula*.

Treatment with neomycin, by altering bacterial flora and possibly by combining with bile salts and so disrupting micelles, can cause a similar syndrome.

Many intestinal bacteria require vitamin B_{12} for metabolism, and *megaloblastic anaemia* is common in the syndrome. Bile salts may be metabolised to bile acids with resultant *steatorrhoea* and malabsorption of all those substances associated with fat absorption (p. 264). Protein and carbohydrate are usually normally absorbed.

Biliary Obstruction (Malabsorption of Fat)

Bile salts are synthesised and secreted by the liver. In biliary obstruction these cannot reach the intestinal lumen in normal amounts and *steatorrhoea* results, with the usual consequences (p. 264). Because of extreme jaundice there is rarely any difficulty in the differential diagnosis.

Differential Diagnosis of Steatorrhoea

The malabsorptive conditions described so far are all associated with steatorrhoea. As has been mentioned, biliary obstruction is rarely a

TABLE XXIV

DIFFERENTIAL DIAGNOSIS OF STEATORRHOEA

	Upper Small Intestinal Disease	Pancreatic Disease	Blind-Loop Syndrome
Xylose absorption	Reduced	Normal	Usually normal
Anaemia	Mixed megaloblastic and iron deficiency common	Rare	Megaloblastic common
Intestinal Biopsy	May show flattened villi or other cause	Normal	Normal
Glucose Tolerance Test	May be flat (except postgastrectomy)	Normal or diabetic	Normal

problem of diagnosis. The important points in laboratory findings in differential diagnosis of the other conditions are summarised in Table XXIV.

Local Disease or Surgery

Local disease or resection may cause selective malabsorption of substances whose absorption occurs predominantly at these sites. The lower ileum is concerned with vitamin B_{12} absorption and with reabsorption of bile salts, and resection or disease of this area (for example, Crohn's disease or tuberculosis), may cause megaloblastic anaemia.

Pernicious Anaemia

Vitamin B_{12} cannot be absorbed in the absence of intrinsic factor (p. 265). In pernicious anaemia, when antibodies to both parietal cells and intrinsic factor are present, the stomach cannot secrete this substance and significant deficiency may occur after total gastrectomy or with extensive malignant infiltration of the stomach.

The Schilling test.—In such subjects malabsorption of vitamin B_{12} can be demonstrated if a small dose of the radioactive vitamin is given and its excretion in the urine measured (compare xylose absorption test, p. 282). If the malabsorption is due to pernicious anaemia, administration of the labelled vitamin together with intrinsic factor results in normal absorption: if it is due to intestinal disease malabsorption persists. A "flushing" dose of non-radioactive vitamin B_{12} is given parenterally at the same time as, or just after, the labelled dose, to ensure quantitative urinary excretion. Haematological tests such as examination of blood and marrow films should have been completed before the vitamin B_{12} is given.

Disaccharidase Deficiency

Because disaccharidases are localised in the brush border, generalised disease of the intestinal wall usually causes a non-selective disaccharidase deficiency. This is relatively unimportant compared with general malabsorption and tests for this syndrome are therefore useful only in the absence of steatorrhoea when a selective rather than a generalised malabsorption of carbohydrate may be present.

The *symptoms* of disaccharidase deficiency are those of the effects of unabsorbed, osmotically active sugars in the intestinal tract, and include faintness, abdominal discomfort and severe diarrhoea after ingestion of the offending disaccharide. (Compare the "dumping syndrome", p. 272.)

Lactase deficiency.—*Acquired lactase deficiency* is much more common than the congenital form and is probably the commonest type of disaccharidase deficiency: it may first present in adults. It may be due to a sensitivity reaction, perhaps in subjects with a genetic predisposition.

Lactase deficiency associated with prematurity.—In premature infants lactase may not be present in normal amounts in intestinal cells. Initially these cases resemble the congenital ones. However, the sensitivity to milk usually disappears within a few days of birth.

Congenital lactase deficiency.—This is very rare. Infants present soon after birth with severe diarrhoea. Stools are typically acid because of bacterial production of lactic acid from lactose. The syndrome is cured by removal of milk and milk products from the diet.

Sucrase and isomaltase deficiency usually coexist.

Congenital sucrase-isomaltase deficiency is more common than congenital lactase deficiency.

Acquired sucrase-isomaltase deficiency and *maltase deficiency* of any kind are very rare.

The *diagnosis* of disaccharidase deficiency is most reliably made by estimation of the relevant enzymes in intestinal biopsy tissue.

The relevant disaccharide may be given orally and blood glucose estimated as in the glucose tolerance test. If the disaccharide cannot be hydrolysed the constituent monosaccharides (glucose, or those converted to glucose in the intestinal cell) cannot be absorbed, and the curve is flat. The result should usually be compared with a glucose tolerance curve. If the patient experiences typical symptoms, or if a child excretes disaccharides in the urine when the offending sugar is given, the comparison need not be made.

Radiological examination may assist in the diagnosis of disaccharidase deficiency. Barium is administered first without, and then with, the disaccharide. When the sugar is not absorbed the flocculation pattern of barium is altered by its osmotic effect.

Protein-losing Enteropathy

This is a very rare syndrome in which the intestinal wall is abnormally permeable to large molecules (as the glomerulus is in the nephrotic syndrome). This is not strictly speaking a malabsorption syndrome, but is due to excessive loss of protein from the body into the gut. It occurs in a variety of conditions in which there is ulceration of the bowel, lymphatic obstruction and intestinal lymphangiectasis, or hypertrophic lesions of the bowel. It is only rarely associated with steatorrhoea, and the clinical picture, like that of the nephrotic syndrome (p. 327), is due to hypoalbuminaemia: protein is not usually found in the urine.

In this syndrome the hypoproteinaemia is probably not entirely due to loss from the body, because some of the protein which passes into the intestinal tract is digested and the amino acids and small peptides reabsorbed. However, the rate of protein breakdown probably exceeds the rate at which the reabsorbed amino acids can be resynthesised into protein.

Diagnosis can be made by measuring the loss into the bowel after intravenous injection of a substance of a molecular weight approximating to that of albumin: substances that have been used are radioactive polyvinylpyrrolidone (PVP), dextran or albumin. Typically the electrophoretic pattern is similar to that found in the nephrotic syndrome: as in that syndrome the relatively high molecular weight α_2 fraction (on cellulose acetate, p. 313) is retained in the blood stream while all other fractions are lost from the body.

GASTRIC FUNCTION

The important components of gastric secretion by the parietal cells are *hydrochloric acid, pepsin* and *intrinsic factor*. All these factors have already been mentioned as being of importance in digestion and absorption, and loss of hydrochloric acid in pyloric stenosis has been discussed as a cause of metabolic alkalosis (p. 102).

Stimulation of gastric secretion occurs by two main pathways:

through the *vagus nerve*, which in turn responds to stimuli from the cerebral cortex, normally resulting from the sight, smell and taste of food. Hypoglycaemia can stimulate this pathway and this fact can be used to assess the completeness of vagotomy;

by *gastrin*, a hormone normally produced by G cells in the gastric antrum in response to the presence of food; it is carried by the blood stream to the parietal area of the stomach where it stimulates secretion, perhaps through the mediation of histamine. Acid in the pylorus, in turn, inhibits gastrin secretion, providing feedback control.

Histamine acts after binding to receptors on the surface of cells. There are probably two types of such receptors:

those on which antihistamines compete for the sites with histamine (H_1 receptors). These are on smooth muscle cells;

those on which antihistamines have no effect. These H_2 *receptors* are found on *gastric parietal* cells.

HYPERSECRETION

Hypersecretion of gastric juice may be associated with *duodenal ulceration*, when it may be neurogenic in origin. However, there is overlap between acid secretion in normal subjects and in those with duodenal ulceration, and the estimation is of very limited diagnostic value in this condition.

In the *Zollinger-Ellison syndrome* (p. 434) acid secretion by the stomach is very high. The consequent ulceration of the stomach and upper small intestine may cause severe diarrhoea: the low pH, by inhibiting lipase activity, may cause steatorrhoea.

HYPOSECRETION

Hyposecretion of gastric juice occurs in *pernicious anaemia* (p. 413) and is probably the result of antibodies to the parietal cells of the gastric mucosa: this is by far the commonest cause, and the achlorhydria is usually pentagastrin "fast" (p. 283). In extensive *carcinoma of the stomach* and in *chronic gastritis* there may also be gastric hyposecretion. However, in none of these conditions is estimation of gastric acidity of help in diagnosis: the diagnosis of pernicious anaemia should be made on haematological grounds, and on the result of the Schilling test, and results in the other two conditions are too variable to be useful.

TESTS OF GASTRIC FUNCTION

Routine tests of gastric function involve measurement of the acid secretion by the stomach, either at rest or in response to stimuli. Passage of a tube into the stomach is unpleasant for the patient: moreover, valuable results will only be obtained when the operator is skilled and able to recognise when the specimens obtained are incomplete, either because of blockage or malpositioning of the tube. For these reasons the tests should only be carried out when really necessary for diagnosis and management, and only by a skilled operator.

Resting Juice or Overnight Secretion

Gastric contents may be aspirated overnight and the total night secretion measured, or a timed specimen may be obtained without a stimulus and the hourly secretion determined. Both these secretions will usually be high in volume and acid content in duodenal ulceration. In the Zollinger-Ellison syndrome (p. 434) diagnosis partly depends on finding a very high rate of acid secretion in a one-hour basal collection (more than 15 mmol/hour) (see Appendix).

Stimulation of Gastric Secretion

Direct stimulation of parietal cells is used to demonstrate achlorhydria, which is typically pentagastrin "fast" in pernicious anaemia: as already pointed out, the test is of limited use in the diagnosis of this condition, and is unpleasant for the patient. *Pentagastrin* is a pentapeptide consisting of the physiologically active part of the gastrin molecule. It acts directly on the parietal cells of the stomach and can be used to stimulate secretion.

Vagal stimulation of gastric secretion is used before vagotomy to test the therapeutic prospects of this operation in treatment of peptic ulcer, and after vagotomy to test completeness of the section of the nerve. The stimulus used is insulin-induced hypoglycaemia (compare the use of insulin to stimulate cortisol secretion). If vagotomy is complete there should be no acid secretion even when the blood glucose level falls below 2·2 mmol/l (40 mg/dl), and when there is clinical evidence of hypoglycaemia.

Treatment of Hypersecretion

Until recently only palliative therapy with antacids and carbenoxolone, or surgery, was available for peptic ulceration due to hypersecretion of acid. Recently the drug *cimetidine* ("Tagamet") has been shown to compete with histamine for H_2 receptors (p. 277), so directly reducing acid and pepsin secretion. Trials of this treatment are still in progress.

SUMMARY

Intestinal Absorption

1. Normal intestinal absorption depends on adequate digestion of food (and therefore on normal pancreatic function), and on a normal area of functioning intestinal cells.

2. Normal digestion and absorption of fat depends on the presence of bile salts as well as of lipase.

3. Absorption of cholesterol, phospholipids and fat-soluble vitamins depends on normal neutral fat absorption.

4. In *intestinal* malabsorption there is malabsorption of small molecules, usually due to a reduced absorptive area.

5. In *pancreatic* malabsorption there is malabsorption of fats, proteins and polysaccharides, but small molecules are usually absorbed normally.

6. An *abnormal bacterial flora* may cause steatorrhoea (because of competition for bile salts), and megaloblastic anaemia (because of competition for vitamin B_{12}).

7. There may be steatorrhoea in biliary obstruction (because of lack of bile salts).

8. Selective malabsorption of vitamin B_{12} occurs in pernicious anaemia (lack of intrinsic factor).

9. Selective disaccharidase deficiencies cause malabsorption of disaccharide. These deficiencies are more commonly acquired than congenital in origin.

Pancreatic Function

1. Direct tests for pancreatic hypofunction are unsatisfactory unless performed in special centres.

2. Acute pancreatitis is associated with a transient rise in plasma amylase levels. This enzyme can also reach high concentrations in many acute abdominal emergencies and in renal failure.

Gastric Function

1. Hypersecretion of acid occurs in:
   duodenal ulceration;
   the Zollinger-Ellison syndrome.
2. Hyposecretion of acid (pentagastrin "fast") occurs in:
   pernicious anaemia;
   extensive gastric infiltration.
3. The parietal cells can be directly tested by stimulation with pentagastrin.
4. The vagal stimulation of gastric secretion can be tested by producing insulin-induced hypoglycaemia.

FURTHER READING

Course in Gastroenterology. (1973). *Proc. Mayo Clin.*, **48**, 605.

LOSOWSKY, M. S., WALKER, B. E., and KELLEHER, J. (1973). *Malabsorption in Clinical Practice*. Edinburgh: Churchill-Livingstone.

ROMMEL, K., and GOEBELL, H. (Eds.) (1976). *Lipid Absorption: Biochemical and Clinical Aspects*. Lancaster: Medical & Technical Publishing Co.

The Pancreas. Articles by Wormsley, K. G., Hatfield, A. R. W., Hermon-Taylor, J., Mallinson, C. and Go, V. L. W. (1977). *Brit. J. hosp. Med.*, **18**, 518, 528, 546, 553 and 567.

BOUCHIER, I. A. D. (1975). Acid secretion, peptic ulceration and H_1 and H_2 receptors. *11th Symposium on Advanced Medicine*, p. 1 (Ed. A. F. Lant). Tunbridge Wells: Pitman Medical.

BARON, J. H. (1977). Are gastric secretion tests worthwhile? *Proc. roy. Soc. Med.*, **70**, 223.

APPENDIX

TESTS FOR MALABSORPTION

COLLECTION OF SPECIMENS FOR FAECAL FAT ESTIMATION

Estimation of faecal fat output measures the difference between the fat absorbed and that entering the intestinal tract from the diet and from the body (p. 260). Absorption of fat and addition of fat to the intestinal contents occurs throughout the small intestine. A single 24-hour collection of faeces will usually give inaccurate results for two reasons:

the transit time from the duodenum to the rectum is variable;

rectal emptying is variable and may not be complete.

It has been shown that consecutive 24-hour collections yield answers which may vary by several hundred per cent, and which are therefore useless as an estimate of daily excretion from the body into the gut. The longer the period of collection, the nearer does the calculated daily mean value approach the "true" one.

For obvious reasons patients cannot be kept in hospital indefinitely. A usual compromise is to collect *at least a 3-day* and preferably a 5-day specimen of stools. Precision may be increased by collecting between "markers"— usually dyes which can be taken orally and which colour the stool.

Procedure

Day 0.—The first "marker" (usually two capsules of carmine) is given.

As soon as the "marker" appears in the stool the collection is started. This "marker" will gradually disappear.

Day 5.—The second "marker" is given.

As soon as the second "marker" appears in the stool the collection is stopped.

The estimation is made on all the specimens passed between the appearance of the two "markers" and including one of the marked stools.

Important.—1. It is very important that *all* stools should be collected during this period. To ensure that none is missing it is best to label each specimen with the following information:

Name of Patient
Ward
Date of Specimen
Time of Specimen ⎱ A record of these should also be kept on the
Number in the series ⎰ ward.

The patient should not be allowed to go to the toilet, and should be impressed with the importance of a complete collection.

2. The time for collection of specimens will be about a week (including the time for appearance of the marker). Before starting the test try to make sure that during this time the patient is *not to be discharged*, is *not going to be*

operated on (except in emergency), and does *not receive enemas or aperients*: any patient requiring aperients to keep his bowels open is most unlikely to have significant steatorrhoea at that time. *Neither barium enemas nor barium meals* should be performed during this time as the barium interferes with the estimation.

The collection and estimation of faecal fat excretion is time-consuming and unpleasant for all concerned (including the patient). It is important that specimen collection is carefully controlled so that the answer may be meaningful.

Interpretation

A *mean* daily fat excretion of more than 18 mmol (5 g) indicates steatorrhoea. Since, in the normal person, almost all the faecal fat is of endogenous origin, it is not affected by diet within very wide limits.

XYLOSE ABSORPTION TEST

An oral dose of 25 g of xylose has been given for this test. However, the absorption of xylose is relatively inefficient and the presence of this large dose in the intestine may, by its osmotic effect, cause abdominal discomfort and diarrhoea, and further interfere with absorption. For this reason a 5 g dose is preferable.

Warning. In the presence of poor renal function the test is invalid. Oedema, because the volume through which the xylose is distributed is increased, also invalidates the test.

Procedure

The patient is fasted overnight.

8 a.m.—The bladder is emptied and the *specimen discarded*. 5 g of xylose dissolved in a glass of water is given orally.

All specimens passed between 8 a.m. and 10 a.m. are put into Bottle 1.

10 a.m.—The bladder is emptied and *the specimen put into Bottle 1* which is now complete.

All specimens passed between 10 a.m. and 1 p.m. are put into Bottle 2.

1 p.m.—The bladder is emptied and *the specimen is put into Bottle 2*, which is now complete.

Both bottles are sent to the laboratory for analysis.

Interpretation

In the normal subject more than 23 per cent of the dose (1·15 g) should be excreted during the five hours of the test. Fifty per cent or more of the total excretion should occur during the first two hours. In mild intestinal malabsorption the total 5-hour excretion may be normal, but delayed absorption is reflected in a 2- to 5-hour excretion ratio of less than 40 per cent.

In pancreatic malabsorption the result should be normal (p. 271).

TESTS OF GASTRIC FUNCTION

PENTAGASTRIN STIMULATION

Procedure (applies to all types of gastric secretion test)

The patient fasts from 10 p.m. on the evening before the test. On the morning of the test a radio-opaque Levin tube is passed into the stomach (preferably under x-ray control) until it lies in the gastric antrum: the tube is attached to the side of the face with plaster.

8 a.m.—All the gastric juice is aspirated from the stomach and is put in a bottle marked *"Resting Juice"*.

8 a.m. to 9 a.m.—Gastric juice is aspirated continuously for the next hour and put in a bottle marked *"Basal Secretion"*.

9 a.m.—**Pentagastrin 6 μg/kg body weight is injected intramuscularly.**

9 a.m. to 9.15 a.m.
9.15 a.m. to 9.30 a.m. ⎱ Four 15-minute samples are aspirated and put
9.30 a.m. to 9.45 a.m. ⎰ into bottles marked with the appropriate times.
9.45 a.m. to 10 a.m.

All specimens are sent to the laboratory for analysis.

Interpretation

1. In pentagastrin-fast *achlorhydria* no specimen has a pH as low as 3·5.

2. A *basal* acid secretion of greater than 15 mmol of hydrogen ion in the hour, with no further response to stimulation, is suggestive of the *Zollinger-Ellison syndrome*.

3. If any specimen is of pH 3·5 or less the hydrogen ion secretion is checked by titration. In *hyperchlorhydria* the highest mean acid secretion (calculated by adding the results of the two consecutive specimens of highest acidity and by multiplying by two, to give an hourly secretion), is greater than 40 mmol/hour.

The resting juice is inspected for *blood*. A large quantity of altered blood is suggestive of carcinoma. Small flecks may be due to trauma during aspiration.

INSULIN STIMULATION OF GASTRIC SECRETION (p. 279)

This test is usually carried out at about six months after vagotomy. Insulin induced hypoglycaemia stimulates gastric secretion via the vagus nerve.

Procedure

The preparation of the patient and positioning of the tube is described above.

8 a.m.—The stomach is emptied and the specimen discarded.

8 a.m. to 9 a.m.—A 1-hour "basal secretion" is collected.

9 a.m.—Soluble insulin (0·15 units/kg body weight) is injected intravenously.

9 a.m. to 11 a.m.—Specimens are collected for blood glucose estimation every 30 minutes for two hours. During this time gastric juice is aspirated every 15 minutes.

The specimens are sent to the laboratory for analysis.

Warning.—1. The test is potentially dangerous and should be done only under *direct medical supervision*. *Glucose* for intravenous administration should be *immediately available* in case severe hypoglycaemia develops. At the conclusion of the test the patient should be given something to eat.

2. If it is necessary to administer glucose, *continue with the sampling*. The stress has certainly been adequate.

Interpretation

Analysis of gastric samples is only useful if the blood glucose has fallen to levels below 2·2 mmol/l (40 mg/dl) and if clinical hypoglycaemia is present.

If more than 2 mmol of hydrogen ion is present in any 60-minute sample the response is positive, suggesting the presence of intact vagal fibres.

Chapter XIII

LIVER DISEASE AND GALL STONES

LIVER DISEASE

OUTLINE OF FUNCTIONS OF THE LIVER

THE liver plays an important role in many metabolic processes. Because it receives blood from the portal vein, all nutrient from the gut except fat (p. 264), reaches the liver before entering the systemic circulation.

Carbohydrate metabolism.—*Glycogen is synthesised* and stored in the liver during periods of carbohydrate availability. During fasting, blood glucose levels are maintained within normal limits by breakdown of stored glycogen (*glycogenolysis*). The formation of glucose from such substrates as amino acids (*gluconeogenesis*) occurs mainly in the liver (see Chapter IX).

Lipid metabolism.—*Synthesis* of almost all lipoproteins, phospholipids, cholesterol and endogenous triglyceride occurs in the liver and the breakdown products of cholesterol are excreted in bile. *Fatty acids* reaching the liver from fat stores are *metabolised* in the tricarboxylic acid cycle. Any excess is incorporated into endogenous triglyceride, or is converted into ketones (see Chapter X).

Protein synthesis.—Many of the plasma proteins, including special carrier proteins and most of the coagulation factors, but with the important exception of the immunoglobulins, are synthesised in hepatic cells. Prothrombin and factors VII, IX and X require vitamin K for their synthesis (see Chapter XIV).

Vitamin D metabolism is discussed on p. 235.

Storage functions.—In addition to glycogen, *vitamins* A, D and B_{12} are stored in the liver and it is quantitatively the most important site of *iron* storage (p. 378).

Excretion and detoxication.—*Bile pigments* and *cholesterol* (p. 299) are excreted in the bile. Many *drugs* are detoxicated by the liver and some are excreted in bile. *Ammonia*, derived from amino acid metabolism or produced in the bowel by bacteria, is converted to *urea* and so rendered non-toxic. *Steroid hormones* are inactivated by conjugation with glucuronate and sulphate in the liver and are excreted in the urine. Many of the clinical findings of advanced liver failure, such as gynaecomastia, testicular atrophy, spider naevi and "liver palms" are thought to be due to failure to detoxicate steroids.

Reticulo-endothelial function.—The Kupffer cells lining the sinusoids of the liver form part of the reticulo-endothelial system.

BILIRUBIN METABOLISM

Disturbances of bile pigment metabolism occur in many hepato-biliary disorders. The consequences may be clinically obvious or may be detected by simple tests. These changes can be used in diagnosis and in monitoring progress. Abnormalities are best discussed in the context of the normal sequence of events outlined below and in Fig. 29.

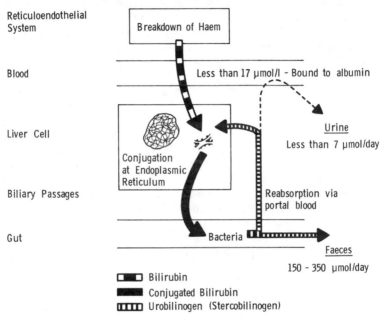

FIG. 29.—Bile pigment metabolism.

At the end of their life circulating red cells are broken down in the reticulo-endothelial system, mainly in the spleen. The released haemo-globin is split into globin, which enters the general protein pool, and haem, which is converted to bilirubin after removal of iron. The iron is re-utilised.

Bilirubin formed by this process is carried to the liver and accounts for about 80 per cent of the bilirubin metabolised daily. Other sources include the breakdown of immature red cells in the bone marrow and of compounds chemically related to haemoglobin, such as myoglobin and cytochromes. In all about 500 μmol (300 mg) of bilirubin reaches the liver daily. Healthy hepatic cells are capable of handling much greater loads than this, and only very excessive haemolysis will cause jaundice in the adult.

Bilirubin in transit to the liver (*unconjugated bilirubin*, or simply *bilirubin*) is bound to plasma albumin. Some *drugs* displace bilirubin from albumin. Free bilirubin enters the brain more easily than that bound to protein: in neonatal jaundice such drugs may therefore increase the danger of cerebral damage (p. 298). Normally most plasma bilirubin is unconjugated, and is not excreted in the urine because it is protein-bound and insoluble in water. There is no bilirubinuria in jaundice due to a rise in unconjugated, without conjugated, bilirubin ("acholuric jaundice").

At the hepatic cell membrane bilirubin is split off from albumin and passes into the cell where it is accepted by the specific binding proteins Y protein (ligandin) and Z protein. Many other organic anions (including certain drugs) are similarly bound and compete with bilirubin. Bilirubin is then transported within the cell to the smooth endoplasmic reticulum. Here it is *conjugated* with glucuronic acid to form bilirubin glucuronide, or *conjugated bilirubin*, by the enzyme uridyl diphosphate (UDP) glucuronyl transferase. The activity of this enzyme may be increased (enzyme induction), for example, by phenobarbitone. Conjugated bilirubin is transported out of the liver cell into the bile canaliculi and *excreted in the bile*. Elimination is nearly complete and the plasma concentration is normally negligible. Interference with excretion may cause jaundice due to high plasma levels of water-soluble conjugated bilirubin, which *can be excreted in the urine*. *Bilirubinuria is always pathological* because it reflects a high circulating conjugated bilirubin level.

The conjugated bilirubin enters the gut in bile. In the colon it is broken down by bacteria to a group of products known collectively as *stercobilinogen* (or *faecal urobilinogen*). A small amount is absorbed into the portal circulation and most of this is re-excreted in bile: a very small fraction appears in the urine as *urobilinogen* which in turn can be oxidised to *urobilin*.

Urobilin(ogen), by contrast with bilirubin, *is* often detectable in normal urine by routine tests: this is especially common if concentrated urine is passed. Urinary urobilin(ogen) is increased if:

    haemolysis is excessive, when large amounts of bilirubin enter the bowel and are converted to stercobilinogen. Much of the reabsorbed urobilinogen is passed in the urine;

    liver damage impairs re-excretion of normal amounts of urobilinogen into the bile.

Urinary urobilinogen excretion is very variable in the normal subject, and only very high urinary levels are clinically significant. Such levels are most commonly found during acute haemolysis.

Unabsorbed stercobilinogen is oxidised to stercobilin, a pigment which contributes to the brown colour of the faeces. Pale stools may, therefore, be evidence of biliary obstruction.

Bilirubin, urobilin and *stercobilin* are *coloured* ("bile pigments");
Urobilinogen and stercobilinogen are *colourless*.

Conjugated and *unconjugated* bilirubin are sometimes referred to as *direct* and *indirect* bilirubin respectively. These terms reflect the fact that the water-soluble conjugated bilirubin reacts *directly* with a colour reagent, whilst the unconjugated fraction reacts only after addition of methyl alcohol or another solubilising reagent (*"indirectly"*).

Bile acids, end products of cholesterol metabolism, are excreted in bile, reabsorbed in the ileum, and re-excreted by the liver (enterohepatic circulation, see p. 264).

JAUNDICE

A clinically detectable increase in plasma bilirubin levels is called jaundice, and may be due to abnormalities of any of the steps outlined above.

Jaundice may be the result of:

a predominant *rise in unconjugated bilirubin,* due most frequently to excessive *haemolysis,* in which bilirubin production may exceed the hepatic conjugating capacity. There is an excess of urobilinogen, but no bilirubin in the *urine,* which is therefore of *normal colour*: bilirubin which has been conjugated reaches the gut, and *faecal colour* is also *normal*;

a *rise in both fractions* due to hepatocellular damage or to interference with bilirubin excretion, whether due to intra- or extrahepatic causes. In most cases both urobilin(ogen) and bilirubin are present in the urine, and the *urine is dark.* If biliary obstruction is nearly complete no bilirubin enters the gut and the *faeces* are *pale*: urobilinogen is then absent from the urine because it is not formed in the intestine.

BIOCHEMICAL TESTS IN LIVER DISEASE

There are many tests for liver cell damage and for liver function. In this chapter we discuss those commonly used and which we have found useful. Most parameters are affected by diseases other than those of the liver. Selection of tests depends on the clinical picture, and especially on the likelihood of such other disease.

Basic Processes in Liver Disease

Although many agents impair hepatobiliary function, there are only a limited number of ways in which they can do so. The first purpose of investigation is to identify the underlying pathological change. These changes fall into two main groups:

liver cell damage, with or without demonstrable liver dysfunction;
biliary tract involvement.

Liver cell damage.—This may vary from areas of focal damage to destruction of most of the liver leading to liver failure. Examples of the main causes are:

acute hepatitis, which is most commonly viral but may occur in the course of other diseases such as infectious mononucleosis, or septicaemia;

the action of *toxins* on the liver, especially *paracetamol* (acetaminophen) and chlorinated hydrocarbons such as carbon tetrachloride. Phosphorus is a hepatotoxin;

chronic hepatitis due to the continuing action of an infective or toxic agent, or that associated with an autoimmune process (*chronic aggressive hepatitis*);

secondarily to prolonged *biliary obstruction*;

cellular destruction of known or unknown aetiology may be followed by *cirrhosis,* and the process may persist in the cirrhotic liver;

hepatic congestion and/or hypoxia such as occur in congestive cardiac failure or "shock".

The *consequences* of liver cell damage vary in degree. If mild, biochemical tests may provide the only evidence. With more severe damage there is usually jaundice and demonstrable failure of synthetic function. Extensive destruction is incompatible with life.

Biliary tract involvement.—Involvement of the biliary tract is often associated with obstruction to bile flow (*cholestasis*), and may therefore present as obstructive jaundice. Intrahepatic obstruction due to involvement of bile canaliculi is not amenable to surgery, but if the obstruction is in the extrahepatic bile ducts it may be. Distinction between the two is therefore important.

Intrahepatic cholestasis.—This is frequently associated with liver cell destruction. The main causes are:

viral hepatitis (cholangiolytic hepatitis);

chlorpromazine;

steroids, whether administered (methyl testosterone and oral contraceptives) or endogenous, as in some subjects in late pregnancy: these patients also tend to develop cholestasis in subsequent pregnancies, or when taking oral contraceptives;

cholangitis (some cases of cholangitis are not jaundiced);

biliary cirrhosis;
cirrhosis (some cases);
infiltrations of the liver (e.g. Hodgkin's disease, malignancy);
intrahepatic biliary atresia.

Extrahepatic cholestasis.—The main causes are:
gall-stone in the common bile duct;
carcinoma of head of pancreas, ampulla of Vater or, rarely, bile duct;
fibrosis of the bile duct;
external pressure from tumour or glands;
extrahepatic atresia of bile duct.

The *consequences* of cholestasis depend largely on its duration. There is accumulation in the blood of substances normally excreted in the bile, including conjugated bilirubin, bile salts and cholesterol. Retention of bilirubin leads to jaundice. The high levels of cholesterol found in some cases, especially in biliary cirrhosis, may cause xanthomatosis. In chronic cases, with prolonged absence of bile salts from the gut, absorption of fat and fat-soluble vitamins is impaired (p. 264). Retention of bile salts may be the cause of the itching experienced by patients with severe cholestatic jaundice.

Tests for Liver Disease

Tests may be classified according to the underlying pathology.

Hepatocellular damage can be demonstrated by detecting release of intracellular constituents, usually enzymes (such as *transaminases*), into the extracellular fluid. These tests are very sensitive and may be abnormal with minimal damage, and without demonstrable loss of function. Jaundice only occurs when cell damage is extensive.

Involvement of the biliary tract is characterised by an increased production, and therefore raised plasma levels, of hepatobiliary enzymes such as *alkaline phosphatase* (ALP) and γ-glutamyltransferase (GGT). There is usually, but not always, intra- or extrahepatic obstruction of the biliary tract with jaundice.

Impaired hepatocellular function affects synthesis of albumin and of coagulation factors, as well as bilirubin metabolism. Hypoalbuminaemia and a prolonged prothrombin time are detectable when damage is extensive and prolonged.

The **aetiology** of hepatic dysfunction may be demonstrated by tests which, for example, suggest autoimmunity or infection.

Tests Indicating Liver Cell Damage

Damage to liver cells, with or without necrosis, causes the acute release of intracellular constituents into the blood stream. This is detected by measuring plasma enzymes.

Transaminases.—Raised levels of transaminases are found whatever the cause of liver cell damage. Mildly raised levels occur in cholestasis and some cases of cirrhosis. Both aspartate (AST: SGOT) and alanine (ALT; SGPT) transaminases are affected.

AST is present in both mitochondria and cytoplasm whereas ALT is found in the cytoplasm only. In conditions in which there is cytoplasmic damage, but in which relatively few cells are totally destroyed, as in hepatitis, ALT levels are relatively higher than those of AST; in conditions in which damage, however focal, involves the whole cell (as in space-occupying lesions), AST levels are relatively higher than those of ALT. As ALT has a longer half-life than AST, raised levels usually persist for a longer period of time.

Lactate dehydrogenase levels may be raised, but the test is too insensitive and non-specific to be useful in detecting hepatocellular damage.

Tests Indicating Biliary Tract Involvement

Plasma alkaline phosphatase.—As mentioned on p. 342, plasma alkaline phosphatase is derived from at least two sources—the osteoblasts and the cells lining bile canaliculi. If the biliary tract is involved in any disease process, alkaline phosphatase is released from these cells into the extracellular fluid. In *cholangitis* the cells lining the biliary tract are stimulated, and a *raised plasma alkaline phosphatase may be found with normal transaminase levels and without jaundice.* If there is significant biliary obstruction, conjugated bilirubin is regurgitated into the extracellular fluid, and *raised alkaline phosphatase levels with jaundice* generally indicate *cholestasis,* although moderately raised levels, as in hepatitis, may be a result of liver cell damage. The higher the level of alkaline phosphatase the greater the likelihood of extrahepatic cholestasis.

If only one of the hepatic ducts is obstructed, or if there are patchy lesions such as multiple secondary carcinomata, there is local retention of both bilirubin and alkaline phosphatase. Bilirubin, but not alkaline phosphatase, can be excreted by liver cells in the unaffected part of the biliary system, so plasma levels may be only minimally raised. In the absence of bone disease the finding of a raised alkaline phosphatase out of proportion to plasma bilirubin is suggestive of a space-occupying lesion(s), obstruction of one of the radicles of the common bile duct, or cholangitis.

γ **Glutamyltransferase** (*GGT*) (p. 344) is the most sensitive indicator of hepatobiliary disease, especially if there is a cholestatic element. If other tests for hepatocellular damage are normal a high level of this enzyme may suggest liver disease, or indicate that a raised alkaline phosphatase in the presence of normal transaminases is at least partly of hepatic origin: it can *not* indicate whether there is also a contribution from osteoblasts. Many drugs, including barbiturates and alcohol, induce GGT synthesis causing elevated plasma levels despite normal liver function: in such cases 5'-nucleotidase estimation may very occasionally be indicated (p. 342).

Tests Indicating Impaired Hepatocellular Function

The liver, like the kidney, has considerable functional reserve, and tests for impaired hepatic function only become abnormal when there is widespread and prolonged damage.

The most obvious feature of many hepatobiliary diseases is jaundice, due to impaired bilirubin excretion. The differential diagnosis of jaundice is discussed on p. 288. Many other substances, including drugs, share the same excretory pathway. One of these substances, *bromsulphthalein* (*BSP*) has been used to test liver function by measuring the rate of excretion after an injected dose. This test has largely been replaced by sensitive plasma tests of hepatic damage.

It must be emphasised that these tests *do not distinguish* between intra- and extrahepatic cholestasis.

Serum albumin and coagulation factors.—*Albumin* is catabolised at a rate of about 4 per cent of the body pool daily; decreased synthesis is soon reflected in reduced plasma levels. Hypoalbuminaemia may be the only finding in some chronic liver diseases, and is an index of the severity and extent of liver damage.

Many of the coagulation factors are synthesised by the hepatic parenchymal cells. Hepatocellular damage may give rise to a bleeding state, or laboratory tests of coagulation may be abnormal. Four factors, II (prothrombin), VII, IX and X are synthesised only in the presence of vitamin K. Their activities (with the exception of factor IX) are conveniently measured by the *one-stage prothrombin time*. Deficiencies may arise in two ways. In cholestasis the absorption of the fat-soluble vitamins, including vitamin K, is impaired (p. 264). In the presence of parenchymal damage synthesis is impaired despite an adequate supply of vitamin K. These two mechanisms offer a way of distinguishing cholestasis from cellular damage by the response in the former to parenteral vitamin K. Shortening of the prothrombin time after vitamin K injection suggests cholestasis whereas total lack of response indicates liver cell damage. As in many liver function tests, there is considerable overlap of results.

In severe liver disease deficiencies of factors V and I (fibrinogen) may also be encountered. The latter is more often due to fibrinolysis than to deficient synthesis.

Tests Indicating Aetiology

Changes in the *α- and β-globulin* electrophoretic fractions are inconstant and non-specific. *γ-globulin* levels increase in chronic liver disease and in cirrhosis a characteristic, but not specific, pattern often occurs: the β and increased γ fractions fuse due to an increase in fast-moving immunoglobulins.

Estimation of the individual **immunoglobulins** may occasionally be of value. IgA increases in *early cirrhosis* but in most established cases IgG and IgM are also elevated. Predominant elevation of IgG suggests *chronic active hepatitis*. *Primary biliary cirrhosis* often has a marked increase in IgM.

Hepatitis B antigen.—Several antigens may be detectable in the serum of patients with hepatitis B virus infection. HBsAg, an antigen of the surface coat of the virus, is the most usually measured of these. It can be detected before the onset of symptoms, and persists for a few weeks. Its disappearance may be associated with the development of an antibody, anti-HBs. HBsAg may persist for months or years in patients with, for example, leukaemia and chronic renal failure, and in whom the immune response may be impaired. It may also be found in apparently unaffected people (carriers). The disease may be transmitted by blood, or blood-products, and donors should always be screened for HBsAg. The references at the end of the chapter contain more detailed information.

Circulating autoantibodies.—Circulating antibodies to mitochondria and smooth muscle, as well as antinuclear factor, often occur in three related conditions, *chronic active hepatitis, cryptogenic cirrhosis* and *primary biliary cirrhosis*. The finding of such antibodies helps to distinguish these conditions from clinically similar diseases. Mitochondrial (or M) antibody is especially helpful in differentiating between primary biliary cirrhosis (more than 80 per cent of cases positive) and extrahepatic biliary obstruction (usually negative).

α-**Fetoprotein.**—α-Fetoprotein, normally present in fetal serum, is undetectable by *routine immunological techniques* in children and adults. Detectable levels occur in most patients with *primary hepatocellular carcinoma* and some patients with germinal cell tumours of testis and ovary. Transient positive results may be found during acute viral hepatitis, particularly in children.

Biochemical Changes in Individual Liver Diseases and the Selection of Tests

Acute hepatitis (viral or toxic).—The primary pathological abnormality in this condition is cell damage: this may be detectable only by plasma enzyme estimations or may cause detectable disturbance in function. In some cases cholestasis is a dominant feature (cholestatic hepatitis).

The patient gives a history of nausea, anorexia, and tenderness or discomfort over the liver. There is commonly a cholestatic element, and he may notice that he is yellow and that his *stools are pale* (impaired entry of bilirubin into the intestine) and *urine dark* (reflecting a rise in plasma conjugated bilirubin). *Transaminases* (both AST and ALT) rise to *very high* levels as soon as the patient feels ill, and reach their peak at about the time jaundice is detectable: they remain elevated for several weeks.

Plasma *bilirubin* levels rarely exceed 350 μmol/l (about 20 mg/dl). Because jaundice is the result of both cell damage and cholestasis *both bilirubin and conjugated bilirubin levels rise.* The *alkaline phosphatase* activity is *often normal,* but may be *moderately increased.* If cholestasis is predominant the picture *may resemble obstructive jaundice,* with very high plasma alkaline phosphatase and bilirubin levels. Anicteric hepatitis is not uncommon, and the only laboratory finding may be markedly raised transaminases.

The *course* of hepatitis may be followed by plasma transaminase estimation (ALT is the last to return to normal). If relapse is suspected clinically, it may be confirmed by enzyme levels which remain high, or which show a second rise. The development of a chronic phase is suggested by an increase in γ-globulin levels.

The *severity and extent* of liver cell impairment are shown best by the prothrombin time, although this is also affected by bile salt deficiency (p. 292): occasionally, in very severe damage, albumin levels may be unequivocally low. Bilirubin levels are affected by both cholestasis and cell damage and the rise in transaminases reflects the rapidity rather than the degree of destruction of liver cells.

Most cases of hepatitis recover, but a very small percentage die in liver failure and a small number develop cirrhosis, or chronic hepatitis.

Chronic hepatitis (whether chronic aggressive or chronic active). The only abnormal finding in these conditions may be raised transaminases. In this situation *ALT is more sensitive than AST* (p. 291).

Biliary obstruction.—In complete extrahepatic biliary obstruction *plasma bilirubin* levels rise progressively, although they may fluctuate slightly, for several weeks, and may reach values of 850 μmol/l (50 mg/dl) or greater. Levels then flatten out. The rise is predominantly in the conjugated fraction. The urine contains bilirubin but no urobilin(ogen) and the stools are pale.

Plasma alkaline phosphatase levels increase progressively. Cholesterol levels also rise, but are of less use diagnostically. After prolonged obstruction there may be secondary liver damage due to bile necrosis of liver cells or ascending cholangitis and tests of liver cell damage become positive.

Differential diagnosis of prolonged cholestasis.—The main causes to be considered are:

intrahepatic—cholestatic hepatitis
         drug-induced jaundice (for example, chlorpromazine)
         primary biliary cirrhosis
extrahepatic obstruction

Surgery is usually indicated for extrahepatic obstruction: because anaesthesia may aggravate hepatocellular damage, it is important to distinguish between the types.

As outlined above, not only does prolonged extrahepatic obstruction cause liver damage, but cholestasis develops in many cases of hepatitis. *It is most important to perform the relevant tests as early as possible* if the primary process is to be diagnosed. This is second in importance only to an adequate clinical history (with emphasis on possible drug exposure) and examination.

In both conditions the rise of plasma bilirubin is predominantly in the conjugated fraction. Points helpful in differential diagnosis are the following:

Transaminase levels which are very high, or are elevated early in the course of the disease, are in favour of hepatitis. In extrahepatic obstruction the levels of these enzymes may initially be normal, although they tend to increase as cell damage occurs.

Raised *alkaline phosphatase*, or γ-glutamyltransferase levels indicate the presence of cholestasis but not its cause. The higher the alkaline phosphatase the greater the likelihood of extrahepatic obstruction, but there is no level clearly separating the two.

A *positive mitochondrial antibody* (M) test is a strong point in favour of biliary cirrhosis, and therefore against extrahepatic obstruction (p. 293). A predominant increase in IgM also favours a diagnosis of primary biliary cirrhosis.

These laboratory tests must be supplemented by other diagnostic procedures, such as radiological studies and liver biopsy.

Cirrhosis.—In cirrhosis during phases of active cellular destruction there are usually slightly raised transaminase levels, sometimes with jaundice. Many cases show the typical *plasma protein* changes (p. 292) of a low albumin and diffusely elevated γ-globulin with β–γ fusion and it is in this type of case that protein estimations are most valuable. If ascites is present there may be dilutional hyponatraemia (p. 58). Plasma γ-glutamyltransferase is a sensitive indicator of liver disease, and may be the only enzyme elevated, particularly in alcoholic cirrhosis (p. 344).

Hepatocellular failure.—Most cases of hepatocellular failure occur during the course of severe hepatitis or decompensated cirrhosis, or following ingestion of liver toxins. Jaundice is usually present although fulminating cases may die before it develops. Any or all of the biochemical abnormalities of hepatitis may be found, depending on the stage of the disease.

Other features may include:

severe electrolyte disturbances, particularly hypokalaemia, due to secondary hyperaldosteronism;

a prolonged prothrombin time;

a low plasma urea. Normally ammonia from deamination of amino acids is utilised in the synthesis of urea in the liver. Impairment of this process results in deficient urea production and accumulation of amino acids in the blood with consequent overflow aminoaciduria (p. 356);

rarely, hypoglycaemia (p. 199).

Hepatic infiltration.—Hepatomegaly may be a result of infiltration of the liver by carcinoma or lymphoma, or of replacement by granulomata. Transaminase levels are often raised, *AST being more sensitive in this situation than ALT*. If there is biliary tract involvement alkaline phosphatase levels may also rise. This may be the only abnormal finding. In many cases bilirubin levels are normal, and liver function is rarely demonstrably impaired.

Haemolytic jaundice.—Increased erythrocyte destruction, with release of haemoglobin, results in an abnormally high rate of bilirubin production. This may overload the hepatic mechanisms for conjugation and excretion. The reserve capacity of the adult liver is such that jaundice, if present, is mild (bilirubin usually below 70 μmol/l or 4 mg/dl). The rise in plasma bilirubin is confined almost entirely to the unconjugated fraction unless there is associated liver disease. The increased load of excreted bilirubin in the bile, however, results in increased stercobilinogen production in the gut and, after this has been reabsorbed, an increased urinary urobilin(ogen). Bilirubin does not appear in the urine (acholuric jaundice). Other "liver function" tests are unaffected with the exception, in some cases, of raised AST and LD (and HBD) levels, due not to liver damage but to red cell destruction (p. 347).

In the newborn, however, the massive red cell destruction occurring in haemolytic disease of the newborn, coupled with the immature hepatic handling of bilirubin, can produce elevations of unconjugated bilirubin of 400–500 μmol/l (25–30 mg/dl) or greater. Such elevations are associated with the risk of developing *kernicterus* and levels may be reduced by exchange transfusion. In premature infants, too, the poorly developed conjugating mechanism may result in so-called "physiological" jaundice with markedly raised levels of unconjugated bilirubin, also necessitating exchange transfusion.

The effects of certain drugs in neonatal jaundice are considered on p. 298.

The student should read the section on "Investigating Liver Disease" on p. 304.

CONGENITAL HYPERBILIRUBINAEMIA

A group of disorders, apparently of congenital origin, has been described in which there is defective handling of bilirubin. The three best defined entities will be described briefly.

Gilbert's disease.—In patients with this disease there is probably defective transport of bilirubin into the liver cell. In some, reduced glucuronyl transferase (p. 287) activity has been demonstrated. Plasma levels of *unconjugated bilirubin* are mildly raised (20–35 μmol/l or 1–2 mg/dl), and tend to fluctuate. The condition may be noted at any age. It is often discovered when bilirubin levels fail to return to normal after an attack of hepatitis, or during any mild illness which, because of the jaundice, may be misdiagnosed as hepatitis. The condition is harmless but must be differentiated from haemolysis and from hepatitis. It has been suggested that Gilbert's "disease" represents the upper end of the spectrum of "normal" plasma bilirubin levels.

Crigler-Najjar syndrome.—In contradistinction to the other two syndromes discussed the Crigler-Najjar syndrome is *not* harmless. It presents in the first few days of life as jaundice due to a rise in *unconjugated bilirubin* levels. They may often be high enough (350 μmol/l [20 mg/dl] or higher) to cause kernicterus. The probable cause is a deficiency of glucuronyl transferase. In infants who survive, the level of bilirubin tends to stabilise, suggesting the existence of alternative pathways of bilirubin excretion.

Dubin-Johnson syndrome.—This condition is characterised by mildly raised *conjugated bilirubin* levels that tend to fluctuate. There is defective excretion of conjugated bilirubin. Bilirubin is present in the urine. Alkaline phosphatase levels are normal. There may be hepatomegaly and the liver is dark brown due to the presence in the cells of a pigment with the staining properties of lipofuscin. The condition is harmless and the diagnosis may be confirmed by the characteristic staining in the liver biopsy.

There is overlap between these and related conditions.

DRUGS AND THE LIVER

Many drugs are capable of producing jaundice with or without liver damage and a drug history is an essential part of the investigation of a patient with jaundice. This summary is based on the references given at the end of the chapter.

There are several mechanisms whereby drugs can produce jaundice.

1. Drug-induced *haemolysis* rarely causes jaundice. The subject is dealt with in textbooks of haematology.

2. Novobiocin interferes with bilirubin *conjugation*.

3. *Acute overdosage* (usually accidental or suicidal) of *paracetamol*, and, in children, of *ferrous sulphate*, can produce severe hepatic necrosis.

Direct *hepatotoxins*, such as carbon tetrachloride and *Amanita phalloides* (a toadstool) poisoning, produce liver cell necrosis of varying severity. Ingestion is usually accidental or with suicidal intent.

Tetracyclines by intravenous injection, especially during pregnancy, have produced acute fatty liver and death from liver failure.

4. *Hepatitis-like reaction.*—Several drugs may produce a clinical, biochemical and histopathological syndrome closely resembling viral hepatitis.

(*a*) Hydrazine derivatives, e.g. iproniazid, phenelzine and rarely isoniazid.

(*b*) Chlordiazepoxide (rare).

(*c*) Methyldopa (occasionally).

(*d*) Halothane anaesthetics rarely produce severe liver damage and death.

5. *Cholestatic reaction.*—Another group of drugs produces the picture of cholestatic jaundice :

(*a*) 17-α-alkylated steroids such as methyl testosterone, norethandrolone, or norethisterone (components of certain oral contraceptives).

(*b*) Phenothiazines, such as chlorpromazine.

(*c*) Rarely thiouracil, chlorpropamide, or phenylbutazone.

(*d*) Antibiotics such as erythromycin or oleandomycin.

(*e*) Para-aminosalicylic acid (PAS) (rare).

(*f*) Cholestasis may occur as part of a generalised sensitivity reaction (e.g. to penicillin).

Neonatal jaundice and drugs.—In the newborn, particularly if premature, the hepatic handling of bilirubin is immature (p. 296) and jaundice due to a rise in unconjugated bilirubin levels is common. This phase lasts two to three days in full term infants and five to six days in premature infants. The course and severity of unconjugated hyperbilirubinaemia may be influenced by drugs in three ways:

(*a*) Several drugs *displace bilirubin from plasma albumin* and increase the risk of deposition in the brain with cerebral damage (*kernicterus*). These include salicylates and sulphonamides. The most important of these is salicylates.

(*b*) Novobiocin *inhibits the glucuronyl transferase system* and aggravates unconjugated hyperbilirubinaemia.

(*c*) Any drug producing *haemolysis* aggravates the condition.

BILE AND GALL STONES

BILE ACIDS AND BILE SALTS

Four bile acids are produced in man. Two of these, *cholic acid* and *chenodeoxycholic acid*, are synthesised in the liver from cholesterol and are referred to as *primary bile acids*. These are excreted in the bile into the gut where bacterial action converts them to the *secondary bile acids*, *deoxycholic acid* and *lithocholic acid*, respectively. In human bile the bile acids are in the form of sodium salts, all conjugated with the amino acids glycine or taurine (*bile salts*). Some of the secondary bile salts are absorbed and re-excreted by the liver (enterohepatic circulation of bile salts, p. 264). Bile therefore contains a mixture of primary and secondary bile salts.

Bile salts are important because they contain both polar and non-polar chemical groups and *form micelles*. In these micelles the non-polar groups are orientated towards the centre of the molecule and form a small pool of lipid solvent, while the polar (water-soluble) groups are on the outside.

Deficiency of bile salts in the intestinal lumen leads to impaired micelle formation, and malabsorption of fat (p. 274). Such deficiency may be caused by cholestatic liver disease (failure to reach the gut) or by ileal resection or disease (failure of reabsorption, with reduced bile-salt pool). The role of deficient bile salts in cholelithiasis is discussed below.

FORMATION OF BILE

About 1 to 2 litres of bile is produced in the liver daily. This *hepatic bile* contains bilirubin, bile salts, phospholipids and cholesterol as well as electrolytes in similar concentrations to those in plasma. Small amounts of protein are also present. In the gall bladder there is active reabsorption of sodium, chloride and bicarbonate, together with an isosmotic amount of water. The end result is *gall-bladder bile* which is ten times more concentrated than hepatic bile and in which sodium is the major cation and bile salts are the major anions. The concentration of other non-absorbable molecules, conjugated bilirubin, cholesterol and phospholipids also increases.

GALL STONES

Although most gall stones contain all constituents of bile, there are several types differing in the main constituent. Only about 10 per cent of gall stones contain sufficient calcium to be radio-opaque (unlike renal calculi).

Cholesterol gall stones may be single or multiple and are not usually radio-opaque. They are white or yellowish and the cut surface has a crystalline appearance. They may be mulberry shaped.

Pigment gall stones consist largely of bile pigments with organic material and variable amounts of calcium. They are small multiple stones, dark green or black and are hard. Rarely they are radio-opaque.

Mixed gall stones, the commonest form, are, as the name suggests, composed of a mixture of cholesterol, bile pigments, protein and calcium. They are multiple and appear as faceted dark brown stones with a hard shell and softer centre. They may be radio-opaque.

CAUSES OF GALL STONE FORMATION

Pigment stones are found in *chronic haemolytic states* such as hereditary spherocytosis where there is an increase in bilirubin formation and therefore excretion. In all other forms of gall stones, however, the aetiology is uncertain. The role of infection as a precipitating factor in gall stone formation is controversial.

Pathological bile has a greater concentration of cholesterol and a lower concentration of bile salts than normal. As mentioned above, bile salts are required to maintain cholesterol in solution. In addition, the salts of cholic acid (with three OH groups) form more stable micelles than do those of chenodeoxycholic acid and deoxycholic acid (with two OH groups). Stone-forming bile has been shown to have a higher proportion of the dihydroxy bile salts than normal. This combination of bile salts deficient in quantity and quality would favour the precipitation of cholesterol. The cause of such alterations is not clear. It may be due to primary hepatic or gall bladder dysfunction or it may be the result of reduced enterohepatic circulation of bile salts due to biliary stasis or obstruction. It has been suggested that inflammation of the gall bladder wall leads to greater absorption of the more water-soluble bile salts (those with more OH groups).

There is *no* association between hypercholesterolaemia and the formation of cholesterol gall stones.

Routine chemical analysis of gall stones is of little value.

CONSEQUENCES OF GALL STONES

Gall stones may remain silent for an indefinite length of time and be discovered only at laparotomy for an unrelated condition, or on x-ray of the abdomen. They may, however, lead to several clinical consequences:

1. Acute cholecystitis, due to obstruction of the cystic duct by a gall

stone with chemical irritation of the gall bladder mucosa by trapped bile and secondary bacterial infection.

2. Chronic cholecystitis.

3. Common bile duct obstruction, if a stone lodges in the bile duct. This may present as biliary colic, obstructive jaundice (usually intermittent) or acute pancreatitis if the pancreatic duct is also occluded.

4. Extremely rarely, carcinoma of the gall bladder occurs.

SUMMARY

LIVER DISEASE

1. The liver has a central role in many metabolic processes.

2. Bilirubin derived from haemoglobin is conjugated in the liver and excreted in bile. Conversion to stercobilinogen (faecal urobilinogen) takes place in the bowel. Some reabsorbed faecal urobilinogen is excreted in the urine.

3. Bilirubin metabolism may be assessed by plasma levels of total and conjugated bilirubin, and by visual inspection of the stool and urine.

4. Jaundice is due to a raised plasma bilirubin. It may be classified initially as that due to a raised unconjugated bilirubin only, and that in which both fractions are increased. The majority of cases of jaundice fall into the latter group.

5. The two basic processes in liver disease are liver cell damage and cholestasis. Tests help to distinguish the underlying process, but not necessarily its cause.

6. Tests of liver function are considered in the following categories:
   (*a*) tests indicating liver cell damage;
   (*b*) tests indicating biliary tract involvement;
   (*c*) tests indicating impaired function;
   (*d*) tests indicating aetiology.

7. The selection of tests is governed by the particular problem. The main indications are:
   (*a*) differential diagnosis of jaundice;
   (*b*) assessment of the severity and progress of liver disease;
   (*c*) detection of liver disease.

8. A group of congenital conditions exist, characterised by hyperbilirubinaemia. Most are relatively harmless, but the Crigler-Najjar syndrome may lead to kernicterus.

9. Drugs may produce jaundice in several ways and a drug history is important in the assessment of the jaundiced patient. In the newborn, drugs may increase the risk of kernicterus.

BILE AND GALL STONES

Bile secreted by the liver is concentrated in the gall bladder before passing into the gut. Cholesterol in bile is held in solution by bile salt micelles. A change in either the concentration or type of bile salts may lead to defective micelle formation and consequent precipitation of cholesterol with gall stone formation.

FURTHER READING

SHERLOCK, SHEILA (1975). *Diseases of the Liver and Biliary System*, 5th edit. Oxford: Blackwell Scientific Publications.

ZIMMERMAN, H.J., Ed. (1975). Symposium on diseases of the liver. *Med. Clin. N. Amer.,* **59,** 4.

MELNICK, J. L., DREESMAN, G. R., and HOLLINGER, F. B. (1977). Viral hepatitis. *Sci. Amer.,* **237** (July), 44.

STENGER, R. J. (1977). Liver disease. *Hum. Path.,* **8,** 603.

BURKE, M. D. (1975). Liver function. *Hum. Path.,* **6,** 273.

KNELL, A. J. (1977). Acute liver failure. *Practitioner,* **218,** 230.

APPENDIX

Handling of Blood Samples from Patients with Possible Hepatitis

Samples from all patients with viral hepatitis, undiagnosed jaundice, or positive HBsAg tests, as well as at-risk patients from dialysis units, should be considered *infective*. Anyone handling such a sample (medical and nursing staff, porters and laboratory staff) is at risk. It is *the duty of the clinician sending the blood to identify it clearly as potentially dangerous*. The sample should be *sent to the laboratory in a sealed plastic bag*.

Rapid Tests for Urinary Bile Constituents

Urine containing bilirubin is usually dark yellow or brown, whereas fresh urine containing urobilinogen only is initially of normal colour.

Reagent strips (Ames) are available for the detection of these substances. *Fresh* urine is essential.

Ictostix includes stabilised diazotised 2, 4-dichloraniline which reacts with *bilirubin* to form azobilirubin.

The test will detect about 3 μmol/l (0·2 mg/dl) of bilirubin. Drugs (such as large doses of chlorpromazine) may give *false positive* reactions.

Urobilistix includes paradimethylaminobenzaldehyde which reacts with *urobilinogen*. It does *not* react with porphobilinogen.

This test will detect urobilinogen in some normal urines. *False positive* results may occur with drugs such as *p*-aminosalicylic acid and certain sulphonamides. Positive results should be confirmed by a test for *urobilin* (below).

Both these reagent strips must be stored, and the test performed, strictly according to the instructions of the manufacturers.

Alternative test for urobilinogen

1. Mix about 2 ml of *fresh* urine with an equal volume of Ehrlich's reagent and allow to stand for 5–10 minutes.

2. Add 4 ml of saturated sodium acetate solution and mix.

A red colour denotes the presence of urobilinogen. Distinction from porphobilinogen (p. 403) is made by shaking the coloured solution with *n*-butanol. The colour due to urobilinogen is extracted into butanol (upper layer) whereas that of porphobilinogen is not.

Urobilin.—This test is preferable to the above if the urine is not fresh, as urobilinogen is converted to urobilin on standing.

1. Mix 5 ml of urine with 2 drops of alcoholic iodine solution (converts urobilinogen to urobilin).

2. Add 5 ml of zinc acetate suspension, mix and allow to settle.

3. A greenish fluorescence in the supernatant is due to a zinc-urobilin complex. This is best seen in darkened surroundings by shining the light from a pencil torch through the fluid.

INVESTIGATING LIVER DISEASE

The following are the most useful tests in individual conditions.

Suspected acute hepatitis
    Plasma bilirubin
    ALT *or* AST
    HBsAg.

Suspected chronic hepatitis
    ALT (AST may be normal)
    In some cases HBsAg and antibodies
    Serum protein electrophoresis.

Suspected cirrhosis
    Serum albumin, protein electrophoresis and immunoglobulin levels.
In this condition any or none of the biochemical findings may be abnormal;
in addition ALT, AST, alkaline phosphatase, bilirubin, and, *if all these are
normal*, GGT. An AST level relatively higher than that of ALT favours a
diagnosis of cirrhosis rather than chronic hepatitis.

Suspected cholestasis or cholangitis
    Alkaline phosphatase *or* GGT
    Bilirubin, with special reference to the conjugated fraction
    Transaminases are less useful, but may help to differentiate extra- from
intrahepatic cholestasis.

Suspected hepatic deposits or infiltration
    AST (ALT may be normal)
    Alkaline phosphatase *or* GGT.

Suspected primary hepatocellular carcinoma
    α-fetoprotein
    AST
    Alkaline phosphatase *or* GGT.

Chapter XIV

PLASMA PROTEINS AND IMMUNOGLOBULINS: PROTEINURIA

PLASMA PROTEINS

PLASMA contains a complex mixture of proteins at a concentration of about 70 g/l. These include simple proteins as well as those incorporating carbohydrates and lipids. These different proteins have different functions and originate from several different cell types.

Functions

The following is an outline of the main functions of the plasma proteins.

Antibodies (p. 315), **the complement system** (p. 314) **and protease inhibitors** (p. 313) are proteins. They are components of the immune system.

Control of extracellular fluid distribution (ECF).—Distribution of water between the intra- and extravascular compartments is affected by the concentration of plasma proteins. Albumin is the most important in this respect (p. 36).

Transport.—Plasma proteins transport many hormones (for example, cortisol, p. 135 and thyroxine, p. 162) and vitamins, lipids (p. 216), calcium, trace metals and some drugs. Combination with proteins renders them soluble (for example, lipids), or physiologically inactive (for example, calcium and drugs).

Nutrient.—Circulating protein, particularly albumin, can be a source of nutrient for tissues.

Buffering.—Proteins form a minor part of the plasma buffering system (p. 90),

Blood clotting factors, enzymes and many **hormones** are physiologically important proteins which are not discussed in this chapter. Only a few enzymes are functional in the circulation, most arising from leakage from cells.

This list is by no means complete. The function of many of the proteins which have been identified in plasma is unknown.

Many plasma proteins are synthesised in the *liver*, but the *plasma cells* and *lymphocytes* of the immune system (immunocytes) synthesise immunoglobulins.

Abnormal plasma levels of hepatic proteins may be due to hepatocellular dysfunction, but are more commonly a non-specific response to other disease. Abnormal plasma immunoglobulin levels may be a response to an immune challenge, or may be due to primary immunocyte dysfunction: they may cause symptoms directly.

Total protein concentration is easy to measure, but of limited diagnostic value. *Electrophoresis*, which separates the proteins into groups, may sometimes provide useful information. Measurement of *individual proteins* may be indicated on clinical grounds, or after inspection of the electrophoretic pattern.

TOTAL PROTEIN

Albumin is quantitatively the major single contributor to the plasma total protein, and *hypoproteinaemia* is almost always due to *hypoalbuminaemia*. True hyperalbuminaemia probably does not occur, and *hyperproteinaemia* is usually due to a major *increase in one or more of the immunoglobulins*.

Total protein levels may be misleading, and may be normal in the face of quite marked changes in the constituent proteins. For example:

a fall in albumin may roughly be balanced by a rise in immunoglobulin levels. This is quite a common combination;

most individual proteins, other than albumin, make a relatively small contribution to total protein: quite a large *percentage* change in the concentration of one of them may not be detectable as a change in total protein.

Examples of conditions in which total protein concentration may be abnormal are listed below, the proteins most affected being indicated in brackets.

Causes of Raised Total Protein Concentration

Changes due to relative water deficiency (all fractions). These are concentration changes only and *do not indicate alterations in absolute amounts of protein*.

Dehydration.

Artefactual—stasis during venepuncture. Stasis induced by keeping the tourniquet on too long during venepuncture causes fluid to escape into the extravascular compartment, leading to haemoconcentration and a falsely raised protein concentration. *This is the commonest cause of a high total protein level.*

Paraproteinaemia (paraprotein, p. 321).

Certain chronic diseases (immunoglobulins).

Chronic inflammatory conditions (p. 313).

Cirrhosis of the liver.

Autoimmune disease, for example, systemic lupus erythematosus (SLE).
Sarcoidosis.

Causes of Low Total Protein Concentration

Changes due to relative water excess (all fractions). These are concentration changes only and *do not indicate alterations in absolute amounts of protein.*
Overhydration.
Artefactual—blood taken from the "drip" arm.
Excessive loss of protein (mainly albumin).
Through the kidney in the nephrotic syndrome.
From the skin after severe burns.
Through the intestine in protein-losing enteropathy.
Decreased synthesis of protein.
Severe dietary protein deficiency, for instance in kwashiorkor (all fractions but mainly albumin).
Severe liver disease (mainly albumin).
In both of the above there may be no fall in total protein because of a rise in immunoglobulins.
Severe malabsorption (all fractions).
Total protein estimation may be helpful in the assessment of hydration. Extracellular proteins, because of their large size, are mostly confined to the vascular compartment: changes in plasma volume will cause a change in total protein *concentration* (and in haematocrit). Serial estimations may be useful in following changes in hydration.

ELECTROPHORESIS

Electrophoresis, using the differences in electrical charge on different proteins to separate them, is routinely performed by applying a small amount of serum to a strip of cellulose acetate and passing a current across it for a standard time. In this way five main groups of protein may be distinguished after staining, and may be compared with a normal serum. Alternatively, it may be passed through a light path (*scanned*); the absorption or reflection of light by each fraction is graphically represented as the electrophoretic "scan" (Fig. 30).

Each of the five fractions, albumin and the α_1-, α_2-, β- and γ-globulins, includes many proteins (Fig. 30). Changes in electrophoretic pattern are most obvious when the level of a protein normally present in high concentration (for instance, albumin) is abnormal, or when there are parallel changes in several proteins in the same fraction (for instance, in the immunoglobulins of the γ fraction).

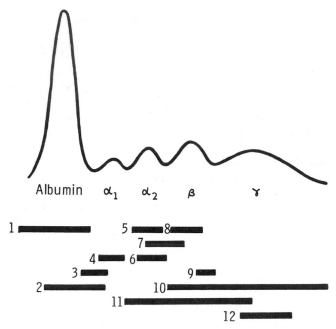

FIG. 30.—The plasma protein electrophoretic pattern and the twelve main plasma proteins (excluding fibrinogen).

1. Albumin	5. α_2-macroglobulin	9. C_3 component of complement
2. α-lipoprotein (HDL)	6. haptoglobin	10. IgG
3. α_1-acid glycoprotein	7. β-lipoprotein (LDL)	11. IgA
4. α_1-antitrypsin	8. transferrin	12. IgM

CHANGES IN INDIVIDUAL PLASMA PROTEIN FRACTIONS

Albumin

There are two rare congenital anomalies of albumin synthesis. There are a variety of *bisalbuminaemias*, in which, because the subject carries alleles for more than one albumin, two albumin types are present. These are curiosities only, as there are no clinical consequences. In the other, *analbuminaemia*, there is deficient synthesis of the protein. Clinical consequences are slight, and oedema, though present, is surprisingly mild.

An abnormally high albumin level is found only with dehydration, or, artefactually, in a sample taken with prolonged venous stasis (p. 306), and only low albumin levels are of clinical interest.

Consequences of a low serum albumin.—The role of plasma proteins in fluid distribution has been discussed. As albumin is the most important fraction in this respect, one of the consequences of a reduced level is *oedema*. The critical level varies, but oedema is almost always present if the plasma albumin is below 20 g/l. The surprisingly mild oedema of analbuminaemia has been mentioned above.

About half the plasma calcium is bound to albumin (p. 232) and hypoalbuminaemia is accompanied by *hypocalcaemia*. As this involves only the protein bound (physiologically inactive) half, symptoms of tetany do not develop and calcium or vitamin D supplements are contra-indicated. As a rough guide there is about 0·02 mmol/l (0·8 mg/dl) of calcium per g/l albumin.

Albumin also binds *bilirubin, free fatty acids*, and a number of *drugs* such as salicylates, penicillin and sulphonamides. The albumin bound fractions are physiologically and pharmacologically inactive. A marked reduction in serum albumin, by reducing the binding capacity, may increase free levels of these substances and may cause toxic effects if drugs are given in their normal dosage. Drugs which are albumin bound, if administered together, may compete for binding sites, also increasing free concentrations: an example of this is simultaneous administration of salicylates and the anticoagulant warfarin, with potentiation of the effect of the latter.

Causes of hypoalbuminaemia.—In normal subjects albumin accounts for almost half the total protein concentration. Significant changes in albumin levels therefore usually affect the latter and the causes of hypo-albuminaemia are very similar to those listed on p. 307 for low total protein levels. In addition, mild changes, such as are found non-specifically, may not significantly affect total protein levels.

Decreased synthesis.—Albumin is catabolised at a steady rate of about 4 per cent of the body pool daily, and any impairment of synthesis soon causes hypoalbuminaemia. Albumin is synthesised in the liver, and *liver disease*, especially if chronic, causes low levels.

Loss.—Albumin is a relatively small molecule with a molecular weight of about 70 000. It is the predominant fraction lost in protein-losing states such as the *nephrotic syndrome* (p. 327), *protein-losing enteropathy* (p. 276) or exudation from the skin after *extensive burns* or with extensive *psoriasis* or other exudative skin diseases. The lowest levels are usually reached in this category of disease.

Increased catabolism.—In many illnesses the rate of albumin cata-bolism is increased, and this may aggravate the effects of decreased synthesis, or loss from the body.

Deficient intake.—*Malabsorption* or *malnutrition* due to severe dietary deficiency causes hypoalbuminaemia because of the poor supply of amino acids.

Non-specific.—One of the commonest reasons for a low serum albumin is less easily explained. In many acute conditions, including apparently *minor illnesses* such as colds and boils, the serum albumin level falls. The answer cannot lie entirely in decreased synthesis or increased breakdown as there is a rapid return to normal on recovery. It could be due partly to a switch to synthesis of other fractions, because the total protein level often remains the same. There is some evidence that there is a redistribution of albumin from the plasma. This phenomenon is seen commonly in hospitalised patients, and changes of concentration of up to 5–10 g/l may be a non-specific finding. *Posture* influences albumin levels which may be 5–10 g/l higher in the upright than in the recumbent position.

Albumin levels fall briefly at birth, but soon reach adult levels.

Haemodilution.—In the late stages of pregnancy, or in overhydrated subjects, low concentrations of albumin and all protein fractions may be found due to dilution effects. This is *not* an abnormality of albumin metabolism.

Artefactual.—Dilution may result artefactually if blood is taken from a "drip" arm (p. 457). Bilirubin and drugs may cause an apparently low albumin if dye-binding methods of assay are used. These methods depend on measuring the amount of dye bound to albumin, and bilirubin and drugs may compete for binding sites.

α_1- and α_2-Globulins

The proteins making up this group of globulins all tend to increase when there is active tissue damage (*"acute phase reaction"*). *Raised levels of both α_1- and α_2-globulins are therefore a non-specific finding in the serum of most ill patients.* They occur:

in inflammatory conditions;
associated with malignancy;
after trauma;
post-operatively;
in autoimmune disease.

α_1-**Antitrypsin** is the only α_1-globulin which stains well on routine electrophoresis. *Low levels* occur in α_1-*antitrypsin deficiency* (p. 313): such deficiency can be excluded if an α_1-band is visible.

α_2-*Macroglobulin* and *haptoglobin* make up most of the α_2-globulin. α_2-Macroglobulin is a very large molecule (MW approximately 900 000) and is retained in the blood stream in the *nephrotic syndrome*, when there may be loss of all other protein fractions: levels may even *increase* in this condition, possibly due to feedback stimulation of synthesis. β-lipoprotein (LDL) overlaps the α_2 and β fractions on cellulose acetate electrophoresis; a rise in this protein also occurs in the nephrotic syndrome, and contributes to the raised α_2-globulin. *A relatively or absolutely raised α_2-globulin, with reduction of other fractions is characteristic of the nephrotic syndrome and other protein-losing states.*

β-Globulins

The main components of β-globulin are part of the β-lipoprotein (LDL, p. 217) and transferrin (p. 382).

High levels of β-globulin are very occasionally found in *pregnancy* and in *biliary obstruction*.

γ-Globulins

This group of proteins is considered more fully in the section on immunoglobulins. Although all γ-globulins are immunoglobulins, not all immunoglobulins are found in the γ region of the electrophoretic strip: some may be found in the β and $α_2$ regions.

Immunoglobulins are antibodies, and a general increase in levels occurs in chronic inflammatory states and in some autoimmune diseases. The most important use of electrophoresis is to distinguish this diffuse increase from a sharp paraprotein band (p. 321).

A diffusely raised γ-globulin occurs in:

chronic infections. In the tropical disease kala-azar and in lymphogranuloma venereum very high values may be found;

cirrhosis of the liver;

sarcoidosis;

autoimmune disease

systemic lupus erythematosus (SLE);

rheumatoid arthritis.

Decreased γ-globulin levels may be due to:

increased loss

nephrotic syndrome and other protein-losing states;

decreased synthesis

severe malabsorption and malnutrition;

primary immune deficiency (p. 320);

secondary suppression of synthesis

secondary immune deficiency (p. 321).

ABNORMAL ELECTROPHORETIC PATTERNS

The commonest abnormal protein patterns seen on routine electrophoresis are illustrated in Fig. 31. This figure represents "scanned" strips (p. 307).

Parallel changes in all fractions (not shown in Fig. 31). This is a *normal pattern with an abnormal total protein concentration*. An *increase* in all protein fractions (including immunoglobulins) may be found in *dehydration* and due to *stasis* during venepuncture, and a reduction in *overhydration* (p. 307) or in specimens taken from a "drip arm". A reduction also occurs in severe protein *malnutrition* and *malabsorption*, unless accompanied by infection.

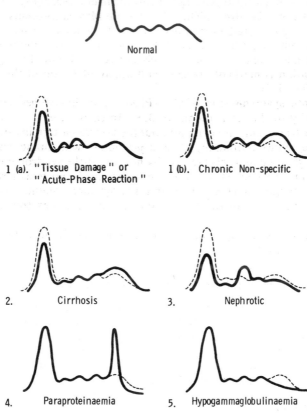

Normal

1 (a). "Tissue Damage" or
"Acute-Phase Reaction"

1 (b). Chronic Non-specific

2.   Cirrhosis

3.   Nephrotic

4.   Paraproteinaemia

5.   Hypogammaglobulinaemia

FIG. 31.—Electrophoretic patterns in disease (refer to corresponding
numbers in text).

Dotted line = normal pattern

Non-specific patterns.—These occur in a variety of illnesses and are
of no more importance in diagnosis than a raised erythrocyte sedimentation rate (ESR).

A low albumin level, without changes in other fractions, occurs
in many acute illnesses (not shown in Fig. 31).

A low albumin and raised α_1- and α_2-globulin concentrations occur
in more severe acute infections, inflammation and neoplasia ("tissue
damage" pattern or *"acute phase reaction"*, Fig. 31/*1a*).

A low albumin with diffusely raised γ- and often raised α_2-globulin *levels* occur with chronic "tissue damage", as in chronic infections and some autoimmune diseases (for example, rheumatoid arthritis). This pattern may also be found in chronic liver disease (Fig. 31/*1b*).

Cirrhosis of the liver.—The changes in plasma proteins in liver disease are considered more fully on p. 292. They are usually "non-specific", but in cirrhosis a characteristic pattern is sometimes seen. Albumin and often α_1-globulin levels are reduced and the γ-globulin concentration is markedly raised, with apparent fusion of the β and γ bands (Fig. 31/*2*).

Nephrotic syndrome (Fig. 31/*3*).—Plasma protein changes depend on the severity of the renal lesion (p. 328). In early cases a low albumin level may be the only abnormality, but the typical pattern in established cases is a reduced albumin, α_1- and γ-globulin with an increase in α_2-globulin. β-globulin levels are usually normal. If the syndrome is due to systemic lupus erythematosus the γ-globulin may be normal or raised.

Paraproteinaemia (Fig. 31/*4*) and hypogammaglobulinaemia (Fig. 31/*5*) are discussed in the section on immunoglobulins (pp. 321 and 320).

Although the changes in electrophoretic pattern usually indicate disease, they are rarely pathognomonic.

SPECIFIC PROTEINS OF CLINICAL IMPORTANCE

Concentrations of many of the individual plasma proteins may be measured by immunochemical or chemical methods. Of these, albumin is discussed on p. 308, the lipoproteins in Chapter X, transferrin in Chapter XVIII, and the immunoglobulins on p. 315. Some other proteins of clinical value are discussed here. Concentration changes in some, but not all, are visible on electrophoretic strips.

α_1-Antitrypsin

The α_1-antitrypsins are important protease inhibitors (Pi). Because of their relatively low molecular weight (54 000) they can diffuse through the capillary wall, and are found in interstitial fluid as well as in plasma. There are more than 20 genetic variants. In several of these (most of them combinations of the alleles called Z, S and null) plasma levels of α_1-antitrypsin are so *low* that the α_1 band cannot be seen on the electrophoretic strip (it is unlikely that α_1-antitrypsin is deficient if this band is visible). The patient with such α_1-*antitrypsin deficiency* may present in the neonatal period or during childhood with cirrhosis of the liver, or as a young adult with emphysema. The α_1-antitrypsin response to inflammation and tissue damage is subnormal, and it may be that the unopposed proteolytic action of inflammation causes this permanent pulmonary and hepatic damage. Although there is no specific therapy for the deficiency, pulmonary infections must always be treated in an attempt to reduce tissue damage. Family studies should be carried out with a view to genetic counselling.

High levels may be a non-specific response to tissue damage, or may be found in women taking oral contraceptives.

Haptoglobin

Haptoglobin minimises urinary loss of haemoglobin (and its important constituent, iron) by binding any free haemoglobin released by haemolysis. The haptoglobin/haemoglobin complex is catabolised faster than haptoglobin alone. *Low levels* are therefore found in haemolytic conditions.

Raised levels account in part for the increased α_2 fraction seen on the electrophoretic strip after tissue damage. This response (like all those of the acute-phase reaction) is non-specific.

Caeruloplasmin.

Caeruloplasmin is a copper-containing protein, with enzymatic activity, but of unknown function. Estimation is most useful in suspected *Wilson's disease* (hepatolenticular degeneration) in which levels are low. Other causes of low levels are malnutrition and the nephrotic syndrome: these conditions are unlikely to cause diagnostic confusion because of the clinical picture, and the other changes on the electrophoretic strip.

High values are not pathognomonic, and occur in active liver disease and tissue damage (as part of the acute-phase response), and during pregnancy and oral contraceptive therapy.

α-Fetoprotein

Fetal plasma levels of α-fetoprotein are high. Postnatal levels are very low, but they increase during pregnancy. The use of α-fetoprotein assays to detect fetal *neural tube defects* is considered on p. 422.

Relatively insensitive screening tests for plasma α-fetoprotein are often positive in patients with *primary hepatocellular carcinoma* or *germinal tumours of the gonads*. There may be a transient positive reaction in some cases of hepatitis, but, more commonly, the slight increase occurring in many non-malignant liver diseases is too small to be detectable by the usual screening methods.

Complement

The complement system plays an important part in the body's defence against infection. Invading organisms, after identification by antibodies, are lysed by activated complement, and are then phagocytosed.

Macrophages are the most important cells synthesising the inactive proteins of the complement system. The most important factors initiating activation of these proteins are antigen-antibody complexes, and bacterial lipopolysaccharides or toxins; a chain of reactions follows in which plasma complement is utilised. Products of the pathway increase vascular permeability, attract phagocytes, and, by causing local thrombosis, limit spread of infection.

In *immune-complex disease* circulating antigen-antibody complexes become locally attached to blood vessel walls, such as those of the renal glomerulus. Under these circumstances the normally beneficial process of complement activation may damage the blood vessels.

C_3 is the most commonly measured plasma complement component and its estimation is used to detect immune-complex disease. *Low levels* suggest that this inactive component is being utilised in the activation pathway faster than it can be synthesised. The finding is of particular value in the differential diagnosis of *glomerulonephritis*. In acute post-strepococcal cases the low levels found in the acute phase rise during recovery, but they fail to do so in the more serious mesangio-capillary glomerulonephritis. *Systemic lupus erythematosus* (SLE) is another immune-complex disease in which low C_3 levels correlate with the activity of the disease: C_3 estimation can be used to assess the effectiveness of therapy for SLE.

High C_3 levels are part of the non-specific acute-phase reaction found with the tissue damage of many inflammatory states.

A full and critical discussion of the complement system will be found in the reference quoted at the end of the chapter.

IMMUNOGLOBULINS

It has long been known that the γ-globulin fraction of plasma proteins has antibody activity. The terms antibody and γ-globulin are often used synonymously. However, antibodies also occur in the β and α_2 fractions (Fig. 30), and the term immunoglobulin (Ig) is preferable. Bence Jones protein (light chain) and heavy chain are naturally occurring subunits of the Ig molecule.

STRUCTURE OF THE IMMUNOGLOBULINS

The immunoglobulin unit is a Y-shaped molecule depicted schematically in Fig. 32.

The following points should be noted:

1. Usually, four polypeptide chains are linked by disulphide bonds. There are two heavy (H) and two light (L) chains in each unit. The *H*

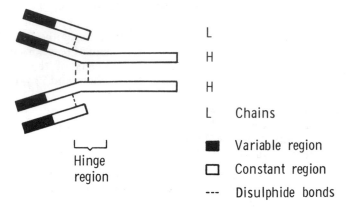

L

H

H

L   Chains

■   Variable region

Hinge region   ☐   Constant region

---   Disulphide bonds

FIG. 32.—Schematic representation of Ig subunit.

chains in a single unit are similar and determine the immunoglobulin *class* of the protein. H chains γ, α, μ, δ and ε occur in IgG, IgA, IgM, IgD and IgE respectively. *L chains* are of two *types* κ or λ. In a single molecule the L chains are of the same type, although the Ig class as a whole contains both types.

2. There are two antigen-combining sites per unit. These lie at the ends of the arms of the Y: both H and L chains are necessary for full antibody activity. The amino acid composition of this part of the chain varies in different units (variable region). When both of these sites combine with antigen, conformational changes are transmitted through the hinge region (Fig. 32) and the molecule becomes activated.

3. The rest of the H and L chains are less variable (constant region). The constant region of the H chains is responsible for such properties of the Ig unit as the ability to bind complement or actively to cross the placental barrier. The H chains are associated with a variable amount of carbohydrate; IgM has the highest content.

NOMENCLATURE OF THE IMMUNOGLOBULINS

Two forms of nomenclature are in use.

Based on Immunochemical Studies

The division of immunoglobulins into classes based on the structure of the H chains and into types based on the structure of the L chains has already been discussed. The chains are identified by use of specific antisera. Several subclasses of the major Ig classes exist, but in routine work IgG, IgA and IgM are the most commonly measured.

Based on Ultracentrifuge Studies

The simplified account of the immunoglobulin unit given above ignores the fact that each molecule may contain more than one unit. The IgM molecule, for example, consists of 5 basic units. This variation in size, and therefore in density, enables the classes to be separated in the ultracentrifuge. Classification is by Svedberg coefficient (S), the S value of a protein increasing with increasing size. IgG, which circulates as a single unit, sediments as 7S, IgM as 19S and IgA, which can contain a variable number of units, as between 7S and 11S. By comparison, albumin sediments as 4·5S.

FUNCTION OF THE IMMUNOGLOBULINS

The immune response mechanism of the body consists of a cellular and a humoral component. Although we are concerned here only with the humoral component—the immunoglobulins—the student should remember that both are necessary: he should consult a textbook of immunology for details of the cellular mechanisms.

The specific functions of each class of immunoglobulin will be discussed briefly, and followed by an outline of changes in disease.

IgG (MW 160000)

IgG accounts for about 75 per cent of circulating immunoglobulins and contains most of the normal plasma antibodies. It is a relatively small molecule, and is present in very low concentration throughout the extracellular compartment: its most important function seems to be protection of the tissue spaces. IgG synthesis is stimulated by soluble antigens such as bacterial toxins. *IgG deficiency* is characterised by *recurrent pyogenic infections* of tissue spaces by toxin-producing organisms such as staphylococci and streptococci: pulmonary and subcutaneous infections are common.

IgG can bind complement and cross the placental barrier. In the first few months of life endogenous IgG levels are very low. Maternal IgG which has been transported across the placenta provides antibody cover during this period. Adult concentrations of this class of protein are not reached before the age of 3 to 5 years.

IgA (MW 160000. Polymers also occur)

About 20 per cent of circulating immunoglobulins consist of this class. However, most IgA is synthesised beneath the mucosa of the gastrointestinal and respiratory tracts. A "secretory piece", synthesised by epithelial cells, binds two IgA units to form secretory IgA. The latter appears in intestinal and bronchial secretions, sweat, tears and colos-

TABLE XXV

PROPERTIES AND FUNCTIONS OF PLASMA IMMUNOGLOBULINS

	IgG	IgA	IgM	IgE	IgD
Molecular weight	160 000	160 000 and polymers	1 000 000	200 000	190 000
Sedimentation coefficient	7 S	7 S, 9 S, 11 S	19 S	8 S	7 S
% total serum immunoglobulin	73	19	7	0·001	1
Able to fix complement	Yes	No	Yes	No	No
Able to cross placenta	Yes	No	No	No	No
Approximate mean normal adult concentration: g/l	9 to 12	2·5	1·0	0·0003	0·03
IU/ml	140	150	140	90	
Adult levels reached by	3–5 years	15 years	9 months	? 15 years	? 15 years
Major function	Protects extra-vascular tissue spaces. Secondary response to antigen. Neutralises toxins.	Protects body surfaces as secretory IgA (11 S)	Protects blood stream. Primary response to antigen. Lyses bacteria.	Tissue bound antibodies of immediate hypersensitivity reactions.	Not known

trum; it protects body surfaces, particularly against viral infection. The secretory piece impairs destruction by digestive enzymes and enables IgA to perform an important protective function in the gut. *Deficiency of IgA* may be associated with mild *recurrent respiratory tract infections* and *intestinal disease*, but may be symptomless.

IgA does not bind complement or cross the placenta. Detectable levels are found a few months after birth, but adult levels may not be reached before the age of 15 years.

IgM (MW 1 000 000)

Because of its very large size, IgM (macroglobulin) is almost entirely intravascular. It accounts for about 7 per cent of circulating immunoglobulins. IgM antibodies are usually the first to be synthesised in response to infection: particulate antigens, such as circulating organisms, are particularly effective in stimulating synthesis. In *IgM deficiency septicaemia* is common.

The major blood group isohaemagglutinins belong to this group of proteins.

Although IgM can bind complement it does not cross the placenta. Synthesis occurs even in the fetus, and elevated IgM values at birth indicate intra-uterine infection. Adult levels are reached by about 9 months of age.

IgE (MW 200 000)

IgE is synthesised by plasma cells beneath the mucosae of the gastrointestinal and respiratory tracts and by those in the lymphoid tissue of the nasopharynx. It is present in nasal and bronchial secretions. Circulating IgE is rapidly bound to cell surfaces, particularly to those of mast cells and circulating basophils, and plasma levels are therefore very low. Combination of antigen with this cell-bound antibody results in the cells releasing mediators and accounts for immediate hypersensitivity reactions such as occur in hay fever. Desensitisation therapy of allergic disorders aims at stimulating production of circulating IgG against the offending antigen, to prevent it reaching cell-bound IgE. *Raised concentrations* are found in several diseases with an *allergic* component such as some cases of eczema, asthma and parasitic infestations.

IgD (MW 190 000)

The function and physiological importance of plasma IgD are still unknown.

IMMUNOGLOBULIN RESPONSE IN DISEASE

Most infections produce a general immunoglobulin response and within a few weeks raised levels of all three major Igs are detectable. In such circumstances immunoglobulin estimation adds little to the observation of an increase in γ-globulin on routine electrophoresis. In certain conditions, some of which are listed in Table XXVI, one or more immunoglobulin classes predominate. Although there is considerable overlap, individual Ig estimation may help in diagnosis of such cases.

TABLE XXVI

Predominant Ig	Examples of Clinical Conditions
IgG	Autoimmune diseases, such as SLE or chronic aggressive hepatitis
IgA	Disease of intestinal tract, e.g. Crohn's disease Diseases of respiratory tract, e.g. tuberculosis, bronchiectasis Early cirrhosis of the liver
IgM	Primary biliary cirrhosis Viral hepatitis Parasitic infestations, especially when there is parasitaemia At birth, indicating intra-uterine infection

Immunoglobulin Deficiency

Susceptibility to recurrent infections may be the result of quantitative or functional deficiency of immunoglobulins, complement, or the cells of the immune system. It is important to remember that *normal serum immunoglobulin concentrations do not exclude an immune deficiency state.*

Primary immunoglobulin deficiency is less common than deficiency secondary to other disease.

Several classifications have been proposed for these deficiencies. We will, very briefly, discuss three categories presenting with Ig deficiency.

Transient immunoglobulin deficiency.—In the newborn infant circulating IgG is derived from the mother by placental transfer. Levels decrease over the first 3 to 6 months of life, and then gradually rise as endogenous IgG synthesis increases. In some subjects onset of synthesis is delayed, and "physiological hypogammaglobulinaemia" may persist for several more months.

Most of the IgG transfer across the placenta takes place in the last 3 months of pregnancy. Severe deficiency may therefore develop in very premature babies as maternal IgG falls before endogenous levels rise.

Primary immunoglobulin deficiency.—Primary *IgA deficiency* in plasma, saliva and other secretions is relatively common, with an incidence of about 1 in 500 of the population. It is often symptomless, but may present with intestinal, respiratory tract or renal infection. Several rare syndromes, usually familial, have been described. In one, *infantile sex-linked agammaglobulinaemia* (Bruton's disease), which occurs only in males, there is almost complete absence of circulating immunoglobulins, while cellular immunity seems normal. Other syndromes have varying degrees of immunoglobulin deficiency and impaired cellular immunity, and can occur in either sex.

Secondary immunoglobulin deficiency.—Low serum immunoglobulin levels are most common in patients with malignant disease, particularly of the haemopoietic and immune systems, often being precipitated by chemo- or radiotherapy; they are an almost invariable finding in patients with myelomatosis. In severe protein-losing states, such as the nephrotic syndrome, low immunoglobulin levels (especially of IgG) are partly due to the loss of relatively low molecular weight Igs, and partly due to increased catabolism.

Investigation of Immunoglobulin Deficiency

Although severe immunoglobulin deficiency may be obvious on electrophoresis, determination of individual immunoglobulin classes should be performed on all suspected cases. Diagnosis during the first year of life may be difficult because of the "physiologically" low IgG levels: the condition should be suspected if IgA and IgM concentrations fail to rise normally.

PARAPROTEINAEMIA

Immunoglobulin-producing cells (plasma cells and lymphocytes) are very specialised in function. Each cell produces only immunoglobulins of a single class and type. Cells producing the same class and type form a *clone*. Immunocyte proliferation may involve many cell types synthesising a range of Ig classes and types (*polyclonal response*), or a single cell type producing a single class and type (*monoclonal response*).

The normal response of immunocytes to infection is polyclonal, resulting in a diffusely raised γ-globulin on routine electrophoresis. More rarely a narrow, dense band is seen, usually in the γ region, but which may be anywhere from the a to γ regions inclusive. If this is shown to consist of a single class and type of immunoglobulin it is termed a *paraprotein*, and is probably due to proliferation of a single clone of immunocytes. The protein is usually structually normal: the monoclonal proliferation, however, is pathological, producing abnormally high levels of protein. Although paraproteins can occur in benign conditions, paraproteinaemia usually indicates malignancy of immunocytes.

Confusing terms are "M band", "M globulin" or "M protein". "M" originally denoted "myeloma" but now means "monoclonal". It does *not* mean IgM.

Bence Jones Protein

Bence Jones protein (BJP) is found in the urine of many patients with malignant immunocytomata. It consists of free monoclonal light chains, or fragments of them, which have been synthesised in excess of H chains. Its presence implies a degree of dedifferentiation of immunocytes. Because of its low molecular weight (20 000 to 40 000) the protein is filtered at the glomerulus and only accumulates in the plasma if there is glomerular failure. BJP may damage renal tubular cells and may form large casts, producing the "myeloma kidney". It may also pass into tissue spaces and is associated with amyloid formation.

BJP appears as a sharp band after electrophoresis of urine; if the protein concentration is low the urine should be concentrated 100- to 300-fold. The presence of BJP should be confirmed, and the protein typed, by immunological studies, which will show it to be light chains of *either* κ or λ type. Simpler "screening" tests are considered in the Appendix.

Causes of Paraproteinaemia

Although the presence of a paraprotein is strongly suggestive of a malignant process, this is not always so. Paraproteins may be found in the following conditions:

Malignant paraproteinaemia

Myelomatosis (this accounts for the majority of paraproteinaemias).

Macroglobulinaemia.

Some of the heavy chain diseases.

Malignancy of the lymphoreticular system (including chronic lymphatic leukaemia).

Benign paraproteinaemia

Associated with conditions normally producing a polyclonal response, for example,

autoimmune diseases, including rheumatoid arthritis;

severe chronic infections;

cirrhosis of the liver.

"Essential" paraproteinaemia (including "essential" cryoglobulinaemia).

Benign paraproteinaemias may be transient.

Myelomatosis (*Synonyms*: multiple myeloma; plasma cell myeloma)

Myelomatosis is a condition occurring with increasing frequency after the age of 50; it is very rare before the age of 30. It occurs equally in both sexes. In the commonest form there is malignant proliferation of plasma cells throughout the bone marrow. In such cases the clinical and laboratory features are due to:

malignant proliferation of plasma cells;

disordered immunoglobulin synthesis and/or secretion from the cell.

Malignant proliferation of plasma cells

Bone pain, which may be severe, is due to pressure from the proliferating cells. *X-rays* may show discrete *punched-out areas* of radiotranslucency, most frequently in the skull, vertebrae, ribs and pelvis. There may be generalised osteoporosis. Histologically, there is little osteoblastic activity around the lesions.

Pathological fractures may occur.

Anaemia, the next most common presenting feature, is due to many factors, marrow replacement being a minor one. It may sometimes be aggravated by haemolysis, or by haemorrhage.

Disordered immunoglobulin synthesis

Infections are common due to reduced synthesis of immunoglobulins other than those produced by the malignant cells.

The "hyperviscosity syndrome", although more common in macroglobulinaemia than in myeloma, can occur: production of the paraprotein may be so excessive that the circulating protein levels are very high. Consequent sluggish blood flow leads to sludging and thrombotic episodes in small vessels. This may result in:

retinal vein thrombosis with impairment of vision;

cerebral thrombosis;

peripheral gangrene.

Spurious hyponatraemia is due to the "space-occupying" effect of the paraprotein (p. 37). The sodium concentration in plasma *water* is usually normal.

Haemorrhages may be due to complexing of coagulation factors by the myeloma protein.

Cold sensitivity and Raynaud's phenomenon occur if the paraprotein is a cryoglobulin (p. 325).

Renal failure, one of the causes of which is damage by BJP, is a complication.

Amyloidosis may occur especially in cases with BJP production.

Laboratory findings and diagnosis.—The first clue to the diagnosis may be noticed during venepuncture. The blood may be very viscous and may clot in the syringe. Preparation of blood films can be extremely difficult.

Serum protein changes.—The total protein concentration is often raised, although it may be within normal limits. On electrophoresis there is usually a narrow band, most commonly in the γ region (paraprotein), with reduced levels of normal γ-globulins. The raised protein is usually IgG, less commonly IgA (about 3:1) and rarely Bence Jones protein (if renal failure is present). Occasionally IgD, IgM or IgE are found, the last two being very rare.

Rarely, no abnormal serum protein band is seen. This may occur if the malignant cells are very undifferentiated and fail to synthesise immunoglobulins, or if they are only producing Bence Jones protein, which is excreted in the urine. In either case synthesis of other immunoglobulins is usually depressed and there is often hypogammaglobulinaemia. In IgD myeloma γ-globulin increase may not be apparent on routine electrophoresis.

Bence Jones proteinuria occurs in about 70 per cent of cases of myeloma; light chains are produced in excess of heavy chains.

Hypercalcaemia may be present (p. 244). High levels usually suppress with cortisone (see cortisone suppression test, p. 248) and steroids can be used to treat such hypercalcaemia.

As there is little osteoblastic activity, the *alkaline phosphatase level is normal* unless there is liver involvement; in this case the raised level is accompanied by a raised level of GGT. A normal alkaline phosphatase level in cases with bone lesions helps to differentiate the picture from bony metastases.

If there is significant renal damage, the biochemical *features of renal glomerular failure* (p. 12) such as a raised plasma urea, urate and phosphate concentration may be present. Renal failure is an indication of a poor prognosis.

The *erythrocyte sedimentation rate* (ESR) is usually raised and may be very high. This may be the first indication of the disease and is probably due to the abnormal protein pattern.

Haematological abnormalities.—There is usually anaemia. The leucocyte and platelet counts are usually normal, but may be low. Rarely, large numbers of plasma cells appear in the peripheral blood (plasma cell leukaemia).

Bone marrow.—In myelomatosis malignant plasma cells ("myeloma cells") are seen in a bone marrow aspirate or biopsy. In the early stages the distribution of these cells may be patchy, and puncture at several sites may be necessary to confirm the diagnosis. Inspection of the bone marrow is an essential step in the diagnosis of myelomatosis.

Soft-tissue plasmacytoma.—Rarely myeloma involves soft tissues, without marrow changes (extramedullary plasmacytoma). Although the protein abnormalities of myeloma are often found in these cases, their behaviour and prognosis are different. Spread is slow and tends to be local. Local excision of a solitary tumour is often effective.

Waldenström's Macroglobulinaemia

Like myeloma, macroglobulinaemia occurs in the older age group (between 60 and 80 years), but is more common in males than in females. Also like myeloma, it is due to malignancy of immunocytes, but the malignant cells resemble lymphocytes rather than plasma cells. Symptoms of the "hyperviscosity syndrome" are commoner than in myeloma, probably because of the large size of the IgM molecule, but skeletal manifestations are rare. There is anaemia and lymphadenopathy.

Laboratory findings and diagnosis.—*Serum protein changes.*—There is usually a raised total protein concentration and electrophoresis shows a paraprotein in the γ-region. This can be identified as monoclonal IgM. Serum IgA concentration is usually reduced, but that of IgG *may* be raised.

Bence Jones protein can be detected in urine by electrophoresis in most cases, if the urine has first been concentrated 100- to 300-fold.

Haematological findings.—Anaemia is common and is due to several factors including bleeding, marrow replacement and haemolysis.

As in myeloma, the ESR is usually markedly raised.

The bone marrow aspirate or lymph node biopsy contains atypical lymphocytoid cells.

Heavy Chain Diseases

This is a rare group of disorders characterised by the presence of an abnormal protein identifiable as part of the H chain (α, γ or μ). The clinical picture is that of generalised lymphoma (γ-chain disease), intestinal lymphomatous lesions, with malabsorption (α-chain disease) or chronic lymphatic leukaemia (μ-chain disease). In some cases, a paraprotein is detectable in the serum.

Cryoglobulinaemia

Proteins which precipitate when cooled below body temperature are called cryoglobulins. Most commonly they are monoclonal IgG or IgM, or a mixture of the two. Their presence may be associated with a number of diseases in which there is a disturbance of immunoglobulin production; like other paraproteinaemias, monoclonal cryoglobulins may be an isolated finding ("essential cryoglobulinaemia"). They may only be detected by laboratory testing. Occasionally, especially if they precipitate at temperatures above 22°C, they give rise to symptoms of hyperviscosity syndrome after exposure to cold, the temperature at which the particular protein precipitates and its concentration being factors determining the severity of the symptoms.

Cryoglobulins have been described as occurring in all the conditions which may be associated with paraproteinaemia (p. 322), heavy-chain disease being the exception, and in immune complex diseases such as SLE.

Diagnosis.—A paraprotein is often detectable, but its absence, even in a properly collected specimen, does not exclude cryoglobulinaemia. If the diagnosis is suspected blood should be collected in a syringe *warmed to 37°C, and maintained at this temperature until it has been tested.* Failure to observe this precaution may result in false negative findings, as the cryoprecipitate is incorporated into the blood clot on cooling.

"Benign" and "Essential" Paraproteinaemia

The only difference between these two syndromes is that in "benign" paraproteinaemia the paraprotein is found in association with a disease normally producing a polyclonal response, while in "essential" para-proteinaemia the subject is apparently healthy. Such cases have been reported to account for between 10 per cent and 30 per cent of all paraproteinaemias.

The diagnosis of either "benign" or "essential" paraproteinaemia should be made provisionally; the patients should always be followed up because they may later be found to have had early myeloma or macroglobulinaemia. The following points strongly suggest that the condition is malignant.

1. Bence Jones proteinuria.
2. Reduced levels of other immunoglobulins ("immune paresis").
3. An initial level of paraprotein above 20 g/l of IgG, or 10 g/l of IgA or IgM.
4. Increasing levels of paraprotein. In malignant paraproteinaemia the concentration usually doubles in less than 2 years.

If after 5 years there is no evidence of malignancy the diagnosis of benign paraproteinaemia can probably be made.

PROTEINURIA

Small amounts of the relatively low molecular weight plasma proteins are filtered at the glomerulus. Most of these are reabsorbed by the renal tubular cells, and normal subjects excrete up to 0·08 g of protein a day in the urine, amounts undetectable by usual screening tests. Proteinuria of more than 0·15 g a day almost always indicates disease.

Proteinuria may be due to renal disease or more rarely may occur because large amounts of low molecular weight proteins are circulating.

It is important to remember that blood and pus in the urine give positive tests for protein.

Renal Proteinuria

Glomerular proteinuria is due to increased glomerular permeability (*nephrotic syndrome*): this is discussed more fully below. *Albumin* is usually the predominant protein in the urine.

"*Orthostatic (postural) proteinuria*".—Most proteinuria is more severe in the upright than in the prone position. The term "orthostatic" or "postural" has been applied to proteinuria, often severe, which disappears at night. It appears to be glomerular in origin and is commonest in adolescents and young adults. Although often harmless, evidence of renal disease may occur after some years.

Tubular proteinuria may be due to renal tubular damage from any cause, especially pyelonephritis (p. 16). If glomerular permeability is normal, proteinuria is usually less than 1 g a day, and consists mainly of the relatively low molecular weight α_2- *and* β-*globulins*, especially the small β_2-microglobulin.

Proteinuria with Normal Renal Function

Proteinuria can be due to production of *Bence Jones protein* (p. 322), to severe haemolysis with *haemoglobinuria*, or to severe muscle damage with *myoglobinuria*. In the latter two cases the urine will be red or brown in colour.

Bence Jones proteinuria can be inferred by inspection of the zone electrophoretic pattern of urinary proteins. BJP is the only protein of a molecular weight lower than albumin likely to be found in significant amounts in unconcentrated urine (in the absence of haemoglobinuria or myoglobinuria). The presence in the urine of a band denser than that of albumin, especially if it is not present in the serum, suggests such a protein.

NEPHROTIC SYNDROME

In the nephrotic syndrome *increased glomerular permeability* causes protein loss, by definition of *more than 5 g a day*, and often massive, with consequent hypoalbuminaemia and oedema, and with hyperlipoproteinaemia. The renal disease may be primary, or secondary to other pathology. It has been reported:

in most types of glomerulonephritis (about 80 per cent of nephrotics). In children "minimal change" glomerulonephritis is the commonest cause.

secondary to:
  diabetes mellitus
  systemic lupus erythematosus (SLE)
  inferior vena caval or renal vein thrombosis
  amyloidosis
  malaria due to *P. malariae* (in malarial districts).

Laboratory Findings

Protein abnormalities.—*Proteinuria* in the nephrotic syndrome ranges from 5 to 50 g per day. The proportion of different proteins lost is an index of the severity of the glomerular lesion. In mild cases albumin (MW about 70 000) and transferrin (MW about 80 000) are the predominant urinary proteins, and α_1-antitrypsin (MW 45 000) is also present. With increasing glomerular permeability IgG (MW 160 000) and larger proteins appear.

The *serum protein* electrophoretic picture reflects the pattern of loss and has been described on p. 313.

Electrophoresis of urine on cellulose acetate gives some idea of the severity of the lesion. The *differential protein clearance* is a more precise measure of the selectivity of the lesion. The clearance of a low MW protein, such as transferrin or albumin, is compared with that of a larger one, such as IgG. The result is usually expressed as a ratio, obviating the need for timed collections. A ratio of IgG to transferrin clearance of less than 0·2 indicates high selectivity (predominant loss of small molecules), with a more favourable prognosis than those cases in which the ratio is higher: such cases usually respond well to steroid or cyclophosphamide therapy.

The consequences of the protein abnormalities are:

oedema due to hypoalbuminaemia (p. 55);

reduction in the concentration of protein-bound substances due to loss of carrier protein. It is important not to misinterpret low total levels of calcium, thyroxine, cortisol and iron.

Lipoprotein abnormalities.—In mild cases β-lipoprotein (LDL) increases, with consequent *hypercholesterolaemia*. In more severe cases a rise in VLDL and its associated *triglycerides* may cause turbidity of the plasma. The cause of these changes is not fully understood.

Fatty casts may appear in the urine.

Renal function tests.—In the early stages glomerular permeability is high, and the plasma urea concentration normal. Later glomerular failure may develop, with uraemia. At this stage, protein loss is reduced and plasma levels of protein and lipid may revert to normal. In the presence of uraemia this does *not* indicate recovery.

The student should read the indications for protein estimations on p. 333.

SUMMARY

Plasma Proteins

1. Total protein estimations are of limited use in the diagnosis of protein abnormalities, although they may be used to assess hydration.

2. Serum proteins are usually fractionated by zone electrophoresis on

cellulose acetate. Five fractions are recognised by this method—albumin, α_1-, α_2-, β-and γ-globulins. These fractions are affected in different ways by disease.

3. A non-specific pattern is seen in many inflammatory diseases, including liver disease and in conditions characterised by tissue damage. Diagnostic patterns may be seen in the nephrotic syndrome, the paraproteinaemias, hypogammaglobulinaemias, α_1-antitrypsin deficiency and some cases of cirrhosis of the liver.

4. Estimation of one or more specific proteins is of value in selected conditions.

Immunoglobulins

1. The immunoglobulins (Ig) are a group of proteins that are structurally related. Five classes are described. The main ones are IgG, IgA and IgM. Paraproteins are immunoglobulins and Bence Jones protein is related to them.

2. The functions and properties of the Ig classes and their response to antigenic stimuli differ. Estimation of IgG, IgA and IgM may be helpful in diagnosis of a few conditions.

3. Immunoglobulin deficiency is only one aspect of immunological deficiency. *Normal Ig levels do not exclude immunological deficiency.* Ig deficiencies may be primary or secondary and may involve one or all Ig classes.

4. A paraprotein is a narrow band, usually found in the γ-globulin region of the electrophoretic strip. It usually, but not always, indicates malignant proliferation of immunocytes.

5. Paraproteins are most commonly associated with myeloma. Myeloma presents clinically in a variety of ways, reflecting bone marrow replacement and abnormal plasma protein levels. The laboratory diagnosis is made by bone marrow examination, and by finding protein abnormalities in serum and/or urine.

6. Bence Jones protein (usually found only in the urine) consists of free light chains. Its presence usually indicates malignancy of immunocytes.

7. Cryoglobulins are proteins which precipitate when cooled below body temperature. They cause symptoms on exposure to cold. They may occur in any of the diseases associated with paraproteinaemia.

Proteinuria

1. Proteinuria may be due to glomerular or tubular disease. Glomerular proteinuria is the commoner form: massive proteinuria is always of glomerular origin.

2. Proteinuria may occur with normal renal function if abnormally large amounts of low molecular weight proteins are being produced.

3. The nephrotic syndrome is characterised by proteinuria of at least 5 g a day, with a low serum albumin, oedema and hyperlipoproteinaemia. The proteinuria is glomerular in type. The severity of the lesion may be assessed by differential protein clearance.

FURTHER READING

Review Articles on immunoglobulins and immune response. *Brit. J. hosp. Med.*, 1970, **3**, 669 *et seq.* and 1972, **8**, 648 *et seq.*

ROITT, I. M. (1977). *Essential Immunology*, 3rd edit. Oxford: Blackwell Scientific Publications. (This is a concise, readable, comprehensive account of immune mechanisms).

Immunology in Medical Practice (1976). *Aust. Fam. Phycn.*, **5**, Special Issue. A comprehensive account of plasma proteins and immunological mechanisms and their disorders.

SHARP, H. L. (1976). The current status of α-1-antitrypsin, a protease inhibitor, in gastrointestinal disease. *Gastroenterology*, **70**, 611.

WHICHER, J. T. (1978). The value of complement assays in clinical chemistry. *Clin. Chem.*, **24**, 7.

APPENDIX

BLOOD SAMPLING FOR PROTEIN ESTIMATION

Blood for protein estimation, including that for immunoglobulins, should be taken with a *minimum of stasis*, or falsely high results may be obtained.

Clotted blood is essential for routine electrophoresis, because the presence of fibrinogen may mask, or be interpreted as, an abnormal protein.

Samples for *complement* estimations must be taken and processed so as to minimise *in vitro* activation. *It is essential to contact your laboratory for details before taking the blood.*

TESTING FOR URINARY PROTEIN

Several rapid screening tests are in routine use. There are two common *limitations:*

1. Most of the tests were developed to detect albumin and may be negative in the presence of even large amounts of other proteins such as B.JP.

2. As these tests depend on protein *concentration*, negative results may be found with very dilute urines despite significant proteinuria.

It is *essential* that the sample should be *fresh*.

Albustix (Ames).—The test area of the reagent strip is impregnated with an indicator, tetrabromphenol blue, buffered to pH 3. At this pH it is yellow in the absence of protein; because protein forms a complex with the dye, stabilising it in the blue form, it is green or bluish-green if protein is present. The colour after testing is compared with the colour chart provided, which indicates the approximate protein concentration. The strips should be kept in the screwtop bottles, in a cool place. The instructions on the container should be carefully followed.

False positives occur:

1. In strongly alkaline (infected or stale) urine, when the buffering capacity is exceeded. A green colour in this case is a reflection of the alkaline pH.

2. If the urine container is contaminated with disinfectants such as chlorhexidine.

False negatives occur if acid has been added to the urine as a preservative (for example, for estimation of urinary calcium).

Salicylsulphonic acid test.—The urine is filtered if turbid. 2 to 3 drops of 25 per cent salicylsulphonic acid are added to about 1 ml of clear urine. The appearance of turbidity or flocculation indicates the presence of protein.

False positives may be due to:

    radio-opaque media such as are used in intravenous pyelograms;
    the presence of metabolites of tolbutamide;
    administration of large doses of penicillin;
    the presence of urate in high concentration.

False negative results may occur in very alkaline (stale) specimens.

Boiling test.—The urine is filtered if turbid. About 5 ml of clear urine is boiled in a test tube. If turbidity appears 33 per cent acetic acid is added drop by drop, while mixing thoroughly. Turbidity due to phosphates will redissolve on acidification, while that due to protein remains.

These three tests detect protein present in concentrations above about 0·2 g/l. Rough quantitation of the salicylsulphonic acid and heat tests can be made by noting the degree of turbidity.

Appearance	Approximate protein concentration g/l
+Turbidity	0·1–0·3
+ +Turbidity with granular precipitation	0·4–1·0
+ + +Heavy turbidity with flocculation	2·0–5·0
+ + + +Marked flocculation. Coagulation of specimen	More than 5·0 More than 20

Bradshaw's test for globulins, including Bence Jones protein.—This test is a better side-room test for BJP than the heat test. *It must be stressed that the presence of BJP can only be confirmed or excluded in the laboratory.*

Urine is layered gently on a few ml of concentrated hydrochloric acid in a test tube. A thick, white precipitate at the interface indicates the presence of *globulin* in the urine (most commonly BJP). Although albumin does not give a positive reaction, in severe nephrotic syndrome the presence of globulins may cause a positive result.

INDICATIONS FOR PROTEIN ESTIMATIONS

We have discussed the changes that may occur in serum protein levels in disease. Demonstrating these changes does not always aid in diagnosis. Before making a request, some assessment should be made as to whether the result of the estimation will aid diagnosis or treatment.

The following are the commonest indications for protein estimation.

To Assess Changes in Hydration

Changes in *serum total protein or albumin* levels over short periods of time are almost certainly due to changes in hydration (p. 42).

To Evaluate Apparently Abnormal Levels, or Changes in Level, of a Protein-Bound Substance

Serum total protein or albumin estimation must always accompany that of calcium (p. 232). If changes in other protein-bound substances parallel those of the *total protein*, they are probably due to changes in protein levels.

To Investigate Oedema (see p. 55)

Very low *serum albumin* levels suggest that hypoalbuminaemia is the cause of the oedema.

In the presence of hypoalbuminaemia a *serum electrophoretic pattern* typical of nephrotic syndrome (p. 313) or hepatic cirrhosis (p. 292) points to the cause. If oedema is due to causes other than hypoalbuminaemia the dilution leads to a fall in level of albumin and all the other protein fractions.

To Investigate Suspected Immune Complex Disease

Serum C_3 determination is indicated if immune complex disease is suspected. If such disease is in an active phase, levels will be low: if it is high it suggests another cause for tissue damage.

The *electrophoretic strip* may show a diffuse rise in γ-globulins in immune complex disease. It should be remembered that this finding may be due to a number of other types of disease.

To Investigate the Cause of Recurrent Infections

Serum immunoglobulin concentrations should be measured as part of the assessment of the adequacy of the immune system.

Serum electrophoresis may indicate the cause of low immunoglobulins (for example, there may be a protein-loss pattern [p. 313] or a paraprotein).

Estimation of the 24-hour *urinary protein loss* may be indicated if this cause is suspected.

To Investigate Suspected Myeloma or Macroglobulinaemia

Electrophoresis should be carried out on *serum*, to detect a paraprotein, *and* on *urine*, to detect Bence Jones protein.

Serum immunoglobulin levels should be measured to assess the degree of immune paresis.

To Investigate Liver Disease

In chronic liver disease *serum albumin* levels may be low.

Serum electrophoresis may show β–γ fusion in cirrhosis or a diffuse increase in γ-globulin in chronic aggressive hepatitis.

Serum immunoglobulin levels may help in differential diagnosis. In cirrhosis there is a predominant increase of IgA, in biliary cirrhosis of IgM, and in chronic aggressive hepatitis of IgG.

Protein estimations are usually unhelpful in acute liver disease.

To Evaluate Proteinuria

Electrophoresis of serum and urine may help to assess the severity of glomerular damage.

Measurement of the differential protein clearance (p. 328) is a more precise index of severity.

A diffuse increase in γ-globulin on the electrophoretic strip suggests an autoimmune origin (such as SLE) for the nephrotic syndrome.

Low *serum C_3* levels (p. 315) suggest active immune complex disease.

Analysis of Individual Proteins

Analysis of the following proteins, for example, may be helpful in the diagnosis of specific disorders.

a-fetoprotein (primary hepatocellular carcinoma)

caeruloplasmin (Wilson's disease)

transferrin (some disorders of iron metabolism)

The changes of acute tissue damage (for instance, due to inflammation) are so non-specific as to be unhelpful in differential diagnosis.

Chapter XV

PLASMA ENZYMES IN DIAGNOSIS

MOST enzymes are present in cells at much higher concentrations than in plasma: some enzymes occur predominantly in certain types of cell. "Normal" plasma levels reflect the balance between the *synthesis* and *release of enzymes* during ordinary cell turnover and their *catabolism and excretion*. *Proliferation of cells*, an *increase in the rate of cell turnover, cell damage* or *enzyme induction* usually result in *raised plasma levels*; these are readily demonstrable, because very low concentrations produce easily measurable activity *in vitro*. Hence enzymes can be used as "markers" to detect and localise cell damage or proliferation.

Very occasionally plasma activities may be *lower than normal*, due either to *reduced synthesis* or to *congenital deficiency or variants*; an example is plasma cholinesterase.

Assessment of Cell Damage and Proliferation

Plasma enzyme levels depend on:
  the rate of release from cells, in turn dependent on the rate of cell damage;
  the extent of cell damage, or the *degree of proliferation* or *induction of enzyme synthesis*.
These two factors are balanced against
  the rate of enzyme excretion and catabolism.

The rapid damage of relatively few cells as, for example, in many cases of viral hepatitis, may produce very high enzyme levels, which fall as excretion and catabolism take place. By contrast, the liver may be much more extensively involved in advanced cirrhosis, but the rate of cell breakdown is often low and enzyme levels may be only slightly raised, or may even be "normal". In very severe liver disease, if many cells have been destroyed, fewer are left to undergo further damage and plasma enzyme levels may even *fall* terminally.

The knowledge that enzymes have different half-lives in the circulation may help to assess the stage of an *acute* illness. After a myocardial infarction, for example, plasma levels of creatine kinase and aspartate transaminase (with short half-lives) fall to normal before those of lactate dehydrogenase (with a longer half-life).

If the fall of levels during recovery depends largely on renal excretion, as in the case of the relatively small α-amylase molecule, renal disease may delay this fall; it may even be the cause of high levels in the absence of pancreatic disease.

For these reasons it is unwise to predict the extent of cell damage from single or serial enzyme estimations without taking the clinical picture and other findings into account.

Localisation of Damage

An increase in cell numbers (for instance in osteoblasts with calcium deficiency or with metastatic bony deposits) increases plasma levels of associated enzymes (in this example of alkaline phosphatase).

Many enzymes occur in nearly all cells, although some are associated predominantly with certain tissues. Even if the enzyme is relatively specific for a particular tissue, elevated levels indicate only damage to the cells, and not the type of disease process. For example, following *major surgical procedures* or after extensive *trauma*, levels of lactate dehydrogenase, aspartate transaminase and creatine kinase may be raised as the enzymes leak out of traumatised cells, mainly of skeletal muscle. If *circulatory failure* is present, due to cardiac failure or shock, and especially after cardiac arrest, very high levels of several enzymes may occur: results of assays cannot be used to detect the cause of the arrest. When unexplained raised levels (especially of lactate dehydrogenase) are found, the possibility of *malignancy* should be considered: some malignant cells contain very high concentrations of such enzymes. *Haemolysis* is often associated with increases in transaminases and other enzymes liberated from damaged erythrocytes.

Specificity may be improved in two ways:

by isoenzyme determination.—Measured enzyme activity may be due to closely related but slightly different molecular forms of the enzyme. These different forms (isoenzymes) can be identified by physical or chemical means and assume clinical significance when isoenzymes have different tissues of origin: for instance, different lactate dehydrogenase isoenzymes predominate in heart and liver;

by estimation of more than one enzyme.—Although many enzymes are widely distributed, their *relative concentrations in different tissues* may vary. For instance, both alanine and aspartate transaminases are abundant in liver, while in heart muscle there is much more aspartate than alanine transaminase. This difference is reflected in the markedly higher plasma aspartate than alanine transaminase level after myocardial infarction, whereas both are usually elevated if there is liver involvement.

The *distribution of enzymes within the cell* may differ. Some, such as alanine transaminase and lactate dehydrogenase, occur only in the *cytoplasm*. Others are found only in *mitochondria*, while aspartate transaminase, for example, occurs in both of these cellular compartments. Different relative levels may indicate different types of disease (p. 291).

NON-SPECIFIC CAUSES OF RAISED ENZYME LEVELS

Before attributing enzyme changes to a specific disease process, it is advisable to consider more generalised causes. Some of these have already been mentioned.

Small rises in aspartate transaminase are common as a non-specific finding in a variety of illnesses, some of which may be minor. Moreover, even only moderate exercise or large intramuscular injections may elevate the plasma levels of such muscle enzymes as creatine kinase.

Other causes which should be considered are:

physiological
newborn.—The levels of some enzymes such as aspartate transaminase are moderately raised in the neonatal period;
childhood.—Alkaline phosphatase levels are high until after puberty;
pregnancy.—In the last trimester levels of alkaline phosphatase are raised. During and immediately after labour there are moderate rises in several enzymes, such as transaminases and creatine kinase, probably due to the uterine contractions.

enzyme induction by drugs
Some drugs, particularly diphenylhydantoin and barbiturates may stimulate the production of microsomal enzymes (induction). Raised levels of, for example, γ-glutamyltransferase in patients taking them do not necessarily imply a pathological process, but should be investigated further.

artefactual elevations of enzyme levels
haemolysed samples in general are *unsuitable* for enzyme estimation. Even if the red cell does not contain the enzyme being measured, other released enzymes may interfere with the estimation to produce a falsely elevated result.

ENZYME UNITS

The total amount of enzymes (as protein) in the blood is less than 1 g/l. Results are not expressed as concentrations, but as *activities*. With the development of clinical enzymology many methods of estimation were introduced, each with its own units (for example, Reitman-Frankel units, King-Armstrong units, spectrophotometric units). In an attempt to achieve standardisation international units were introduced, but the use of these is still controversial. Moreover, results differ with variations in methodology, particularly as there is no standardised temperature at which the tests are performed. *Enzyme activities must be interpreted in relation to the normal range of the issuing laboratory*, and "normal ranges" will not be given in this chapter.

ALTERED PLASMA ENZYME LEVELS

The enzymes we will discuss, and the abbreviations recommended by the panel quoted on p. 349 are:

Aspartate transaminase	AST
Alanine transaminase	ALT
Lactate dehydrogenase	LD
Creatine kinase	CK
α-Amylase	AMS
Alkaline phosphatase	ALP
Acid phosphatase	ACP
5'-nucleotidase	NTP
γ-Glutamyltransferase	GGT

TRANSAMINASES

The transaminases are enzymes that are involved in the transfer of an amino group from an α-amino to an α-oxo acid. They are widely distributed in the body.

Aspartate Transaminase (AST)

AST (serum glutamate oxaloacetate transaminase, SGOT) is widely distributed, with high concentrations in the heart, liver, skeletal muscle, kidney and erythrocytes, and damage to any of these tissues may cause raised levels.

Causes of Increased AST

Artefactually raised levels occur in haemolysed specimens.
Physiological—newborn (approximately 1½ times adult "normal").
Markedly raised levels (10–100 times normal).
   Myocardial infarction (p. 345).
   Viral hepatitis (p. 293).
   Toxic liver necrosis (p. 293).
   Circulatory failure, with "shock" and hypoxia.
Moderately raised levels.
   Cirrhosis (up to twice normal).
   Cholestatic jaundice (up to 10 times normal).
   Malignant infiltrations of the liver.
   Skeletal muscle disease (p. 346).
   After trauma or surgery (especially cardiac surgery).
   Severe haemolytic anaemia.
   Infectious mononucleosis (liver involvement).

Alanine Transaminase (ALT)

ALT (serum glutamate pyruvate transaminase, SGPT) is present in high concentration in the liver and to a lesser extent in skeletal muscle, kidney and heart.

Causes of Raised ALT

Markedly raised levels
   Viral hepatitis (p. 293).
   Toxic liver necrosis (p. 293).
   Circulatory failure, with "shock" and hypoxia.
Moderately raised levels
   Cirrhosis (p. 295).
   Cholestatic jaundice (p. 295).
   Liver congestion secondary to cardiac failure.
   Infectious mononucleosis (liver involvement).
   Extensive trauma and muscle disease (much less than AST).

LACTATE DEHYDROGENASE (LD)

LD catalyses the reversible interconversion of lactate and pyruvate. It is widely distributed with high concentrations in the heart, skeletal muscle, liver, kidney, brain and erythrocytes.

Causes of Raised LD Levels

Artefactually raised levels are found in haemolysed specimens.
Marked increase (more than five times normal)
   Myocardial infarction (p. 345).
   Haematological disorders.
In blood diseases such as pernicious anaemia and leukaemias, very high values (up to 20 times normal) may be found. Lesser increases occur in other states of abnormal erythropoiesis, such as thalassaemia and myelofibrosis.
   Circulatory failure, with "shock" and hypoxia.
Moderate increase
   Viral hepatitis.
   Malignancy anywhere in the body.
   Skeletal muscle disease.
   Pulmonary embolism.
   Infectious mononucleosis.
   Acute haemolysis.
   Cerebral infarction ⎫
   Renal disease   ⎬ occasionally.
                        ⎭

Isoenzymes of LD

Five isoenzymes are normally detectable by electrophoresis and are referred to as LD_1 to LD_5. LD_1 is the fastest moving fraction on electrophoresis and is the form found in heart muscle, erythrocytes, blast cells and kidney. LD_5 occurs predominantly in the liver. The heart isoenzyme may be measured by two further characteristics. It is relatively heat-stable and it is active against the substrate hydroxybutyrate. As *hydroxybutyrate dehydrogenase* (HBD) activity measures mainly heart isoenzyme, it is used in some laboratories in place of LD. Levels are raised in myocardial infarction, megaloblastic anaemias, acute leukaemias, and severe active renal damage (such as rejection of transplants), and much less so in liver disease.

CREATINE KINASE (CK)

CK (CPK) is found in heart muscle, brain and skeletal muscle.

Causes of Raised CK Levels

Artefactually high levels occur in haemolysed samples (certain methods only).

Physiological
>   Newborn—slightly raised.
>   Parturition—for a few days.

Marked increase
>   Myocardial infarction (p. 345).
>   Muscular dystrophies (p. 346).
>   "Shock" and circulatory failure.

Moderate increase
>   Muscle injury.
>   After surgery (for about a week).
>   Physical exertion. There may be a significant rise in plasma levels after only moderate exercise, or even muscle cramps.
>   After intramuscular injection.
>   Hypothyroidism (thyroxine apparently influences the catabolism of the enzyme).
>   Alcoholism (possibly due partly to alcoholic myositis).
>   Acute psychotic episodes.
>   Some cases of cerebrovascular accident and head injury.
>   Some patients predisposed to malignant hyperpyrexia (p. 364).

Isoenzymes of CK

The three CK isoenzymes are combinations of the protein subunits, M and B.

MM is the predominant *muscle* isoenzyme (*skeletal and cardiac*). MM is detectable in normal plasma.

The *myocardium* also contains the *MB* form; this is normally undetectable in the plasma, and its presence suggests myocardial damage. The uses and limitations of CK-MB estimation are considered on p. 346.

BB, is the *brain* isoenzyme and is also rarely detectable in the plasma. Surprisingly, the high plasma CK that may occur in acute psychosis or after cerebrovascular accidents is of the MM type.

α-AMYLASE (AMS)

α-Amylase is the enzyme concerned with the breakdown of dietary starch and glycogen to maltose. It is present in pancreatic juice and saliva as well as in liver, fallopian tubes and muscle. The enzyme is excreted in the urine.

The main use of amylase estimations is in the diagnosis of acute pancreatitis (p. 273), when levels may be very high. Moderately raised levels are found in a number of conditions and are not diagnostic (p. 273).

Causes of Raised Plasma Amylase Levels

Marked increase (5–10 times normal)
    Acute pancreatitis.
    Severe glomerular dysfunction.
    Severe diabetic ketoacidosis.
Moderate increase
    Other acute abdominal disorders
        Perforated peptic ulcer.
        Acute cholecystitis.
        Intestinal obstruction.
        Abdominal trauma.
        Ruptured ectopic pregnancy.
    Salivary gland disorders
        Mumps—not usually required for diagnosis except in mumps encephalitis when it may be raised without obvious salivary gland enlargement.
        Salivary calculi.
        After sialography.
    Morphine administration (spasm of sphincter of Oddi).
    Severe glomerular dysfunction (may be markedly raised).
    Myocardial infarction (occasionally).
    Acute alcoholic intoxication (transient).
    Diabetic ketoacidosis (may be markedly raised).

Macroamylasaemia. In some subjects a high plasma amylase activity is accompanied by low renal excretion of the enzyme, despite normal renal function. The condition is symptomless, and it is thought either that the enzyme is bound to a high molecular weight plasma component, such as a protein, or that the amylase molecules themselves form high molecular weight polymers. This harmless condition may be confused with other causes of hyper-amylasaemia.

Low levels of plasma amylase may be found in infants up to one year and in some cases of hepatitis. They are of no diagnostic importance.

α-Amylase is a relatively small molecule, rapidly cleared by the kidneys. Raised *urinary* levels occur whenever plasma levels are raised, unless these are due to renal failure or macroamylasaemia.

ALKALINE PHOSPHATASE (ALP)

The alkaline phosphatases are a group of enzymes which hydrolyse phosphates at an alkaline pH. The activity measured by routine methods includes that of several isoenzymes. They are found in bone, liver, kidney, intestinal wall, lactating mammary gland and placenta. In bone the enzyme is found in *osteoblasts* and is probably important for normal bone formation.

In *adults*, the normal levels of alkaline phosphatase are derived largely from the *liver*. In *children*, in whom osteoblasts are very active, there is an additional contribution from *bone*, and this accounts for the higher total levels at this age; during the *pubertal growth spurt* activities rise even higher. In both adults and children there is a variable contribution from the intestine. *Pregnancy* raises the "normal" values because of the production of a heat-stable alkaline phosphatase by the *placenta*.

Bone disease (with increased osteoblastic activity) or liver disease (with involvement of the biliary tract) are the commonest causes of a raised plasma alkaline phosphatase activity. If the source of the enzyme is not apparent on clinical and radiological grounds, and if the transaminase levels are normal, *γ-glutamyl transferase* (GGT) (p. 344) may be measured; if this is normal it is very unlikely that the alkaline phosphatase is derived from the hepatobiliary system. However, GGT synthesis may be induced by drugs or alcohol, and raised levels do not always indicate liver involvement. If such induction is suspected *5'-nucleotidase* (NTP) may help to clarify the situation, because this hepatobiliary enzyme is unaffected by drugs. The estimation of 5'-nucleotidase is, however, technically much less satisfactory than that of GGT, and borderline values should be interpreted with caution. *ALP isoenzyme electrophoresis* separates bone, liver and intestinal isoenzymes, but the method is not available in all laboratories. Placental ALP is the most heat resistant of the isoenzymes. *Heat-stable alkaline phosphatase* is a measure of this fraction, and may be useful in evaluating the significance of a high ALP level in pregnancy. A placental-like isoenzyme (the *"Regan isoenzyme"*, named after the first patient in whom it was found) may be produced by malignant tumours, especially of the bronchus.

Causes of Raised Plasma ALP

Physiological:
  Children—until about the age of puberty (up to 2–2$\frac{1}{2}$ times adult normal).
  Pregnancy—in the last trimester, values may be in the upper normal range or moderately increased.
Bone disease (raised ALP; usually normal GGT):
  Osteomalacia and rickets (p. 241).
  Primary hyperparathyroidism with bone disease (p. 239).
  Paget's disease of bone (may be very high).
  Secondary carcinoma in bone.
  Some cases of osteogenic sarcoma.
Liver disease (raised ALP *and* GGT):
  Cholestasis (p. 294).
  Hepatitis (p. 294).
  Cirrhosis of the liver (sometimes).
  Space-occupying lesions—tumours, granulomata, infiltrations.
Malignancy—bone or liver involvement, or direct production.

Low Levels of Plasma ALP

Arrested bone growth:
  Achondroplasia.
  Cretinism.
  Vitamin C deficiency.
Hypophosphatasia.

ACID PHOSPHATASE (ACP)

Acid phosphatase is found in the prostate, liver, red cells, platelets and bone. The main use of the estimation is in the diagnosis of prostatic carcinoma, and the level is usually little, if at all, affected in liver and bone disease. Many methods have been devised to measure the prostatic fraction only in plasma. There is no entirely satisfactory routine technique, but one of the best is based on the fact that the prostatic fraction is inhibited by 1-tartrate ("tartrate-labile").

Normally acid phosphatase in prostatic secretions drains via the prostatic ducts and very little appears in the blood. In prostatic carcinoma, particularly if it has metastasised, plasma acid phosphatase levels rise, probably because of increased numbers of prostatic cells. False negative results may occur when the tumour has not spread, or is too undifferentiated to secrete acid phosphatase.

It is important to remember, in assessing possible prostatic malignancy, to take blood before doing a rectal examination. Following rectal examination, passage of a catheter, or even in a constipated patient, the tartrate-labile acid phosphatase may rise and values above normal may persist for up to a week. In any case in which a marginal rise is found estimations should be repeated on more than one occasion.

Causes of Raised Plasma Acid Phosphatase

Tartrate-labile
> Prostatic carcinoma with extension.
> Following rectal examination.
> Acute retention of urine.
> Passage of a catheter.

Total
> *Artefactually* raised levels occur in a haemolysed specimen.
> Paget's disease and some cases of metastatic malignant disease (only if the ALP levels are very high, and therefore of no diagnostic significance).
> Gaucher's disease (probably from Gaucher cells).
> Occasionally in thrombocythaemias.

Acid phosphatase is unstable, and specimens must be sent to the laboratory without delay.

γ-GLUTAMYLTRANSFERASE (GGT)

γ-Glutamyltransferase (γ-glutamyltranspeptidase), occurs mainly in liver, kidney and pancreas. Plasma levels are *raised* in the same conditions that affect those of liver ALP. It is, however, a more sensitive indicator of hepatobiliary disease, particularly for the detection of cirrhosis, metastatic carcinoma and hepatic infiltrations. Plasma GGT is elevated in chronic alcoholism and roughly correlates with alcohol intake. Whether this represents early liver disease or enzyme induction is uncertain. Values in patients on anticonvulsant therapy and barbiturates are often raised.

ENZYME PATTERNS IN DISEASE

MYOCARDIAL INFARCTION

The enzyme estimations of greatest value in the diagnosis of myocardial infarction are CK, AST and LD (or HBD). The choice of estimation depends on the time interval after the suspected infarction. A guide to the sequence of changes is given in Table XXVII.

TABLE XXVII

| Enzyme | Time after infarction | | |
	Starts to rise (hours)	Peak elevation (hours)	Duration of rise (days)
CK	4–8	24–48	3–5
AST	6–8	24–48	4–6
LD (HBD)	12–24	48–72	7–12

Enzyme elevations are present in about 95 per cent of cases of myocardial infarction, and may reach very high levels. The height of the rise is a very rough index of the extent of damage and as such is of some value in prognosis. It is, however, only one of many factors (p. 335). A second rise in enzyme levels after their return to normal indicates extension of the infarction, or the development of congestive cardiac failure, in which case HBD and CK do not increase. Levels in angina are usually normal, any rise probably being indicative of some myocardial necrosis. If a patient is first seen after the total LD has returned to normal, diagnosis may still be possible on the basis of a raised heart isoenzyme (LD$_1$) as detected by HBD estimation or isoenzyme electrophoresis.

After even a small myocardial infarction there is usually some hepatic congestion, due to right-sided heart failure, and therefore a slight rise in ALT levels. This is not usually a diagnostic problem because the AST is relatively much more elevated than the ALT, and the HBD (LD$_1$) is unequivocally high. If there is primary hepatic dysfunction, congestive cardiac failure without infarction, or pulmonary embolism (which usually causes some hepatic congestion), there is an approximately equal rise in ALT and AST, with normal HBD levels. If the suspected infarction is recent (see Table XXVII), and there is possible hepatic dysfunction, assessment of plasma CK levels may occasionally help. Total CK levels may be misleading because they are often elevated following muscle damage, whether due to *intramuscular injection*, trauma, or surgical operation; the CK-MB isoenzyme is mainly derived from the heart (p. 341), and its presence in plasma is good evidence for myocardial damage. However, most of the released CK is MM, which is common to skeletal and myocardial muscle; the MM has a longer half-life than the MB fraction and, *after about 24 hours*, plasma elevation of MM with undetectable MB does *not* exclude myocardial damage as a cause of elevated total CK levels: at this time the transaminase and HBD pattern is usually characteristic. *In most cases of suspected myocardial infarction AST and HBD (LD_1) estimations are all that are necessary to make a diagnosis: CK isoenzyme estimation is hardly ever indicated.*

LIVER DISEASE

Enzyme changes in liver disease are discussed in context on p. 291.

MUSCLE DISEASE

In the muscular dystrophies, probably because of increased leakage of enzymes from the damaged cells, plasma levels of muscle enzymes are increased. The relevant enzymes are CK, and the transaminases. Of these, CK estimation is the most valuable. Points to consider in interpretation are:

1. levels are highest (up to 10 times normal or more) in the early stages of the disease. Later, when much of the muscle has wasted, they are lower and may even be normal;

2. levels are higher on activity immediately after rest (build up of muscle CK) than after prolonged activity;

3. in detecting affected newborns, it must be remembered that levels at this age are higher than in adults.

Similar, but less marked, changes are found in many subjects with myositis.

Carriers of the Duchenne type muscular dystrophy can often be detected by raised CK levels. As elevations are only moderate, non-specific causes of raised enzyme activity (p. 337) must be excluded. The specimen should be collected

(*a*) at a time when normal levels would be expected to be maximum, i.e.
late in the day after ordinary physical activity (levels may be normal in known carriers in the morning);
NOT during pregnancy (levels may be low in early pregnancy).

(*b*) at a time when release from skeletal muscle is not abnormally high, i.e.
NOT for 48 hours after severe or prolonged exercise;
NOT for 48 hours after an intramuscular injection.

The assay must be performed on a fresh or carefully preserved specimen, and the laboratory should be informed *before* blood is taken. Preferably the estimation should be performed on three separate occasions, to minimise errors of interpretation due to random variations in plasma levels.

Although enzyme levels are usually normal in *neurogenic muscular atrophy*, the number of exceptions make such tests an unreliable index in the differential diagnosis from primary muscle disease.

ENZYMES IN MALIGNANCY

1. Tartrate-labile acid phosphatase in prostatic carcinoma (p. 343) is the only truly specific enzyme for malignancy.

2. Malignancy anywhere in the body may be associated with a non-specific increase in LD, HBD and occasionally transaminases.

3. For follow-up of treated cases of malignancy, alkaline phosphatase and occasionally GGT estimations are of value. Raised levels may indicate secondaries in bone (normal GGT) or liver (raised GGT) (see p. 344). Liver secondaries may, in addition, produce increases in transaminases or LD.

4. Tumours may produce a number of enzymes, of which the commonest are alkaline phosphatase and LD (HBD).

HAEMATOLOGICAL DISORDERS

The extreme elevation of LD and HBD in megaloblastic anaemia and leukaemia has been mentioned (p. 339). Similar elevations may be found in other conditions with abnormal erythropoiesis. Typically there is much less change in the level of AST than in that of LD and HBD. Severe haemolysis produces changes in AST and LD (HBD).

PLASMA CHOLINESTERASE AND SUXAMETHONIUM SENSITIVITY

There are two cholinesterases, one found predominantly in erythrocytes and nervous tissue (acetyl cholinesterase) and the other in plasma. This latter enzyme, cholinesterase ("pseudocholinesterase") is synthesised chiefly in the liver and is the one routinely measured.

Causes of Decreased Plasma Cholinesterase

Hepatic parenchymal disease:
Hepatitis.
Cirrhosis.
Anticholinesterases (organophosphates).
Inherited abnormal cholinesterase.
Myocardial infarction.

Causes of Increased Plasma Cholinesterase

Recovery from liver damage.
Nephrotic syndrome.

Suxamethonium Sensitivity

The muscle relaxant suxamethonium is broken down by cholinesterase and this limits its period of action. In certain subjects administration is followed by a prolonged period of apnoea. Rarely no enzyme is detectable, but most patients have been found to have low levels of plasma cholinesterase, and the enzyme present is qualitatively different from the normal. The abnormal cholinesterase may be classified by the degree of enzyme inhibition produced by dibucaine (a spinal anaesthetic) or by fluoride. Results are expressed as *dibucaine* and *fluoride numbers*. For example, normal cholinesterase is 80 per cent inhibited by dibucaine (dibucaine number 80) while in subjects homozygous for the defective gene the dibucaine number is about 20. Heterozygotes have intermediate numbers.

Ten possible combinations have been described. In general, marked sensitivity is found only in homozygotes. Discovery of a patient with this abnormality requires investigation of the whole family and issuing of appropriate warning cards to all affected individuals.

SUMMARY

1. Enzyme activities in cells are high. Natural decay of cells releases enzymes into the plasma, where activities are measurable, but usually low.

2. The main use of plasma enzyme estimations is the detection of raised levels due to cell damage.

3. Few enzymes are specific for any one tissue, but isoenzyme studies may increase specificity. In general, patterns of enzyme changes together with clinical findings are used for interpretation.

4. Non-specific causes of raised plasma enzyme activity include peripheral circulatory insufficiency, trauma, malignancy and surgery.

5. Enzyme estimations are of value in:

    (*a*) Myocardial infarction—AST, CK, LD and isoenzymes.

    (*b*) Liver disease—transaminases, ALP, and sometimes GGT.

    (*c*) Bone disease—ALP.

    (*d*) Prostatic carcinoma—tartrate labile ACP.

    (*e*) Acute pancreatitis—amylase.

    (*f*) Muscle disorders—CK.

6. Artefactual increases may occur in haemolysed samples.

FURTHER READING

BARON, D. N., MOSS, D. W., WALKER, P. G., WILKINSON, J. H. (1975). Revised list of abbreviations for names of enzymes of diagnostic importance. *J. clin. Path.*, **28**, 592.

WILKINSON, J. H. (1976). *The Principles and Practice of Diagnostic Enzymology.* London: Edward Arnold. (The chapters dealing with myocardial infarction (8), enzymes in disease of skeletal muscle (9) and enzyme tests in diseases of the liver and hepatobiliary tract (10) are particularly useful).

WILKINSON, J. H. (1977). The use of enzymes in diagnosis. *Brit. J. hosp. Med.*, **18**, 343.

SCHMIDT, E., and SCHMIDT, F. W. (1976). *Brief Guide to Practical Enzyme Diagnosis*, 2nd edit. Mannheim: Boehringer Mannheim GmbH.

GALEN, R. S. (1975). The enzyme diagnosis of myocardial infarction. *Hum. Path.*, **6**, 141.

Chapter XVI

INBORN ERRORS OF METABOLISM

THE chemical make-up of an individual is determined by the 20 000 to 40 000 gene pairs transmitted from generation to generation on the chromosomes. Random selection and recombination during meiosis, as well as occasional mutation, introduce individual variations. Such variations may at one extreme be incompatible with life or, at the other, produce biochemical differences detectable only by special techniques, if at all. In the latter category are the genetic variations of plasma proteins such as the transferrins or haptoglobins: such differences are useful in population and inheritance studies but do not necessarily impair function. Between the two extremes there are many variations that do produce functional abnormalities. It is to these variations that the term "inborn errors of metabolism" applies.

GENERAL PRINCIPLES

Because the sequence of bases making up the DNA strands in the genes codes, via RNA, for protein structure, it is not surprising that most, if not all, inherited biochemical abnormalities can be explained by defective synthesis of a single peptide. This abnormality may occur in structural proteins, such as in the abnormal globins of the haemoglobinopathies, or in an enzyme, in which case chemically detectable metabolic consequences may result. The abnormality in the protein may be quantitative or qualitative: the haemoglobinopathies and cholinesterase variants fall, at least partly, into the latter group.

Possible Metabolic Consequences

Deficiency of a single enzyme in a metabolic chain may produce its effects in several ways. Let us suppose that substance A is acted on by enzyme X to produce substance B, and that substance C is on an alternative pathway.

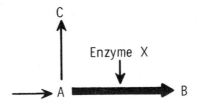

The consequences of a deficiency of X may be due to:

deficiency of the products of the enzyme reaction (B). Examples of this are cortisol deficiency in congenital adrenal hyperplasia (p. 143), and the hypoglycaemia of some forms of glycogen storage disease (p. 202);

accumulation of the substance acted on by the enzyme (A) (for instance, phenylalanine in phenylketonuria, p. 359);

diversion through an alternative pathway; some product of the latter may produce effects (C). The virilisation due to androgens in congenital adrenal hyperplasia falls into this group.

The effects of the last two types of abnormality will be aggravated if the whole metabolic pathway is controlled by feedback from the final product. For instance, in congenital adrenal hyperplasia, cortisol deficiency stimulates steroid synthesis and the accumulation of androgens, with consequent virilisation, is accentuated.

The clinical effects of some inborn errors are evident only in artificial situations. Patients with cholinesterase variants develop prolonged paralysis only if the muscle relaxant suxamethonium is administered (p. 348), and in some subjects with glucose-6-phosphate dehydrogenase deficiency haemolysis occurs only if they ingest drugs such as primaquine. Such cases appear "normal" without the intervention of modern therapeutics.

When to Suspect an Inborn Error of Metabolism

The possibility of an inherited metabolic defect should be considered in the presence of any bizarre clinical or biochemical picture in infancy or childhood, especially if more than one baby in the family has been affected. The following situations, without obvious cause, are particularly suggestive:

failure to thrive; vomiting;

hypoglycaemia;

a peculiar smell or staining of napkins;

hepatosplenomegaly;

retarded mental development; fits; spasticity;

acidosis of obscure origin;

renal calculi;

rickets which is refractory to treatment.

Clinical Importance of Inborn Errors of Metabolism

The recognition of many inborn errors of metabolism is of academic interest only, because the abnormality produces no clinical effect. If no effective treatment is available it may be important to make a diagnosis so that genetic counselling can be undertaken. There is a group of diseases in which recognition in early infancy is vital, since *treatment may prevent irreversible clinical consequences or death.* Some of the more important of these are:

    phenylketonuria (p. 359);

    galactosaemia (p. 202);

    maple syrup urine disease (p. 362).

Examples of conditions which should be sought *in relatives of affected patients,* either because *further ill effects may be prevented,* or because a *precipitating factor should be avoided,* are:

    cholinesterase abnormalities (p. 348);

    glucose-6-phosphate dehydrogenase deficiency (p. 364);

    acute porphyrias (p. 399);

    haemochromatosis (p. 389);

    cystinuria (p. 357);

    Wilson's disease (p. 363).

Other conditions can be *treated symptomatically.* Examples are:

    hereditary nephrogenic diabetes insipidus (p. 48);

    congenital disaccharidase deficiency (p. 275);

    Hartnup disease (p. 358).

Some inborn errors of metabolism are completely, or almost completely, harmless. Their importance lies in the fact that they produce *effects which may lead to misdiagnosis* or which may alarm the patient. Examples of these are:

    renal glycosuria (p. 185);

    alkaptonuria (p. 360);

    Gilbert's disease (p. 297).

Finally, the clinical effects of some inborn errors of metabolism may not appear until after child-bearing age is reached and in these cases *genetic counselling* of relatives is desirable. An example of this type of disease may be:

    Wilson's disease (p. 363).

Laboratory Diagnosis of Inborn Errors of Metabolism

The enzyme deficiency is usually demonstrated *indirectly* by detecting a high concentration of the substance normally metabolised by the deficient enzyme, such as phenylalanine in phenylketonuria. *Direct* enzyme assay is available only in special centres but, if possible, all cases should be confirmed by this method. *Prenatal diagnosis* of some metabolic defects is possible by studying cells cultured from the amniotic fluid in early pregnancy.

Screening for Inborn Errors

Because of the importance of early treatment many countries have instituted programmes of screening all newborn infants for inherited metabolic disorders, particularly for phenylketonuria. Tests may be performed on blood (obtained from a heel prick) or on urine. The sample is often collected on filter paper to facilitate transport to the laboratory.

The timing of the sample collection is important to avoid false negative results. Substances on the metabolic pathway before the enzyme block (for example, phenylalanine in phenylketonuria or galactose in galactosaemia) only accumulate once the infant begins to ingest the precursor (such as protein or milk products). Blood samples for screening apparently well infants are usually collected on the 6th to 9th day of life. Abnormal metabolites may not be found in the urine before 4 to 6 weeks after birth if their "renal threshold" is relatively high.

A positive result on screening should be confirmed by quantitative analysis or by repeat testing. Many abnormalities detected are transient.

The many pitfalls in the interpretation of the results of screening the newborn for inborn errors are reviewed in the reference quoted at the end of the chapter.

Treatment of Inborn Errors

Some inborn errors of metabolism are amenable to treatment by supplying the missing metabolite or by limiting the dietary intake of precursors in the affected metabolic pathway. Accumulated products may occasionally be removed (for example, iron in haemochromatosis).

PATTERNS OF INHERITANCE

The following account is intended for quick revision only. Details are available in books on genetics, a selection of which is listed at the end of the chapter.

Every inherited characteristic is governed by a pair of genes on homologous chromosomes (one received from each of the parents). Different genes governing the same characteristic are called alleles. If an individual has two identical alleles he is *homozygous* for that gene or characteristic; if he has two different alleles he is *heterozygous*. Genes may be carried on the sex chromosomes (X and Y) or on the autosomes (similar in both sexes) and the patterns of inheritance differ.

Autosomal Inheritance

1. Let us suppose that one parent (parent 1 in the example below) carries an "abnormal" gene (A). If N is a normal gene, the possible gene combinations in the offspring are shown in the square.

		Parent 2	
		N	N
Parent 1	A	*AN*	*AN*
	N	NN	NN

It will be seen that, on a *statistical basis*, half the offspring will carry one abnormal gene (*AN*): they are *heterozygous* for this gene, like Parent 1. None will be homozygous for the abnormal gene (*AA*).

2. If both parents are heterozygous, a quarter of the offspring (in a large series) will be homozygous (*AA*) and half heterozygous.

		Parent 2	
		A	N
Parent 1	A	*AA*	*AN*
	N	*AN*	NN

3. If one parent is homozygous and the other "normal" all offspring will be heterozygous.

Since most genes producing clinical abnormalities are rare, example 1 above has the highest statistical likelihood. With consanguineous marriages example 2 becomes more probable, since blood relatives are more likely to carry the same abnormal genes than two unrelated people.

The *consequences* of the carriage of the abnormal gene depend on its potency compared to that of the normal one.

A *dominant* gene produces the abnormality in heterozygotes and homozygotes alike. Thus, in example 1, Parent 1 and half the offspring would be affected, and in example 2, both parents and 75 per cent of the offspring would be affected. Characteristically, cases appear in *successive generations*.

A *recessive* gene produces the abnormality only in homozygous individuals. Thus, in example 1, neither parents nor offspring would be affected, and in example 2 the parents would appear normal but 25 per cent of the offspring would be affected. Characteristically, one or more cases appear in a *single generation* with apparently normal parents.

The terms "dominant" and "recessive" are relative. A dominant gene may fail to manifest itself (*incomplete penetrance*) and so appear to skip a generation. A gene may vary in *expressivity*, that is, in the degree of abnormality it produces. Some genes manifest both in the homozygote and the heterozygote (in a less severe form)—*intermediate* inheritance. Finally, a recessive gene, while it produces disease only in the homozygote, may nevertheless be detectable by biochemical tests in the heterozygote.

Sex-linked Inheritance

Some abnormal genes are carried only on the sex chromosomes, almost always on the X chromosome.

X-linked recessive inheritance.—Females carry two X chromosomes and males one X and one Y. In X-linked recessive inheritance an abnormal X chromosome (Xa) is latent when combined with a normal X chromosome, but active when combined with a Y. If the mother carries Xa she will appear to be normal, but, statistically, half the sons will be affected (*YXa*). Half the daughters will be carriers (*XXa*), but all daughters will appear clinically normal.

	Mother		
	Xa	X	
Father X	*XXa*	XX	←Daughters
Y	*YXa*	YX	←Sons

If the father is affected and the mother carries two normal genes, none of the sons will be affected, but all daughters will be carriers.

	Mother		
	X	X	
Father Xa	*XXa*	*XXa*	←Daughters
Y	XY	XY	←Sons

Inherited disease manifesting only in male offspring, but carried by females, is typical of X-linked recessive inheritance. The female is only clinically affected in the extremely rare circumstance when she is homozygous for the abnormal gene. This would only occur if there were an affected father and carrier mother.

Haemophilia is the classical example of X-linked recessive inheritance.

X-linked dominant inheritance.—In this type of inheritance both XXa and YXa (males and females) are affected. An example is familial hypophosphataemia (p. 246).

Multiple Alleles

Occasionally there may be several alleles governing the same characteristic. Different pair combinations may then produce different disease patterns (e.g. some haemoglobinopathies) or the variation may only be detectable by biochemical testing (e.g. plasma protein variants).

DISEASES DUE TO INBORN ERRORS OF METABOLISM

The diseases discussed below are merely a fraction of the known inborn errors of metabolism. Selection must be biased by what the authors feel to be important, and others might disagree. On p. 352 some of the more clinically important abnormalities have been listed, and in the Appendix a fuller (but by no means complete) list is included under systematic headings; this includes the mode of inheritance, where known. Many of these conditions have been mentioned briefly in the relevant chapters. A few remain, which the authors feel to be of relative importance.

Incidence

All the inborn errors of metabolism are very rare. The approximate incidence of disorders has been established by screening programmes in several countries. Of the disorders discussed below phenylketonuria, Hartnup disease, cystinuria, familial iminoglycinuria and histidinaemia are the commonest (1 in 10 000–20 000); maple syrup urine disease much rarer (about 1 in 350 000).

AMINOACIDURIA

As disturbances of amino acid metabolism or excretion occur in many inborn errors of metabolism this manifestation is one of the first to be sought in suspected cases. It will be discussed briefly before considering specific diseases.

Amino acids are normally filtered at the glomerulus, and reach the proximal renal tubule at concentrations equal to those in plasma: almost all are actively reabsorbed at this site. Aminoaciduria may therefore be of two types:

overflow aminoaciduria in which, because of raised plasma levels, amino acids reach the proximal tubule at concentrations higher than the reabsorptive capacity of the cells;

renal aminoaciduria in which plasma levels are low because of urinary loss due to defective tubular reabsorption.

A further subdivision may be made based on the pattern of the excreted amino acids.

Specific aminoaciduria is the excessive excretion of either a single amino acid or a group of related amino acids. It may be overflow or renal in type and is almost always due to a genetic defect.

Non-specific aminoaciduria is the excessive excretion of a number of unrelated amino acids. It is almost always due to an acquired lesion. It may be overflow in type, such as occurs in severe hepatic disease, when failure of deamination of amino acids causes raised plasma levels; more commonly renal aminoaciduria results from non-specific proximal tubular damage from any cause (p. 14). In the latter, known as the *Fanconi syndrome*, other substances reabsorbed in the proximal tubule are lost in excessive amounts (phosphoglucoaminoaciduria): its occurrence in inborn errors of metabolism is much more commonly due to secondary tubular damage by the substance not metabolised normally (for instance, copper in Wilson's disease—hepatolenticular degeneration) than to a direct primary genetic defect. Acquired lesions will not be discussed further in this chapter.

Inherited Abnormalities of Transport Mechanisms

Groups of chemically similar amino acids are often reabsorbed in the renal tubule by a single mechanism. In several cases similar group-specific mechanisms are involved in intestinal absorption and defects involve both the renal tubule and intestinal mucosa. Inborn errors involving the following group pathways have been identified:

(*a*) the dibasic amino acids (with two amino groups), cystine, ornithine, arginine and lysine (COAL is a useful mnemonic) (*cystinuria*);

(*b*) many neutral amino acids (with one amino and one carboxyl group) (*Hartnup disease*);

(*c*) the imino acids, proline and hydroxyproline, probably shared with glycine (*familial iminoglycinuria*).

Cystinuria

Cystinuria is due to an inherited abnormality of tubular reabsorption of the dibasic amino acids, cystine, ornithine, arginine and lysine, resulting in excessive urinary excretion of these four amino acids. A similar transport defect is present in the intestinal mucosa, but, although cystine absorption is diminished, failure of renal tubular reabsorption results in a high urinary excretion of the endogenously produced amino acid.

Many cases of cystinuria are asymptomatic. Cystine is non-toxic, but is relatively insoluble and the danger is due to its precipitation in the renal tract with calculus formation and its attendant complications. The solubility of cystine is such that only in homozygotes do urinary concentrations reach levels at which precipitation may occur and result in crystalluria and stone formation, although increased excretion can be demonstrated in heterozygotes.

The *diagnosis* of cystinuria is made by the demonstration of excessive urinary excretion of cystine and the other characteristic amino acids. The demonstration of the latter is necessary to distinguish the stone-forming homozygote from heterozygous cystine-lysinurias and from cystinuria occurring as part of a generalised aminoaciduria.

The *management* of cystinuria is aimed at the prevention of the formation of calculi by a high fluid intake, thus reducing the urinary cystine concentration. Alkalinising the urine also increases the solubility of cystine. If these measures are inadequate, administration of D-penicillamine may be tried.

Several genetic forms of cystinuria exist and the condition follows an autosomal recessive pattern of inheritance.

The relatively harmless condition described above must not be confused with **cystinosis**. This is a very rare inherited disorder of cystine metabolism characterised by accumulation of intracellular cystine in many tissues. In the kidney this produces tubular damage and consequently the Fanconi syndrome. The aminoaciduria is non-specific and of renal origin. Death occurs at an early age.

Hartnup Disease

Hartnup disease, named after the first described patient, is a rare but interesting disorder in which there is a renal and intestinal transport defect involving neutral amino acids.

As in cystinuria the defect is present in both the proximal renal tubule and the intestinal mucosa. Most, if not all, the clinical manifestations can be ascribed to the reduced intestinal absorption and increased urinary loss of *tryptophan*. The amino acid is normally partly converted to nicotinamide, this source being especially important if the dietary intake of nicotinamide is marginal (p. 411). The clinical features of Hartnup disease are intermittent and resemble those of pellagra, namely:

a red, scaly rash on exposed areas of skin;

reversible cerebellar ataxia;

mental confusion of variable degree.

The thesis that nicotinamide deficiency is the cause of the clinical picture is supported by the response to administration of the vitamin and the fact that the features of the disease are frequently preceded by a period of dietary inadequacy.

In spite of the generalised defect of amino acid absorption protein malnutrition is not seen: this may possibly be due to absorption of intact peptides by a different pathway.

An associated chemical feature is the excretion of excessive amounts of *indole* compounds in the urine. These originate in the gut from the action of bacteria on the unabsorbed tryptophan.

Hartnup disease has a recessive mode of inheritance.

Diagnosis is made by demonstrating the characteristic amino acid pattern in the urine. Heterozygotes are not readily detectable by present techniques.

Familial Iminoglycinuria

An abnormality of the transport mechanism of the imino acids leads to increased urinary excretion of proline, hydroxyproline and glycine, but with normal plasma levels. The condition is apparently harmless but must be distinguished from other, more serious causes, of iminoglycinuria. It is inherited as an autosomal recessive.

DISORDERS OF AMINO ACID METABOLISM

A number of inborn errors of amino acid metabolism have been described. Most are characterised by raised levels of the relevant amino acid(s) in the blood, with overflow aminoaciduria. Only a few of the better known conditions are described here.

Disorders of Aromatic Amino Acid Metabolism

The main chemical reactions of this pathway are outlined in Fig. 33, together with the site of known enzyme defects. It will be seen that *tyrosine*, normally produced in the body from phenylalanine, is the precursor of several important substances.

The inherited defects in thyroid synthesis are considered on p. 169.

Phenylketonuria.—This condition, if untreated, leads to mental retardation. Extensive screening of newborn infants is carried out in many countries to detect cases early. The problems arising from such surveys are discussed later.

The clinical features are:

irritability, feeding problems, vomiting and fits in the first few weeks of life;

mental retardation developing at between 4 and 6 months with unusual psychomotor irritability;

generalised eczema in many cases;

a tendency to reduced melanin formation. Many patients have a pale skin, fair hair and blue eyes.

The abnormality in phenylketonuria is a deficiency of *phenylalanine hydroxylase*. Phenylalanine accumulates in the blood and is excreted in the urine together with its derivatives, such as phenylpyruvic acid: the disease acquires its name from the recognition of this (phenylketone) in the urine. The cerebral damage is thought to be due to the high circulating levels of phenylalanine or one of its metabolites.

Diagnosis.—1. The *phenylalanine concentration* may be measured in *blood* taken from a heel-prick. The technique is suitable for mass screening and tests are best performed in special centres. The timing of the test is critical (p. 353). It is recommended that tests be performed between 6 and 10 days after birth (just before the infant leaves hospital).

2. Detection of phenylpyruvic acid in the urine with *ferric chloride or Phenistix* (Ames) is the classical method of detecting phenylketonuria. This test may only be positive after about six weeks, when blood phenylalanine levels are very high. Since at this stage the infant has been discharged from hospital and possibly lost to the screening programme,

and because permanent cerebral damage may have already occurred, measurement of blood phenylalanine levels is now the recommended procedure.

3. The absence of the enzyme, phenylalanine hydroxylase, may be demonstrated in liver biopsy material. This is not usually required for diagnosis.

Since the introduction of screening tests it has become apparent that raised blood levels of phenylalanine occur in conditions other than phenylketonuria, particularly in premature babies, when it is possibly due to delayed maturation of enzyme systems. These cases may be distinguished by repeated blood testing and by estimation of blood tyrosine levels (not raised in phenylketonuria, but raised in many of the other conditions).

A variant, persistent hyperphenylalaninaemia, without mental retardation has been described.

A newly recognised problem is *maternal phenylketonuria*. Babies exposed *in utero* to high phenylalanine levels by unrecognised or untreated phenylketonuric mothers are mentally retarded and show other congenital abnormalities, even though they themselves are not phenylketonuric.

Management.—The aim of management is the reduction of blood phenylalanine levels and to this end a low phenylalanine diet is prescribed. Such treatment is difficult, expensive and tedious for the patient and parents, and requires careful biochemical monitoring of blood phenylalanine levels. Phenylalanine deficiency itself has deleterious effects. Tyrosine must also be included in the diet as it is the precursor of many important metabolites (Fig. 33).

Phenylketonuria is inherited as an autosomal recessive. *Heterozygotes* are clinically normal, but may be detected by biochemical tests.

Alkaptonuria.—An inherited deficiency of *homogentisic acid oxidase* results in alkaptonuria. Homogentisic acid accumulates in blood, tissues and urine. Oxidation and polymerisation of this substance produces the pigment "alkapton", in much the same way as polymerisation of DOPA (see Fig. 33) produces melanin. Deposition of alkapton in cartilages, with consequent darkening, is called *ochronosis*: this may cause *arthritis* in later life and may be clinically visible as darkening of the ears. Conversion of homogentisic acid to "alkapton" is speeded up in alkaline conditions and the most obvious abnormality in alkaptonuria is passage of *urine which is black*, or which darkens as it becomes more alkaline on standing. This finding may, however, be absent in a significant number of cases. The condition is often first noticed by the mother who is worried by the black nappies, which only become blacker when washed in alkaline soaps or detergents. The condition is compatible with a normal life span and treatment is unnecessary, but arthritis in middle and

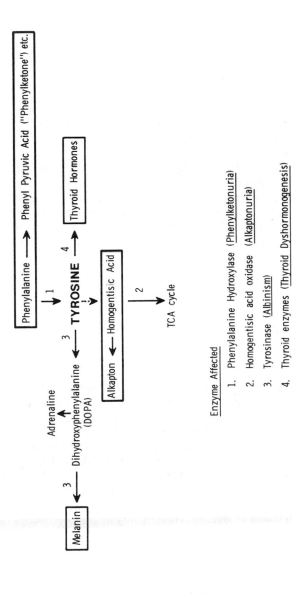

Fig. 33.—Some inborn errors of the aromatic amino acid pathway.

later life is common. Homogentisic acid is a *reducing substance* and reacts with Benedict's solution or Clinitest tablets.

Alkaptonuria is inherited as an autosomal recessive. Heterozygotes are not detectable by clinical or by biochemical findings.

Albinism.—A deficiency of *tyrosinase* in melanocytes causes one form of albinism and is inherited as a recessive character. The patient lacks pigment in skin, hair and iris (the eyes appear pink) and the condition is especially striking in negroes. Acute photosensitivity occurs because of pigment lack in the skin and iris. The tyrosinase involved in catecholamine synthesis is a different enzyme, controlled by a different gene. Adrenaline metabolism is normal in albinos.

Disorders of other Amino Acids

Maple syrup urine disease.—In maple syrup urine disease there is deficient decarboxylation of the oxoacids resulting from deamination of the three *branched-chain amino acids*, leucine, isoleucine and valine. These accumulate in the blood and are excreted in the urine together with their corresponding oxoacids. The odour of the urine, which resembles that of maple syrup, gives the disease its name.

The disease presents in the first week of life and, if untreated, severe neurological lesions develop with death in a few weeks or months. If, on the other hand, the condition is recognised and a diet low in branched-chain amino acids is given, normal development seems possible.

Diagnosis is made by demonstrating the raised levels of branched-chain amino acids in blood and urine. It may be confirmed by demonstrating the enzyme defect in the leucocytes.

The condition has a recessive mode of inheritance.

Histidinaemia.—Histidinaemia is associated with deficiency of the enzyme *histidase* which is required for the normal metabolism of histidine. Blood levels of histidine are raised and histidine and a metabolite, *imidazole pyruvic acid*, appear in increased amounts in the urine. Like phenylpyruvic acid (excreted in phenylketonuria) imidazole pyruvic acid reacts with ferric chloride to give a blue-green colour with ferric chloride or Phenistix (Ames). About half the cases described have shown mental retardation and speech defects but the rest appear normal. The results of dietary therapy are as yet inconclusive.

The condition is probably inherited as an autosomal recessive trait.

DEFECTS OF METAL METABOLISM

Two inherited disorders are associated with an abnormal accumulation of metals in the body. Iron overload (*idiopathic haemochromatosis*) is discussed in Chapter XVIII. Copper accumulates in Wilson's disease.

Wilson's Disease (Hepatolenticular Degeneration)

Some plasma copper is loosely bound to albumin, but most is normally firmly incorporated in the protein *caeruloplasmin*. Copper is mainly excreted in the bile.

There are two defects of copper metabolism in Wilson's disease:

impaired biliary excretion leads to *copper deposition in the liver*;

deficiency of caeruloplasmin results in *low plasma copper* levels: most of this copper is in the loosely bound form, and is *deposited in tissues* and *filtered at the glomerulus* more readily than normal.

Excessive deposition of copper in the basal ganglia of the brain, liver, renal tubules and the eye produces:

neurological symptoms due to basal ganglia degeneration;

liver damage leading to cirrhosis;

renal tubular damage with any or all of the biochemical features of this condition, including aminoaciduria (Fanconi syndrome);

Kayser-Fleischer ring at the edge of the cornea due to copper deposition in Descemet's membrane.

Urinary copper excretion is increased.

Measurement of plasma caeruloplasmin levels is the first step in confirming the clinical diagnosis. They may, however, be normal and in most patients there is poor correlation between the levels of the protein and the severity of the disease.

In assessing caeruloplasmin concentration it is important to remember that *low levels* may also occur during the *first few months of life*, with *malnutrition* and in the *nephrotic syndrome* due to loss in the urine. *Raised levels* are found in *active liver disease* (this may account for some "normal" levels in patients with Wilson's disease), in the *last trimester of pregnancy*, in women taking *oral contraceptives* and non-specifically in states of *tissue damage* (for example, inflammation and neoplasia).

In some cases copper estimation on a liver biopsy specimen may be needed for diagnosis.

The clinical condition has a recessive mode of inheritance, but heterozygotes may have reduced caeruloplasmin levels. Distinction between presymptomatic homozygotes and heterozygotes is important, because the former should be treated.

Treatment with agents chelating copper, such as D-penicillamine, aims to reduce tissue copper concentration.

DRUGS AND INHERITED METABOLIC ABNORMALITIES

The variation in individual response to identical drugs may be partly due to genetic variation. There are, however, a number of well-defined inherited disorders that are aggravated by, or which only become apparent on, administration of certain drugs. These may be classified into two groups.

Disorders Resulting from Deficient Metabolism of a Drug

The muscle relaxant suxamethonium (succinylcholine) normally has a very brief action as it is rapidly broken down by plasma cholinesterase. In *suxamethonium sensitivity* (p. 348) an abnormal cholinesterase is present, breakdown of the drug is impaired and prolonged respiratory paralysis may occur.

Two other inherited disorders are characterised by defective metabolism of the drugs *isoniazid* and *diphenylhydantoin* respectively. In both, toxic effects appear more frequently and at lower dosages than in normal individuals.

Disorders Resulting from Abnormal Response to a Drug

A form of haemolytic anaemia, common in many parts of the world, is due to a deficiency of *glucose-6-phosphate dehydrogenase* (G-6-PD) in the erythrocyte. This is the first enzyme in the hexose monophosphate shunt and is required for the formation of NADPH. This, in turn, is probably essential for the maintenance of an intact red cell membrane. Numerous variants of G-6-PD deficiency have been described. In many cases haemolysis is precipitated by drugs, notably certain antimalarial drugs, such as primaquine, sulphonamides and vitamin K analogues.

In the inherited *hepatic porphyrias* (p. 397) acute attacks may be precipitated by a number of drugs, particularly barbiturates.

Some people react to general anaesthesia (most commonly halothane with suxamethonium) with a rapidly rising temperature, muscular rigidity and acidosis; the majority die as a result (*malignant hyperpyrexia*). Many, but not all, susceptible subjects in affected families have an elevated creatine kinase (CK) level.

This short section should serve to remind readers to consider the possibility of an inborn error when an abnormal reaction to a drug is encountered. A reference to this growing field of *pharmacogenetics* is given at the end of the chapter.

SUMMARY

1. Inborn errors of metabolism are diseases due to inherited defects of protein synthesis. Most cases presenting with clinical symptoms are due to abnormalities of enzyme synthesis.

2. Inborn errors of metabolism may produce no clinical effects, may only produce them under certain circumstances (for example, cholinesterase variants) or, at the other extreme, may produce severe disease. Some are incompatible with life.

3. Recognition of some inherited abnormalities is of academic interest only. Diagnosis is important if the condition is serious but treatable, if precipitating factors can be avoided, or if confusion with other diseases is possible.

4. Inheritance may be autosomal or sex-linked, dominant or recessive. In diseases producing severe clinical effects inheritance is most commonly autosomal recessive, and they are most common in the offspring of consanguineous marriages.

5. In many cases in which the clinical disease is inherited in a recessive manner, lesser degrees of the abnormality can be detected by chemical testing.

6. Some inborn errors of metabolism not mentioned elsewhere in the book are discussed in this chapter.

FURTHER READING

EMERY, A. E. H. (1975). *Elements of Medical Genetics*, 4th edit. Edinburgh: Churchill Livingstone.

ROBERTS, J. A. F., and PEMBREY, M. E. (eds.) (1978). *An Introduction to Medical Genetics*, 7th edit. London: Oxford University Press.

Many similar books on the principles of inherited disease exist. The above two are comprehensive and readable.

RAINE, D. N. (1972). Management of inherited metabolic disease. *Brit. med. J.*, 2, 329.

This includes a useful summary of the clinical features of many inborn errors of metabolism.

PRICE EVANS, D. A. (1968). Clinical Pharmacogenetics. In *Recent Advances in Medicine*, 15th edit., p. 203. Eds. Baron, D. N., Compston, N., and Dawson, A. M. London: Churchill.

VESELL, E. S. (1973). Advances in Pharmacogenetics. In *Progress in Medical Genetics*, *IX*, p. 291. Eds. Steinberg, A. G., and Bearn, A. G. New York: Grune and Stratton.

These two references deal with drugs and inherited abnormalities.

HILL, A., CASEY, R., and ZALESKI, W. A. (1976). Difficulties and pitfalls in the interpretation of screening tests for the detection of inborn errors of metabolism. *Clin. chim. Acta*, 72, 1.

RAINE, D. N. (ed.) (1974). *Molecular Variants in Disease*. Published for the Royal College of Pathologists by the *J. clin. Path.* (London).

This contains general articles and reviews of some diseases, such as the haemoglobinopathies and methaemoglobinaemia, not discussed in this chapter.

REFERENCE

STANBURY, J. B., WYNGAARDEN, J. B., and FREDRICKSON, D. S., Eds. (1978). *The Metabolic Basis of Inherited Disease*, 4th edit. New York: McGraw-Hill.

APPENDIX

The following list of inborn errors of metabolism is far from complete. It is meant for reference only, and the student should not attempt to learn it. Most of the abnormalities have been discussed in this book, and a page reference is given. Where it is known the mode of inheritance is given, unless the heading applies to a group of diseases of different modes of inheritance.

D = Autosomal Dominant
R = Autosomal Recessive
X-linked D = X-linked Dominant
X-linked R = X-linked Recessive

	Inheritance	Page
I. DISORDERS OF CELLULAR TRANSPORT		
Most of these are recognised as renal tubular transport defects, and in some defective intestinal transport can also be demonstrated.		
Generalised Proximal Tubular Transport		
Phosphoglucoaminoaciduria	?	356
Amino Acids		
Dibasic amino acids—Cystinuria	R	357
Neutral amino acids—Hartnup disease	R	358
Familial iminoglycinuria	R	358
Glucose		
Renal glycosuria	D	185
Water (failure to respond to ADH)		
Hereditary nephrogenic diabetes insipidus	X-linked R	48
Sodium (failure to respond to aldosterone)		
Pseudo-Addison's disease	?	50
Potassium (all cells)		
Familial periodic paralysis	D	70
Calcium (failure to respond to PTH)		
Pseudohypoparathyroidism	X-linked D	242
Phosphate		
Familial hypophosphataemia	X-linked D	246
Hydrogen Ion		
Renal tubular acidosis	D	98
Bilirubin (liver cells)		
Congenital hyperbilirubinaemias	—	297
II. DISORDERS OF AMINO ACID METABOLISM		
Aromatic Amino Acids		
Phenylketonuria	R	359
Alkaptonuria	R	360
Albinism	R	362
Thyroid dyshormonogenesis	All R	169

Chapter XVII

PURINE AND URATE METABOLISM

HYPERURICAEMIA AND GOUT

HYPERURICAEMIA may be asymptomatic, or may give rise to the clinical syndrome of gout. In either case it should be treated; the relative insolubility of urate means that there is the danger of precipitation in tissues. If this takes place in the kidney renal damage can result (compare the danger of hypercalcaemia, p. 236). Hyperuricaemia may be due to a primary lesion of purine metabolism or be secondary to a variety of other conditions. The primary syndrome has a familial incidence.

Plasma urate is in the form of the monosodium salt. Uric acid is less soluble than its sodium salts.

NORMAL URATE METABOLISM

Urate is the end product of purine metabolism in man. In most other mammals it is further broken down to the soluble compound, allantoin, and it is the poor solubility of urates which makes man prone to clinical gout and renal damage by urate. The purines, adenine and guanine, are constituents of both types of *nucleic acid* (DNA and RNA). The purines used by the body for nucleic acid synthesis may be derived from the breakdown of ingested nucleic acid (mainly in meat which is rich in cells), or may be synthesised in the body from small molecules *de novo*.

Synthesis of Purines

The synthetic pathway of purines is complex, and involves the incorporation of many small molecules into the relatively complex purine ring. The upper part of Fig. 34 summarises some of the more important steps in this synthesis. Cytotoxic drugs such as 6-mercaptopurine and folic acid antagonists inhibit various stages in this pathway, so preventing DNA formation and cell growth.

The following stages in Fig. 34 should be especially noted.

Step (a) is the first one in purine synthesis. It involves condensation of pyrophosphate with phosphoribose to form phosphoribosyl pyrophosphate (PRPP).

In Step (b) the amino group of glutamine is incorporated into the ribose phosphate molecule and pyrophosphate is released. Amidophosphoribosyl transferase catalyses this *rate-limiting* or controlling step in

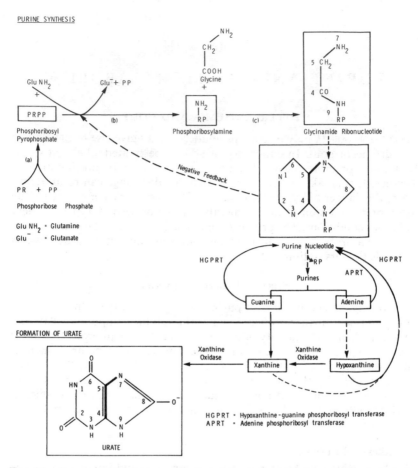

FIG. 34.—Summary of purine synthesis and breakdown to show steps of clinical importance. (See text for explanation of small letters.)

purine synthesis. It is subject to feedback inhibition from increased levels of purine nucleotides: thus the rate of synthesis is slowed when its products increase. This step may be at fault in primary gout.

Step (c) shows how the *glycine* molecule is added to phosphoribosylamine. Labelled glycine can be used to study the rate of purine synthesis. The atoms in the glycine molecule have been numbered in the diagram to correspond with those of the purine and urate molecules, and the heavy lines further indicate the final position of the amino acid in these molecules. By the use of labelled glycine it has been shown that purine synthesis is increased in primary gout.

After many complex steps purine ribonucleotides (purine ribose phosphates) are formed and, as has already been stated, the level of these controls Step (b). Ribose phosphate is split off, thereby releasing the purines.

Fate of Purines

Purines synthesised in the body, those derived from the diet, and those liberated by endogenous breakdown of nucleic acids may follow one of two pathways:

    they may be synthesised into new nucleic acid;

    they may be oxidised to urate.

Formation of urate from purines.—As shown in the lower part of Fig. 34, some of the adenine is oxidised to hypoxanthine, which is further oxidised to xanthine. Guanine can also form xanthine. Xanthine, in turn, is oxidised to form urate. The oxidation of both hypoxanthine and xanthine is catalysed by the liver enzyme *xanthine oxidase*. Thus the formation of urate from purines depends on xanthine oxidase activity, a fact of importance in the treatment of gout.

Reutilisation of purines.—Some xanthine, hypoxanthine and guanine can be resynthesised to purine nucleotides by pathways involving, amongst other enzymes, hypoxanthine–guanine phosphoribosyl transferase (HGPRT) and adenine phosphoribosyl transferase (APRT).

Excretion of urate.—75 per cent of the urate leaving the body is excreted in the urine and 25 per cent passes into the intestine, where it is broken down by intestinal bacteria (*uricolysis*). The urate filtered at the renal glomerulus is probably completely reabsorbed in the tubules and the urinary urate is derived from active tubular secretion: urinary excretion may be enhanced by various drugs used in the treatment of gout.

Renal excretion of urate is inhibited by such organic acids as lactic and oxoacids.

CAUSES OF HYPERURICAEMIA

Figure 35 summarises the factors which may contribute to hyperuricaemia. These are:

    increased rate of urate formation;

        increased synthesis of purines (Step a);

        increased intake of purines (Step b);

        increased turnover of nucleic acids (Step c);

    reduced rate of urate excretion (Step e).

Steps (b), (c) and (e) are causes of secondary hyperuricaemia. Increased synthesis is probably the basic fault in primary hyperuricaemia.

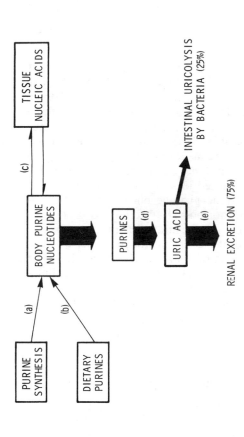

FIG. 35.—Origin and fate of urate in normal subjects.

Causes of Hyperuricaemia

Pathway (a) Increased in Primary Hyperuricaemia

Pathway (b) Affected by diet

Pathway (c) Increased in malignancy, infection, cytotoxic therapy, psoriasis etc.

Pathway (e) Decreased in renal failure, thiazide diuretic therapy, some cases of primary hyperuricaemia and acidosis

Treatment of Hyperuricaemia

Pathway (d) Reduced by Xanthine Oxidase inhibitors (eg. Allopurinol)

Pathway (e) Increased by Uricosuric Drugs (eg. Benemid)

DANGERS OF HYPERURICAEMIA

The solubility of urate in plasma is limited. Precipitation in tissues may be favoured by a variety of local factors of which the most important are probably tissue pH and trauma. Crystallisation in *joints*, especially those of the foot, produces the classical picture of gout, first described by Hippocrates in 460 B.C. It is thought that local inflammation due to urate precipitation produces an increase in leucocytes in the area, and that lactic acid production by these lowers the pH locally: this reduces the solubility of urate and sets up a vicious circle in which further precipitation occurs. It should be noted that in attacks of acute gouty arthritis local factors are of more importance than the plasma urate levels; the latter may even be normal during the attack.

Precipitation can also occur in other tissues, and the subcutaneous collections of urate, which are especially common in the ear and in the olecranon and patellar bursae and tendons, are called *gouty tophi*.

Attacks of gout are extremely painful and unpleasant and may lead to permanent joint deformity. Tophi are disfiguring but harmless. A serious effect of hyperuricaemia is due to precipitation of urate in the kidney, leading to *renal failure*. For this reason it has been recommended that all cases with plasma urate levels consistently above 0·55 mmol/l (9 mg/dl), even if they are asymptomatic, should be treated.

PRIMARY HYPERURICAEMIA AND GOUT

Familial Incidence

In A.D. 150 Galen said that gout was due to "debauchery, intemperance and an hereditary trait". "Intemperance", as we shall see, may aggravate the condition. The striking familial incidence of hyperuricaemia confirms that there is probably "an hereditary trait", but in this respect we know little more than Galen did, because the mode of inheritance (like that of diabetes mellitus) is still obscure.

Sex and Age Incidence

Gout and hyperuricaemia are very rare in children, and rare in women of child-bearing age. The difference in incidence in males and females is not due to a sex-linked inheritance, because it can be transmitted by males. Plasma urate levels are low in children and rise in both sexes at puberty, more so in males than females. Women become more prone to hyperuricaemia and gout in the post-menopausal period (compare plasma cholesterol and iron levels).

Precipitating Factors

The classical image of the gouty subject is the red-faced, good-living, hard-drinking squire depicted in novels and paintings of the 18th century. Galen mentioned "debauchery and intemperance" as causes of gout. Two factors probably account for the high incidence of clinical gout in this type of subject.

Alcohol has been shown to decrease renal excretion of urate. This may be because it increases lactate production, which inhibits urate excretion.

A high meat diet contains a high proportion of *purines*.

Neither of these factors is likely to precipitate gout in a normal person, but may do so in a subject with a gouty trait.

For reasons which are not clear, hyperuricaemia, and even clinical gout, are relatively common in patients with hypercalcaemia from any cause, and in patients with recurrent calcium-containing renal calculi, even when not accompanied by hypercalcaemia.

Biochemical Lesion of Primary Hyperuricaemia

Use of labelled glycine has shown that *purine synthesis is increased* in about 25 per cent of cases of primary hyperuricaemia. There may be overactivity of the enzyme controlling the formation of phosphoribosyl-amine (Fig. 34). This could be due to failure of normal feedback suppression by nucleotides.

Reduced renal secretion of urate has also been demonstrated in other cases of primary hyperuricaemia. Both increased synthesis and decreased excretion may be present in many subjects.

Principles of Treatment of Hyperuricaemia

Treatment may be based on

reducing purine intake (Step (b) Fig. 35).—This is not very effective by itself;

increasing renal excretion of urate with *uricosuric drugs*, such as probenecid and salicylates in large doses (Step (e) Fig. 35). These are very effective if renal function is normal, but are useless in the presence of renal failure. Fluid intake must be kept high. *Low doses* of salicylate *inhibit* urate secretion;

reducing urate production by drugs which inhibit xanthine oxidase (Step (d) Fig. 35), such as *allopurinol* (hydroxypyrazolopyrimidine). This compound is structurally similar to hypoxanthine and acts as a competitive inhibitor of the enzyme. *De novo* synthesis may also be decreased by this drug;

colchicine, which has an anti-inflammatory effect in acute gouty arthritis, does not affect urate metabolism.

Juvenile Hyperuricaemia (Lesch-Nyhan Syndrome)

This is a very rare inborn error, probably carried on an X-linked recessive gene, in which severe hyperuricaemia occurs in young male children.

A deficiency of the enzyme *hypoxanthine-guanine phosphoribosyl transferase (HGPRT)* has been demonstrated in affected subjects. Hypoxanthine and other purines cannot be recycled to form purine nucleotides, and urate production from them is probably increased. The syndrome is associated with mental deficiency, a tendency to self-mutilation, aggressive behaviour, athetosis and spastic paraplegia.

Glucose-6-Phosphatase Deficiency (p. 202)

The tendency to hyperuricaemia in patients with glucose-6-phosphatase deficiency may be related directly to the inability to convert glucose-6-phosphate to glucose. More G-6-P is available for metabolism through intracellular pathways, including:

the pentose-phosphate pathway, increasing ribose phosphate (phosphoribose) synthesis. This may accelerate Step (a) Fig. 34, with consequent *urate overproduction*;

glycolysis, so increasing lactic acid production (p. 182). Lactate may *reduce renal urate excretion*.

SECONDARY HYPERURICAEMIA

High plasma urate levels may result from:
increased turnover of nucleic acids ((c) in Fig. 35);
rapidly growing malignancy, especially leukaemias and polycythaemia vera;
treatment of malignant tumours;
psoriasis;
increased tissue breakdown in
acute starvation
tissue damage.
reduced excretion of urate (Step (e) in Fig. 35);
glomerular failure;
thiazide diuretics;
acidosis.

Increased turnover of nucleic acids in malignancy can cause hyperuricaemia. *Treatment* of large tumours by radiotherapy or cytotoxic drugs can cause massive release of urate and has been known to cause acute renal failure due to tubular blockage by crystalline uric acid. During such treatment allopurinol should be used, and, if renal function is good, fluid intake should be kept high.

Starvation and tissue damage.—In acute starvation and with tissue damage endogenous tissue breakdown is increased. Increased amounts of urate are produced. In both these conditions acidosis is probably present (due to ketosis in starvation, and due to tissue catabolism in both) and these acids probably inhibit renal excretion of urate, aggravating the hyperuricaemia. Levels may reach 0·9 mmol/l (15 mg/dl) or more in complete starvation. Note that in chronic starvation urate levels, like those of urea (p. 19), tend to be low.

Glomerular failure causes retention of urate as well as retention of other waste products of metabolism. When estimating plasma urate, urea should always be estimated on the same specimen to exclude renal glomerular failure as a cause of hyperuricaemia. It has already been mentioned that hyperuricaemia may *cause* renal failure and, in the presence of uraemia, it may be difficult to know which is cause and which is effect. As a rough guide the plasma urate concentration would be expected to be about 0·6–0·7 mmol/l (10–12 mg/dl) at a urea level of about 50 mmol/l (300 mg/dl); if it is much higher than this, hyperuricaemia should be suspected as the primary cause of the renal failure. Clinical gout is rare in secondary hyperuricaemia due to renal failure.

Increased intestinal secretion and bacterial uricolysis have been claimed to occur in renal failure and may account for the fact that, although plasma urea and urate levels rise in parallel, the rise in urate is less on a molar basis than that of urea.

Clinical gout, is a rare complication of therapy with *thiazide diuretics*, although hyperuricaemia is relatively common. These drugs inhibit renal excretion of urate.

PSEUDOGOUT

Pseudogout, while not a disorder of purine metabolism, produces a similar clinical picture to gout. Calcium pyrophosphate precipitates in joint cavities and calcification of the cartilages is seen radiologically. The crystals may be identified under a polarising microscope. The plasma urate is normal.

HYPOURICAEMIA

Hypouricaemia is rare, except as a result of treatment of hyperuricaemia. It is an unimportant finding in the *Fanconi syndrome* (p. 14) when there is decreased tubular reabsorption of urate. It may occur in severe chronic starvation.

Xanthinuria is a very rare inborn error in which there is a deficiency of liver xanthine oxidase. Purine breakdown stops at the xanthine-hypoxanthine stage. Plasma and urinary urate levels are very low. The increased urinary excretion of xanthine may lead to the formation of xanthine stones (the reason why this does not happen during therapy with xanthine oxidase inhibitors is not clear, but may be due to reduction of synthesis). The mode of inheritance is probably autosomal recessive.

SUMMARY

1. Urate is the end product of purine metabolism.
2. Hyperuricaemia may be the result of:
    increased nucleic acid turnover (malignancy, tissue damage, starvation);
    increased synthesis of purines (primary gout);
    reduced rate of renal excretion of urate (glomerular failure, thiazide diuretics, acidosis).
3. Hyperuricaemia may be aggravated by:
    high purine diets;
    acidosis and a high alcohol intake.
4. Primary hyperuricaemia and gout have a familial incidence and are rare in women of child-bearing age.
5. Because severe hyperuricaemia may cause renal damage it should be treated, even if asymptomatic.
6. Hypouricaemia is rare and usually unimportant. It occurs in the very rare inborn error, xanthinuria.

FURTHER READING

BALIS, M. E. (1976). Uric acid metabolism in man. *Adv. clin. Chem.*, **18**, 213.
WATTS, R. W. E. (1976). Uric acid biosynthesis and its disorders. *J. roy. Coll. Phycns. Lond.*, **11**, 91.

Chapter XVIII

IRON METABOLISM

IN man the circulating iron-containing pigment, haemoglobin, carries oxygen from the lungs to metabolising tissues, and in muscle myoglobin increases the local supply of oxygen. The ability to carry oxygen depends, among other factors, on the presence of ferrous iron in the haem molecule; iron deficiency is associated with deficient haem synthesis, and the symptoms of anaemia are due to tissue hypoxia. Certain enzymes necessary for electron transfer reactions (and therefore, among other things, for oxidative phosphorylation) and the cytochromes also contain iron: it is doubtful whether clinical iron deficiency, unless very severe, affects these.

NORMAL IRON METABOLISM

DISTRIBUTION OF IRON IN THE BODY

Figure 36 represents diagrammatically the distribution of iron in the body. The total body iron is about 50 to 70 mmol (3 to 4 g).

1. About 70 per cent of the total iron is circulating in erythrocyte *haemoglobin*.

2. Up to 25 per cent of the body iron is stored in the reticulo-endothelial system, in the liver, spleen and bone marrow. This storage iron is complexed with protein to form *ferritin* and *haemosiderin*. Ferritin is more readily available than haemosiderin. Haemosiderin (which may be aggregated ferritin) can be seen in unstained sections examined by light microscopy. Both ferritin and haemosiderin (but not iron incorporated in haem) stain with potassium ferrocyanide (Prussian blue reaction): this staining characteristic may be used to assess the size of iron stores.

Iron deficiency is evident when *no stainable iron* is detectable in the *reticulo-endothelial cells* in *bone marrow* films: this is the iron normally drawn upon for haemoglobin synthesis.

Iron overload is evident when *stainable iron* is demonstrable in *liver* biopsy specimens: such parenchymal deposition occurs when storage capacity is exceeded.

Histological assessment is more reliable for iron deficiency than for iron overload (p. 394).

3. *Only about* 50 to 70 μmol (3 to 4 mg: 0·1 per cent) *of the total body iron is circulating in the plasma*, bound to protein. This is the fraction measured in *plasma iron* estimations.

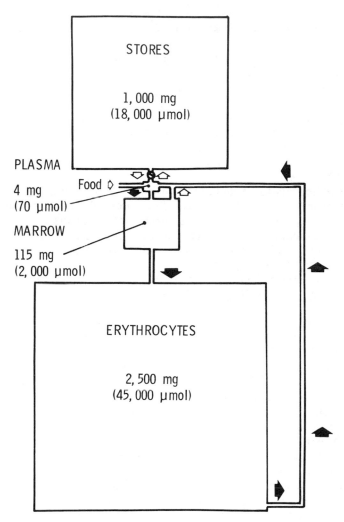

FIG. 36.—Body iron compartments.

4. The remainder of the body iron is incorporated in myoglobin, cytochromes and iron-containing enzymes.

Iron can only cross cell membranes by active transport in the ferrous form: it is in this reduced state in both oxyhaemoglobin and "reduced" haemoglobin. In ferritin and haemosiderin, and when bound to transferrin, it is in the ferric state.

The *control* of iron distribution in the body is poorly understood.

Plasma iron concentrations can vary by 100 per cent or more for purely physiological reasons, and are also affected by a variety of pathological factors other than the amount of iron in the body. These variations in plasma iron are probably due to redistribution between stores and plasma.

IRON BALANCE

As shown in Fig. 36 iron, once in the body, is virtually in a closed system.

Iron Excretion

There appears to be no control of iron excretion and loss from the body probably depends on the iron content of desquamated cells. Negligible amounts appear in the urine, reflecting the fact that it is entirely protein bound in the circulation. Most of the loss is probably into the intestinal tract and from the skin. The total daily loss by these routes is about 18 μmol (1 mg).

In women the mean monthly menstrual loss of iron is about 290 μmol (16 mg) and averages 10–18 μmol (0·5–1 mg) a day over the month above the basal 18 μmol (1 mg) daily, although it can be much higher in those with menorrhagia, who may become iron deficient. During pregnancy the mean extra daily loss to the fetus and placenta is about 27 μmol (1·5 mg).

It should be noted for comparison that a male blood donor losing a unit (pint) of blood every 4 months averages an extra loss of 36 μmol (about 2 mg) daily above the basal loss of 18 μmol (1 mg) (Table XXVIII).

Iron Absorption

The control of the body content of iron depends upon control of absorption.

Iron is absorbed by an active process in the upper small intestine and passes rapidly into the plasma. It can cross cell membranes, including those of intestinal cells, only in the ferrous form. Within the intestinal cell some of the iron is combined with the protein, apoferritin, to form ferritin: this, like ferritin elsewhere, is a storage compound. Ferritin is lost into the intestinal tract when the cell desquamates.

Absorption normally amounts to about 18 μmol (1 mg) of iron a day and just replaces loss. The percentage absorption of dietary iron depends to some extent on the food with which it is taken, but is usually about 10 per cent.

Iron absorption appears to be influenced by oxygen tension, by marrow erythropoietic activity, or by the size of iron stores: all these

TABLE XXVIII

COMPARISON OF IRON LOSSES

	Source of loss	Extra loss	Daily extra loss	Daily total loss
Men and non-menstruating women	Desquamation	—	—	18 μmol (1 mg)
Menstruating women (Mean value)	Desquamation + menstruation	290 μmol (16 mg)/month	9 μmol (0·5 mg)	27 μmol (1·5 mg)
Pregnancy	Desquamation + loss to fetus and in placenta	7000 μmol (380 mg)/9 months	27 μmol (1·5 mg)	45 μmol (2·5 mg)
Male blood donors	Desquamation + 1 unit blood	4500 μmol (250 mg)/4 months	36 μmol (2·0 mg)	54 μmol (3·0 mg)

factors may affect it. From the clinical point of view it is important to note that *iron absorption is increased in many anaemias not due to iron deficiency*.

If an adequate diet is taken most normal women probably absorb slightly more iron than men to replace their greater losses.

The iron requirement for growth in children and adolescents is probably similar to, or slightly higher than, that of menstruating women. This need, too, can probably be met by increased absorption from a normal diet.

Normal iron loss is so small, and normal iron stores are so large, that it would take about 3 years to become iron-deficient on a completely iron-free diet. This period is much shorter if there is any blood loss.

IRON TRANSPORT IN PLASMA

Iron is carried in the plasma in the ferric form, attached to the specific binding protein, *transferrin* (siderophilin), at a concentration of about 18 μmol/l (100 μg/dl). The protein is normally capable of binding about 54 μmol/l (300 μg/dl) of iron, and is therefore about a third saturated. Transferrin-bound iron is carried to stores and to bone marrow: in stores it is laid down as ferritin and haemosiderin, and in the marrow some may pass directly from transferrin into the developing erythrocyte to form haemoglobin.

Little, if any, of the iron in the body is free. In intestinal cells and stores it is bound to protein in ferritin and haemosiderin, in the plasma it is bound to transferrin, and in the erythrocyte it is incorporated in haemoglobin. Free iron is toxic.

FACTORS AFFECTING PLASMA IRON LEVELS

Plasma iron estimation is frequently requested, but is rarely of clinical value, and results are often misinterpreted. As we have seen, plasma iron, like plasma potassium, represents a very small proportion of the total body content, and for this reason alone is likely to be a poor index of it: plasma potassium levels are important to the body (Chapter III) and are normally relatively constant, but those of plasma iron (which is a protein-bound transport fraction) are much less so and can vary greatly even under physiological conditions.

PHYSIOLOGICAL FACTORS AFFECTING PLASMA IRON LEVELS

The causes of physiological changes in plasma iron concentrations are not well understood, but they are very rapid, and almost certainly represent shifts between plasma and stores. The following factors are known to affect levels and some may cause changes of 100 per cent or more.

Sex and Age Differences

Plasma iron levels are higher in men than in women, like those of haemoglobin and the erythrocyte count. This difference is probably hormonal in origin. It is first evident at puberty, before significant menstrual iron loss has occurred, and disappears at the menopause. Androgens tend to increase plasma iron concentration and oestrogens to lower it.

Cyclical Variations

Circardian (diurnal) rhythm.—Plasma iron is higher in the morning than in the evening. If subjects are kept awake at night this difference may be less marked or absent, and it is reversed in night workers.

Monthly variations in women.—Plasma iron may reach very low levels just before or during the menstrual period. This reduction is probably due to hormonal factors rather than to blood loss.

Random Variations

Very large day-to-day variations (occasionally as much as threefold) occur in plasma iron, and these usually overshadow cyclical changes. They may sometimes be associated with physical or mental stress, but more usually a cause cannot be found.

Effect of Pregnancy and Oral Contraceptives

In women taking some oral contraceptives the plasma iron rises to levels similar to those found in men. A similar rise occurs in the first few weeks of pregnancy: if iron deficiency develops in late pregnancy this rise may be masked.

PATHOLOGICAL FACTORS AFFECTING PLASMA IRON LEVELS

1. Iron deficiency and iron overload usually cause low and high plasma iron levels respectively.

Iron deficiency is associated with a hypochromic microcytic anaemia, and with reduced amounts of stainable marrow iron.

Iron overload is associated with increased amounts of stainable iron in liver biopsy specimens.

2. Many illnesses including *infection (acute or chronic, mild or severe), renal failure, malignancy and autoimmune diseases such as rheumatoid arthritis* cause hypoferraemia. Many of these chronic conditions are associated with normocytic, normochromic anaemia. Iron stores are

normal or even increased, and the anaemia does not respond to iron therapy.

3. In conditions in which the *marrow cannot utilise iron*, either because it is hypoplastic, or because some other factor necessary for erythropoiesis (such as vitamin B_{12} or folate) is absent, plasma iron levels are often high. Blood and marrow films may show a typical picture: in, for instance, pyridoxine-responsive anaemia and in thalassaemia, the findings in the blood film may be somewhat similar to that of iron deficiency, but iron stores are increased, and this can be shown on the marrow film.

4. In *haemolytic anaemia* the iron from the haemoglobin of broken down erythrocytes is released into the plasma and reticulo-endothelial system. Plasma iron may be high during the haemolytic episode and is usually normal during quiescent periods. Marrow iron stores are usually increased in chronic haemolytic conditions.

5. In *acute liver disease* disruption of cells may release ferritin iron into the blood stream and cause a transient rise of plasma iron. In *cirrhosis* plasma iron levels may sometimes be high. Although the reasons for this are not clear, it may sometimes be due to increased iron absorption associated with a high iron intake.

TRANSFERRIN AND TOTAL IRON-BINDING CAPACITY (TIBC)

It will be seen that plasma iron levels by themselves give no useful information about the state of iron stores. In the rare situations in which doubt remains about this after haematological investigations have been carried out, diagnostic precision may be improved by measuring the iron-binding capacity of the plasma at the same time as the plasma iron. It is a waste of time and money to estimate only plasma iron.

Transferrin is usually measured indirectly by measuring the iron-binding capacity of the plasma. An excess of inorganic iron is mixed with the plasma and any which is not bound to transferrin is removed, usually with a resin. The remaining iron is estimated on the plasma sample and the result expressed as a total iron-binding capacity (TIBC).

PHYSIOLOGICAL CHANGES IN TIBC

The TIBC is less labile than the plasma iron. However, it rises rapidly in subjects on some *oral contraceptives*, and this point should be remembered when interpreting results in women. It also increases after about the 28th week of *pregnancy*, even in those women with normal iron stores.

PATHOLOGICAL CHANGES IN TIBC

1. The *TIBC rises in iron deficiency* and *falls in iron overload.*
2. The *TIBC falls* in chronic infection, malignancy and the *other pathological conditions associated with a low plasma iron concentration* other than iron deficiency. This includes the nephrotic syndrome, in which the protein is lost in the urine.
3. The TIBC is unchanged in acute infection.

Thus the low plasma iron of iron deficiency is associated with a high TIBC. That of anaemia not due to iron deficiency is associated with a low TIBC.

PERCENTAGE SATURATION OF TIBC

The statement that the TIBC is normally a third saturated with iron is a very approximate one: physiological variations of plasma iron level are rarely associated with much change in TIBC and the saturation of the protein varies widely; the percentage saturation is, of course,

$$\frac{\text{Plasma Iron Concentration} \times 100}{\text{TIBC}}$$

It has been claimed that percentage saturation is a better index of iron stores than plasma iron concentration alone, and that, if it is below 16 per cent, iron deficiency is likely to be present. The first part of the statement is obviously true, because the low plasma iron of iron deficiency is accompanied by a high TIBC; this will result in a lower percentage saturation than with the same level of plasma iron in other conditions. However, saturation as low as 16 per cent can be found, for instance, premenstrually and in acute infections, with no change in TIBC or in iron stores. It is probably more useful to take account of the results of both plasma iron and TIBC rather than to calculate the percentage saturation.

The findings in various conditions which may affect plasma iron levels and TIBC are summarised in Table XXIX.

PLASMA FERRITIN LEVELS

Normal plasma ferritin levels are about 100 μg/l. It has been suggested that circulating ferritin is in equilibrium with that in stores, and that levels less than 10 μg/l indicate iron deficiency, while high levels are found in iron overload or liver disease. This estimation is not yet in general use, and its value is still in doubt.

TABLE XXIX

Changes in Serum Iron (Fe) and Total Iron-Binding Capacity (TIBC)

	Fe	TIBC	% Saturation	Marrow stores
Low Iron Levels Iron deficiency	↓	↑	↓↓	↓ or absent
Chronic illnesses (e.g. infection and malignancy)	↓	↓	Variable	Normal or ↑
Acute illnesses (e.g. infection)	↓	Normal	↓	Normal
Premenstrual	↓	Normal	↓	Normal
High Iron Levels Iron overload	↑	↓	↑↑	↑
Oral contraceptives and late pregnancy	↑ (to male level)	↑	Normal	Normal
Early pregnancy	↑ (to male level)	Normal	↑	Normal
Hepatic cirrhosis	Variable. May be ↑	↓	↑ to ↑↑	May be ↑
Failure of marrow utilisation and haemolysis	↑	Normal or ↓	↑	↑

"IRON DEFICIENCY WITHOUT ANAEMIA"

It has been claimed, but never proven, that the symptoms of iron deficiency can occur when haemoglobin levels are within the "normal" range. Iron deficiency can certainly exist under these conditions, and in most, if not all, such cases the diagnosis may be made on the appearances of the blood film: haemoglobin levels will rise after a short course of oral iron. A low plasma iron level is a poor index of such iron deficiency and subjective, symptomatic response to treatment is extremely difficult to assess in an individual subject. In clinical trials statistically significant symptomatic relief (associated with a rise of haemoglobin concentration) has not been demonstrated unless depleted marrow iron stores could be demonstrated before the trial started. In subjects with low plasma iron levels and vague symptoms, with normal marrow stores, symptomatic improvement occurred as frequently (and in some groups more frequently) when a placebo was given as when iron was administered. Some workers even claim that at haemoglobin levels above 10 g/dl iron therapy, although it may correct the blood picture, does not affect symptomatology. In other words, it is very unlikely that symptomatic iron deficiency exists without anaemia.

The student should read the section on "Investigation of Anaemia" on p. 393.

IRON THERAPY

Because the body does not control iron excretion, and because body content is controlled by absorption, *parenteral iron therapy* may easily lead to iron overload. Repeated blood transfusions carry the same danger, as a unit of blood contains 4·5 mmol (250 mg) of iron. In anaemias other than that of true iron deficiency, stores are normal or even increased, and parenteral iron should not be given unless the diagnosis of iron deficiency is beyond doubt: even when this is so, the oral is preferable to the parenteral route. Repeated blood transfusion may be necessary to correct severe anaemia in, for instance, chronic renal disease and hypoplastic anaemia, but, in such cases, the danger of overload should be remembered and blood should not be given indiscriminately.

Anaemia increases the rate of iron absorption even in the presence of increased iron stores. Treatment of, for instance, chronic haemolytic anaemia with *oral iron supplements* may occasionally cause iron overload; it does not improve the anaemia, which is due to the rate of breakdown of erythrocytes exceeding the rate of production, and not to deficiency of iron. The released iron stays in the body. In other non-iron deficient anaemias the danger is similar.

Although iron absorption is controlled to some extent, this control is inefficient and may rarely be "swamped" by large loads, even in the absence of anaemia. Iron overload has been reported in a non-anaemic woman who continued to take oral iron (against medical advice) over a matter of years.

Iron therapy is potentially dangerous: it should be prescribed with care, and only when iron deficiency is proven.

IRON OVERLOAD

As emphasised on p. 380, the excretion of iron from the body is limited. Iron absorbed from the gastro-intestinal tract, or administered parenterally, in excess of daily loss, accumulates in body stores. If such "positive balance" is maintained over long periods, iron stores may exceed 350 mmol (20 g) (about five times the normal amount).

The main causes of iron overload are increased intestinal absorption of iron, and excessive parenteral administration.

Excessive absorption may be due to
> absorption of a *larger proportion* than usual of a *normal dietary intake* as may occur in
> > idiopathic haemochromatosis;
> > anaemias in which there is increased, but ineffective, erythro-poiesis;
> > alcoholic cirrhosis (rare).
> *high iron intake*
> > usually alcoholic drinks of high iron content;
> > iron therapy in a patient with non-iron deficiency anaemia.

Excessive *parenteral administration* may be in the form of multiple blood transfusions.

A very rare cause of iron overload is genetic transferrin deficiency.

Consequences of Iron Overload

The effect of the accumulated iron depends on the distribution in the body. This in turn is influenced partly by the route of entry. Two main patterns are seen at post-mortem examination or in biopsy specimens.

Parenchymal iron overload is typified by idiopathic haemochromatosis. Iron accumulates in the parenchymal cells of the liver, pancreas, heart and other organs. There is usually associated functional disturbance or tissue damage.

Reticulo-endothelial iron overload is seen after excessive *parenteral administration of iron* or *multiple blood transfusions*. The iron accumulates in the reticulo-endothelial cells of the liver, spleen and bone marrow. There are few harmful effects other than a possible interference with haem synthesis, but under certain circumstances (p. 390), the pattern of distribution may change to that of the parenchymal type. In dietary iron overload both hepatic reticulo-endothelial and parenchymal overload may occur, associated with scurvy and osteoporosis (p. 391). Whatever the cause of *massive* iron overload, there may be parenchymal accumulation and tissue damage.

Two terms in common usage require definition. *Haemosiderosis* is defined as an increase in iron stores as haemosiderin and is a histological definition. It does not necessarily mean that there is an increase in total body iron; for example, many anaemias have reduced haemoglobin iron (less haemoglobin) but increased storage iron. If the increase in storage iron in an organ is associated with tissue damage the condition is called *haemochromatosis*.

SYNDROMES OF IRON OVERLOAD

Idiopathic Haemochromatosis

Idiopathic haemochromatosis is a genetically determined disease in which increased intestinal absorption of iron over many years produces large iron stores of parenchymal distribution. It presents, usually in middle age, as cirrhosis with diabetes mellitus, hypogonadism and increased skin pigmentation. Because of the darkening of the skin, due to an increase in melanin rather than to iron deposition, the condition has been referred to as "bronzed diabetes", although the colour is more grey than bronze. Cardiac manifestations may be prominent, particularly in younger patients, many of whom die in cardiac failure. In about 10 to 20 per cent of cases hepatocellular carcinoma develops.

Increased iron absorption from the gut is not demonstrable in all cases. This is not surprising: the accumulation of 350 mmol (20 g) of iron over a period of 30–40 years requires an increased absorption of less than 36 μmol (about 2 mg) a day, not easily detectable by existing methods because of a wide normal range. In patients who present in the second or third decades, in whom the defect is presumably more severe, an increased rate of absorption is usually demonstrable. A further point is the inhibitory effect on iron absorption of the massive body stores in overt cases (p. 380). Increased absorption is demonstrable after removal of the excess in many cases.

The mode of inheritance is uncertain. Evidence for the genetic nature of the disease is found in the study of families of patients, in whom the role of such environmental factors as a high iron intake can be properly assessed. In addition, evidence of iron overload (see below) has been found in a significant proportion of close relatives of patients with idiopathic haemochromatosis. Factors such as alcohol abuse may hasten the accumulation of iron and development of liver damage.

The distinction between idiopathic haemochromatosis and alcoholic cirrhosis may present difficulty. In both conditions diabetes mellitus and hypogonadism may occur and although the incidence is higher in idiopathic haemochromatosis this does not help in the individual case. Examination of liver biopsy specimens may further confuse the issue. The liver in cases of alcoholic cirrhosis not infrequently contains increased stainable iron. Not only do some alcoholic drinks, notably wines, contain significant amounts of iron, but there is evidence that in cirrhosis there may be increased iron absorption due possibly to an effect of alcohol. However, the majority of patients with cirrhosis do not have demonstrably increased iron stores by chemical testing, and the liver biopsy specimen shows that the iron is mainly present in the portal tracts. The clinical, histological and biochemical changes of cirrhosis are usually more obvious than those of iron accumulation: this contrasts with the picture in haemochromatosis with apparently equivalent iron overload.

Rare cases of cirrhosis may have true iron overload and the distinction between the two conditions may be extremely difficult in the absence of such information. A family history or investigation of near relatives may help in diagnosis. The treatment of iron overload is the same in either case.

The diagnosis of idiopathic haemochromatosis *must* be followed by investigation of other members of the family, and treatment of those in whom increased iron stores are found.

Dietary Iron Overload

A well-described form of dietary iron overload is seen in the African population of Southern Africa. Unlike idiopathic haemochromatosis in which excessive iron is absorbed from a diet of normal iron content, the iron intake is grossly excessive. The main source of this iron is local beer, often brewed in iron containers: daily intakes of 1·4–1·8 mmol (80–100 mg) (at least 5 times normal) are not uncommon. At high oral intakes of iron control of absorption (the only means of controlling body iron, p. 380) is imperfect and daily absorption of 35–55 μmol (2–3 mg) of iron leads to iron overload: many African males over the age of 40 have a heavy deposition of iron in the liver.

Usually the excess iron is confined to the reticulo-endothelial system and the liver (both portal tracts and parenchymal cells), and there is no tissue damage. In a small number of cases, usually those with the heaviest iron deposits and therefore presumably the highest intake of alcohol, cirrhosis develops. In such cases deposition in the parenchymal cells of other organs occurs and the clinical picture now closely resembles that of idiopathic haemochromatosis: it may be distinguished by the high concentrations of iron in the reticulo-endothelial system such as bone marrow and spleen (at autopsy).

Scurvy and osteoporosis may occur in this form of iron overload. The ascorbate deficiency may be due to its irreversible oxidation in the presence of excessive amounts of iron, and osteoporosis sometimes accompanies scurvy. Ascorbate deficiency also interferes with normal mobilisation of iron from the reticulo-endothelial cells; plasma iron levels may be low, and the response to chelating agents poor, despite iron overload. These abnormalities are rapidly corrected by the administration of ascorbate.

OTHER CAUSES OF IRON OVERLOAD

Several types of anaemia may be associated with iron overload. In some, such as aplastic anaemia and the anaemia of chronic renal failure, the cause is multiple blood transfusions, and the iron accumulates in the reticulo-endothelial system. With massive overload (over 100 units of blood), true haemochromatosis may develop with parenchymal overload.

In anaemias characterised by erythroid marrow hyperplasia, but with ineffective erythropoiesis, there is increased absorption of iron from the intestine (p. 387). This may be aggravated by oral iron medication and in a few cases true haemochromatosis develops. It must be stressed again that prolonged iron therapy for anaemia other than that due to iron deficiency carries the risk of overload.

The very rare condition of congenital transferrin deficiency presents as a refractory anaemia and massive iron overload.

The student should read the section on "Diagnosis of iron overload" on p. 394.

SUMMARY

1. No significant excretion of iron can occur from the body. Control of body stores is by control of absorption. For this reason parenteral iron therapy should be given with care.

2. Absorption of iron is increased by anaemia even in the absence of iron deficiency. For this reason oral iron therapy should not be given in anaemia other than that due to iron deficiency.

3. Plasma iron levels vary considerably under physiological circumstances.

4. Plasma iron levels fall in many cases of anaemia not due to iron deficiency.

5. For these two reasons (3 and 4) plasma iron levels alone are a very poor indication of body iron stores.

6. Iron is carried in the plasma bound to the protein transferrin. Transferrin is usually measured indirectly by measuring the total iron-binding capacity (TIBC) of plasma.

7. The TIBC rises in iron deficiency and falls in iron overload.

8. The TIBC falls in many cases of anaemia associated with a low plasma iron, but not due to iron deficiency.

9. A low plasma iron concentration with a high TIBC is more suggestive of iron deficiency than a low plasma iron alone.

10. The quickest, cheapest and most informative tests for iron deficiency are simple haematological ones. These should be performed before requesting plasma iron estimation.

11. The factors governing the distribution of excessive iron are not fully understood. A feature common to all forms of parenchymal overload is a high percentage saturation of transferrin.

12. Iron overload may develop as a result of excessive absorption from a normal diet (idiopathic haemochromatosis) or from a high iron intake; or it may develop as a result of excessive parenteral iron administration. The distribution of the iron in the body differs in the various forms of iron overload.

13. Iron overload can be demonstrated by the response to repeated venesection or to chelating agents. The diagnosis of idiopathic haemochromatosis can only be made if massive iron overload is present.

14. Virtually all cases of parenchymal iron overload show a high plasma iron concentration with a high percentage saturation of transferrin.

15. Although plasma ferritin levels are often raised in iron overload, they may be misleading.

FURTHER READING

JACOBS, A., and WORWOOD, M. (Eds.) (1974). *Iron in Biochemistry and Medicine*. New York: Academic Press. (Especially Chapters 11, 13, 15, 16 and 17).

JACOBS, A., and WORWOOD, M. (1975). Ferritin in serum. Clinical and biochemical implications. *New Engl. J. Med.*, **292**, 951.

KIEF, H. (Ed.) (1975). *Iron Metabolism and its Disorders* (Proc. Third Workshop Conference Hoechst, Schloss Ressenberg, 6–9 April, 1975). Amsterdam: Excerpta Medica.

INVESTIGATION OF DISORDERS OF IRON METABOLISM

Investigation of Anaemia

Anaemia may be due to iron deficiency, or to a variety of other conditions. The subject of the diagnosis of anaemia is covered more fully in textbooks of haematology. However, so that we may see the value of plasma iron estimations in perspective, it is worth considering the order in which anaemia may usefully be investigated.

1. The clinical impression of anaemia should be confirmed by *haemoglobin* estimation. Iron deficiency can, however, exist with haemoglobin levels within the "normal" range.

2. A *blood film* should be examined, or *absolute values* estimated. Iron deficiency anaemia is hypochromic and microcytic in type, and hypochromia may be evident before the haemoglobin level has fallen below the accepted normal range. Normocytic, normochromic anaemia is non-specific and usually associated with other disease; it is not due to iron deficiency unless there has been very recent blood loss. Typical appearances of other anaemias may be seen on the blood film.

In most cases of anaemia, consideration of these findings together with the clinical picture will give the cause. Anaemias such as that of thalassaemia and of the pyridoxine-responsive type, although rare, are most likely to confuse the picture, since they too are hypochromic, but are not due to iron deficiency.

3. A *marrow film* may be required to confirm the diagnosis (for example, of megaloblastic anaemia). If such a film is available, staining with potassium ferrocyanide is by far the best indicator of the state of iron stores, if this information is still required.

If marrow puncture is not felt to be justified, and *in the rare cases* in which diagnosis is not yet clear, biochemical investigations may occasionally help. If these are necessary, plasma iron *and* TIBC should be estimated. Plasma iron estimation alone is uninformative.

Iron absorption test.—If a patient with *proven iron deficiency* does not start to respond to oral iron within a few weeks he is probably not taking the tablets: normal coloured, rather than black, faeces (detectable, if necessary, by rectal examination) confirm this suspicion.

If iron *has* been taken malabsorption, usually part of a general absorption defect, is occasionally the cause. If faecal fat excretion is high, this cause is obvious: if it is normal the adequacy of iron absorption may be assessed by measuring the increment of plasma iron after a standard (300 mg) oral dose of iron elixir. This increment will be very large (up to 40 μmol/l) if absorption is normal *in the patient with proven iron-deficiency anaemia:* it will be very small:

if absorption is deficient;
in the non-anaemic subject (who does not normally absorb much iron);
in the patient with anaemia not due to iron deficiency (the absorbed iron is rapidly deviated to stores).

To avoid misinterpretation the diagnosis of iron deficiency must be made before the test is performed.

Diagnosis of Iron Overload

Plasma iron concentration and TIBC.—The plasma iron concentration is almost invariably high in idiopathic haemochromatosis, often above 36 μmol/l (200 μg/dl). This is associated with a reduced transferrin level (as shown by a lowered TIBC) and the percentage saturation is usually over 80 per cent, and often 100 per cent: in the presence of infection or malignancy, however, the plasma iron level and percentage saturation may be lower than expected; the TIBC remains low. The lowering effect of ascorbate deficiency on plasma iron levels has already been mentioned.

High levels may also be present in cirrhosis.

Demonstration of increased iron stores.—The diagnosis of iron overload can only be made after proof has been obtained of increased iron stores.

Response to venesection.—The lack of response of the patient to a therapeutic course of venesection offers the most convincing proof of increased iron stores, albeit retrospectively. Removal of a unit of blood (4·5 mmol or 250 mg iron) repeated at short intervals produces a rapid fall in plasma iron, soon followed by iron deficiency anaemia, in a subject with normal iron stores. In patients with idiopathic haemochromatosis, however, 350 mmol (20 g) or more of iron may be removed in this way before evidence of iron deficiency develops.

Use of chelating agents.—An alternative method uses the chelating action of substances, such as desferrioxamine, which bind iron and are subsequently excreted in the urine. Following administration of desferrioxamine, subjects with increased iron stores excrete more iron in the urine than do normals.

Plasma ferritin levels are high in most cases of reticulo-endothelial iron overload, but *normal concentrations have been found in cases of idiopathic haemochromatosis.* Raised levels may also occur in many hepatic diseases, including cirrhosis, and the estimation is of doubtful diagnostic value.

Liver biopsy specimens contain large amounts of stainable iron, which may be mainly in parenchymal, or mainly in reticulo-endothelial cells. Chemical estimation of iron is more reliable than histochemical evaluation.

Marrow iron content is *usually normal* in haemochromatosis, in which overload is predominantly parenchymal, but may show greatly increased stores in reticulo-endothelial overload. A similar loading of the reticulo-endothelial cells is found when there is deficient utilisation of marrow iron for haemoglobin synthesis, as in many haematological, neoplastic and chronic inflammatory diseases.

Chapter XIX
THE PORPHYRIAS

THE porphyrias are a group of disorders, usually genetic in origin, caused by abnormalities of the haem synthetic pathway. Although most of them are uncommon and some are very rare, it is important to recognise them and to investigate relatives of known cases. For example, drugs which may precipitate acute attacks, sometimes with fatal consequences, must be avoided in the inherited hepatic porphyrias. It is equally important not to confuse the commoner, secondary, causes of porphyrin abnormalities with true porphyria.

PHYSIOLOGY

Haem is synthesised in most tissues of the body. In the bone marrow it is incorporated into *haemoglobin*. In other cells it is used for the syn-

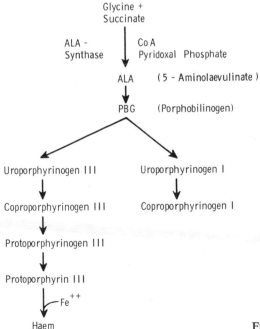

FIG. 37.—Biosynthesis of haem.

thesis of *cytochromes* and related compounds. The cytochromes are constituents of the electron transport chain, which harnesses the energy of metabolic processes. Impaired cytochrome synthesis can therefore have serious consequences. The liver is quantitatively the largest non-erythropoietic haem-producing organ.

Biosynthesis of Haem

The main steps are outlined below and in Fig. 37.

1. *5-Aminolaevulinate* (ALA) is formed by condensation of glycine and succinate. The reaction requires pyridoxal phosphate and is catalysed by the enzyme *ALA synthase*.

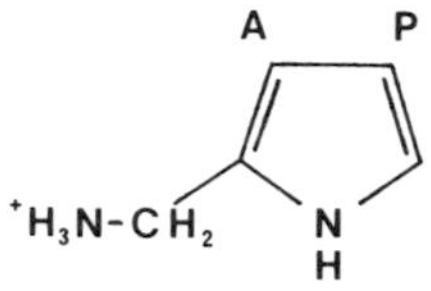

Porphobilinogen

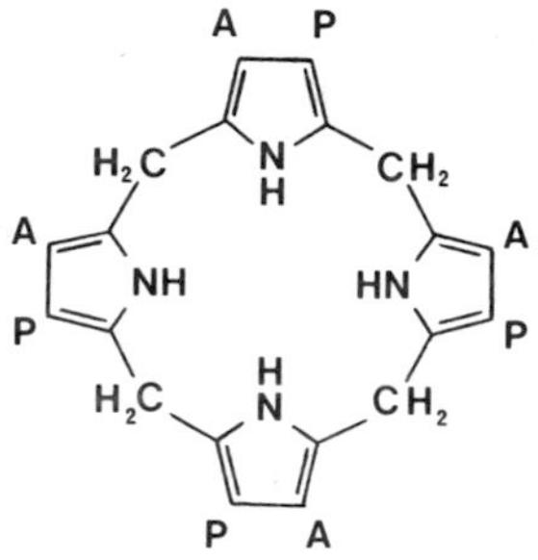

Uroporphyrinogen III

FIG. 38.—Porphobilinogen, and the tetrapyrrole uroporphyrinogen III, which incorporates four porphobilinogen units. Uroporphyrinogen I differs only in the order of side-chains on one of the rings.
(Side-chains: A = acetate; P = propionate.)

2. Two molecules of ALA condense to form a monopyrrole, *porphobilinogen* (PBG).

3. Four molecules of PBG combine to form a tetrapyrrole, *uroporphyrinogen* (Fig. 38). Two isomers are formed, I and III. III is the major pathway and leads to the formation of haem by the successive production of *coproporphyrinogen* and *protoporphyrin*, followed by the incorporation of iron.

Each step is controlled by a specific enzyme. Normally *the rate-limiting step* is that catalysed by *ALA synthase*, and this step is regulated by *feedback inhibition by the final product, haem*.

The *porphyrinogens* and their precursors, *ALA* and *PBG*, are *colourless* compounds. Porphyrinogens, however, oxidise spontaneously to the corresponding *porphyrins* which are *dark red* in colour and which *fluoresce* in ultraviolet light. PBG, too, may spontaneously form uroporphyrin when exposed to air and light. A urine specimen containing large amounts of porphyrinogens or precursors will gradually darken if left standing.

Excretion

Any excess of the intermediates on the haem pathway is excreted. ALA, PBG and uroporphyrin(ogen) are water-soluble and appear in the *urine*. Protoporphyrin is excreted in bile and appears in the *faeces*. Coproporphyrin(ogen) may be excreted by either route. Renal excretion of coproporphyrinogen, like that of urobilinogen, rises with increasing urinary alkalinity.

Normal urine contains ALA, PBG and porphyrin at concentrations undetectable by screening tests (see later). Faeces may contain sufficient porphyrin to impart a slight fluorescence to extracts.

THE PORPHYRIAS

The porphyrias are due to a deficiency, usually inherited, of one of the enzymes on the haem pathway: haem production is therefore impaired. *Reduced feedback inhibition of ALA synthase may maintain adequate haem levels but at the expense of overproduction of porphyrins or their precursors.*

The *symptoms* of porphyria correlate well with the biochemical abnormalities.

Skin lesions, varying from mild photosensitivity to severe blistering, occur when circulating *porphyrin* levels are increased. The lesions typically occur in exposed areas where solar ultraviolet rays may activate the porphyrins to release energy which damages tissue.

Neurological disturbances, such as *peripheral neuritis, abdominal pain*, or both in the serious *acute attack*, occur in those porphyrias in which the precursors, ALA and PBG are produced in excess. It is not known whether the neurological damage is due to haem deficiency in the nervous system or to a direct toxic effect of ALA or PBG.

The porphyrias are classified according to whether the main site of porphyrin accumulation is in the liver or the erythropoietic system.

Hepatic Porphyrias

Acute intermittent porphyria ⎫
Porphyria variegata ⎬ acute porphyrias
Hereditary coproporphyria ⎭
Cutaneous hepatic porphyria
  genetic predisposition
  acquired

TABLE XXX

The Major Clinical and Biochemical Features of the Porphyrias

	Hepatic porphyrias							Porphyrias involving the erythropoietic system	
	Acute intermittent porphyria		Porphyria variegata		Hereditary coproporphyria		Cutaneous hepatic porphyria	Congenital erythropoietic porphyria	Erythrohepatic protoporphyria
	acute	latent	acute	latent	acute	latent			
Clinical Features									
Abdominal and neurological symptoms	+	−	+	−	+	−			
Skin lesions	−	−	+	+	+ rarely	−	+	+	+
Chemical Abnormalities									
Urine PBG and ALA	+	+	+	−	+	−			
Urine porphyrins	+	−	−	−	+	−	+	+ +	
Faecal porphyrins	−	−	+	+	+ +	+	+	−	+
	Acute attacks precipitated by drugs (for instance barbiturates, oestrogens, sulphonamides)						Symptoms may be relieved by venesection	Erythrocyte porphyrins increased	

Erythropoietic Porphyrias

Congenital erythropoietic porphyria
Erythrohepatic protoporphyria
The features of the porphyrias are outlined in Table XXX and discussed briefly below:

THE ACUTE HEPATIC PORPHYRIAS

Acute intermittent porphyria;
Hereditary coproporphyria
Porphyria variegata;

These three *dominantly* inherited disorders have latent and acute phases. The symptoms and biochemical abnormalities of the *latent phases* differ and reflect the nature of the enzyme defect. In the *acute phases*, however, the common biochemical and clinical picture of excessive ALA and PBG excretion, associated with neurological and abdominal symptoms, develops. The similarities and differences are best explained by referring to a simplified scheme of the haem synthetic pathway (Fig. 39).

Latent Phase

The enzyme defect tends to reduce haem levels which in turn increase ALA synthase activity by the negative feedback loop. Increased ALA synthase activity has been demonstrated in all the porphyrias. The increased rate of synthesis maintains haem levels but at the expense of accumulation and excretion of the substance immediately before the block. The biochemical picture is predictable (see Fig. 39).

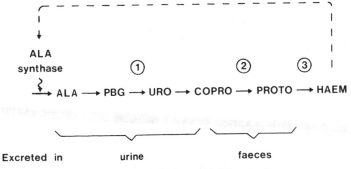

FIG. 39.—Sites of enzyme deficiencies in:
1. acute intermittent porphyria;
2. hereditary coproporphyria;
3. porphyria variegata.

Acute intermittent porphyria—increased *urinary ALA and PBG*
(not detectable in all patients)
Hereditary coproporphyria —increased *faecal coproporphyrin*
Porphyria variegata —increased *faecal protoporphyrin*

In acute intermittent porphyria the latent phase is usually asymptomatic. In porphyria variegata, and less commonly in hereditary coproporphyria, an increase in circulating porphyrins may produce skin lesions.

Acute Phase

ALA synthase activity may be further increased by a number of drugs, particularly barbiturates, oestrogens, sulphonamides and griseofulvin, and by acute illness: this may be a direct effect or may be due to increased demand for haem. This results in a marked increase in ALA and PBG excretion. In acute intermittent porphyria this is due to the block imposed by an inherited deficiency of *uroporphyrinogen-1-synthase*; in hepatic coproporphyria and porphyria variegata, this enzyme becomes rate-limiting and is unable to respond normally to the increased demand. *The increase in urinary ALA and PBG is the hallmark of the acute porphyric attack* and, in hereditary coproporphyria and porphyria variegata, is superimposed on the other biochemical abnormalities. The accelerated activity of the pathway and the spontaneous conversion of the precursors to porphyrin lead to increased urinary porphyrin excretion.

In the acute attack colicky abdominal pain (due to involvement of the autonomic nervous system) and neurological symptoms ranging from peripheral neuritis to quadriplegia are usually the presenting features. Death may result from respiratory paralysis. Electrolyte disturbances, notably hyponatraemia, develop. Acute attacks occur only after puberty and are commoner in women than in men.

The acute attack closely resembles serious acute intra-abdominal conditions, and if the diagnosis is not made the patient may be subjected to surgery, with the use of a barbiturate anaesthetic; barbiturates and the stress of operation may aggravate the condition.

Diagnosis of Latent Porphyria

It is essential to investigate the family of any known porphyric. Carriers of porphyria variegata and hereditary coproporphyria can be identified, after puberty, by demonstration of clearly increased *faecal porphyrin* excretion. Positive screening tests (see Appendix) should be confirmed by quantitative estimation. Screening tests for excess urinary PBG and ALA are inadequate to diagnose acute intermittent porphyria, and even quantitative estimation may fail to detect all carriers. The activity of the enzyme, *uroporphyrinogen-1-synthase*, should be measured in erythrocytes (the most easily available cells). There is often overlap in levels between affected and normal people, but within a family carriers can be shown to have levels about half those of unaffected members. This test will also detect affected children.

CUTANEOUS HEPATIC PORPHYRIA
(Porphyria cutanea tarda)

A symptomatic mild coproporphyrinuria may occur in many liver diseases, but less commonly excessive production and excretion of *uroporphyrin* (mostly of Type I) is associated with severe photosensitivity and skin lesions. This condition, cutaneous hepatic porphyria, is commonest in liver disease associated with alcohol abuse and iron overload, and photosensitivity may be improved for long periods after reducing iron stores by venesection. There is some evidence that only genetically predisposed individuals develop this condition. Hirsutism and hyperpigmentation occur in chronic cases. Cutaneous hepatic porphyria may also be secondary to toxic liver damage. An outbreak of porphyria of this type occurred in Turkey in 1955 and was traced to wheat treated with hexachlorobenzene. The pigmentation and hirsutism earned for the unfortunate victims the title of "monkey children".

ERYTHROPOIETIC PORPHYRIAS

Two rare inherited disorders are associated with accumulation of porphyrins in erythrocytes. Acute porphyric attacks do not occur and ALA and PBG excretion are normal.

Congenital erythropoietic porphyria, unlike all the other porphyrias discussed, is inherited as a *recessive* characteristic. Usually from infancy onwards *erythrocyte* and *plasma uroporphyrin I* levels are very high and there is severe *photosensitivity*. Porphyrin is also deposited in bones and teeth which fluoresce in ultraviolet light. Hirsutism, especially of the face, also occurs and there is a haemolytic anaemia.

Urinary porphyrin levels are grossly increased, and faecal levels less so.

Erythrohepatic protoporphyria.—In this *dominantly* inherited disorder *protoporphyrin* levels are increased in *erythrocytes* and *faeces*. There is mild photosensitivity, and hepatocellular damage may lead to liver failure.

OTHER CAUSES OF EXCESSIVE PORPHYRIN EXCRETION

Porphyria is not the only cause of disordered porphyrin metabolism and positive screening tests must be *confirmed* by quantitative analysis with *identification* of the porphyrin present. Three conditions must be considered.

Lead poisoning inhibits several of the enzymes involved in haem synthesis and eventually produces anaemia. The *urine* contains increased amounts of *ALA* (an early and sensitive test), and *coproporphyrin*. Some of the symptoms of lead poisoning, such as abdominal pain, are similar to those of the acute porphyric attack, and this may cause difficulty in differential diagnosis.

Liver disease may produce an increase in *urinary coproporphyrin*, possibly due to decreased biliary excretion. This is probably the commonest cause of porphyrinuria. Occasionally there is mild photosensitivity (in cutaneous hepatic porphyria there are more severe skin lesions due to uroporphyrin excess).

Ulcerative lesions of the upper gastro-intestinal tract may produce raised levels of *faecal porphyrin* by degradation of haemoglobin. If there is bleeding from the lower part of the tract there is insufficient time for conversion: this may help roughly to localise the site of bleeding.

SUMMARY

1. Porphyrins are by-products of haem synthesis. ALA and PBG are precursors.

2. The porphyrias are diseases associated with disturbed porphyrin metabolism. Most are inherited. The main clinical and biochemical features are outlined in Table XXX.

3. Acute attacks with abdominal or neurological symptoms are a feature of the inherited hepatic porphyrias. Such attacks are potentially fatal and may be provoked by a number of drugs. The diagnosis of porphyria in the acute phase depends on the demonstration of ALA and PBG in the urine.

4. The diagnosis of inherited porphyria must be followed by investigation of all relatives to detect asymptomatic cases. Screening tests may be negative in some types and quantitative estimations are necessary. Both urine and faeces should be examined.

5. Other causes of abnormalities in porphyrin excretion are lead poisoning, liver disease and upper gastro-intestinal bleeding.

6. The very rare erythropoietic porphyrias show excessive red cell porphyrin.

FURTHER READING

ELDER, G. H., GRAY, C. H., and NICHOLSON, D. C. (1972). The porphyrias: a review. *J. clin. Path.*, **25**, 1013. (A full discussion including a list of the drugs known to precipitate acute attacks.)

KRAMER, S., BECKER, D. M., and VILJOEN, J. D. (1977). Enzyme deficiencies in the porphyrias. *Brit. J. Haemat.*, **37**, 439.

APPENDIX

The screening tests for porphyria are usually carried out in the laboratory. In areas of high incidence, such as South Africa, however, simple side-room tests should be available.

The requirements are:
1. A near ultraviolet (Wood's) lamp.
2. An extracting solvent (equal parts of ether, glacial acetic acid and amyl alcohol).
3. Ehrlich's reagent (2 per cent p-dimethylaminobenzaldehyde in 5M HCl).
4. n-Butanol.

Test for Porphobilinogen

(i) Equal parts of *fresh* urine and Ehrlich's reagent are mixed. If PBG is present a red colour develops.

(ii) If a red colour develops, 3 ml of n-butanol is added, the tube shaken and the phases allowed to separate. If the red colour does *not* enter the butanol (upper layer) it is PBG.

It is important to use *fresh* urine as PBG disappears on standing. A number of other substances, especially urobilinogen, may also give a red colour with Ehrlich's reagent. All such known substances are, however, extracted into the butanol layer.

Test for Porphyrins

Urine.—About 10 ml of fresh urine is mixed with 2 ml of the porphyrin extracting solvent (item 2) and the phases allowed to separate. A red fluorescence in the upper layer under ultraviolet light indicates the presence of porphyrin.

Faeces.—A pea-sized piece of faeces is mixed into 2 ml of the solvent. A red fluorescence under ultraviolet light denotes either porphyrin or chlorophyll (with a similar structure and derived from the diet). These may be distinguished by adding 1·5M HCl which extracts porphyrin but not chlorophyll.

Interpretation

The following points are important.

1. It is essential to test the *correct type of sample* (urine or faeces) for the type of porphyria expected.

2. In the *latent phase screening tests are too insensitive* to diagnose *acute intermittent porphyria*. Properly performed, *negative tests for faecal porphyrin* almost certainly *exclude the diagnosis of porphyria variegata*.

3. Tests are *negative before puberty* even in children who will suffer from porphyria.

4. *Positive results must be confirmed* by quantitative analysis and by typing the porphyrins. There are causes other than porphyria for increased porphyrin excretion.

Chapter XX

VITAMINS

VITAMINS, like protein, carbohydrate and fat, are organic compounds which are essential dietary constituents (the name "vitamines" originally meant amines necessary for life): unlike most other constituents they are required only in very small quantities. They must be taken in the diet because the body either cannot synthesise them at all or, under normal circumstances, not in sufficient amounts for its requirements: "vitamin D", for instance, can be synthesised in the skin under the influence of ultraviolet light, but in temperate climates the amount of sunlight is insufficient to provide the required amount, while in hot countries the inhabitants tend to avoid the sun.

A normal mixed diet contains adequate quantities of vitamins and deficiencies are rarely seen in affluent populations except in individuals with intestinal malabsorption, in those on unsupplemented artificial diets, and in food faddists. Vitamin supplementation of a normal diet is unnecessary. Some vitamins (notably A and D) produce toxic effects if taken in excess.

The biochemical function of many vitamins is now understood, but it is not always easy to relate this knowledge to the clinical picture seen in deficiency states.

CLASSIFICATION OF VITAMINS

In 1913 McCollum and Davis first showed that two growth factors are required for normal health, one being fat soluble and the other water soluble. Since then each of these two groups has been shown to consist of many compounds, but classification can still be made on the basis of this solubility. The distinction is important clinically, because steatorrhoea is associated with deficiency of fat-soluble vitamins, but with relatively little clinical evidence of lack of most water-soluble vitamins (the exceptions being vitamin B_{12} and folate, p. 271).

FAT-SOLUBLE VITAMINS

The fat-soluble vitamins are:

A
D
K
(E)

Each of these has more than one active chemical form, but variations in structure are very slight, and in the following discussion we shall refer to each vitamin as a single substance.

VITAMIN A

Source of Vitamin A

Precursors of vitamin A (the carotenes) are found in the yellow and green parts of plants and are especially abundant in carrots: for this reason this vegetable has the reputation of improving night vision, but it is doubtful if it has any effect on a subject eating a normal diet. The vitamin is formed by hydrolysis of the β-carotene molecule, each yielding a possible total of two molecules of vitamin A. In the body this hydrolysis occurs in the intestinal mucosa, but the yield of vitamin is much below the theoretical one, especially in children for whom a diet containing the vitamin itself is essential. It is stored in animal tissues, particularly in the liver, and these tissues are the only source of pre-formed vitamin A.

Stability of Vitamin A

Vitamin A is rapidly destroyed by ultraviolet light and should be kept in dark containers.

Function of Vitamin A

1. Vitamin A is essential for normal *mucopolysaccharide synthesis*.
2. Vitamin A is required for normal *mucus secretion*; deficiency causes drying up of mucus-secreting epithelia.
3. The *retinal pigment*, rhodopsin (visual purple), is necessary for vision in dim light (scotopic vision). Rhodopsin consists of a protein (opsin) combined with vitamin A. In bright light rhodopsin is broken down. It is partly regenerated in the dark, but, because this regeneration is not quantitatively complete, vitamin A is needed to maintain the levels in the retina. The cycle is complex. Vitamin A deficiency is associated with poor vision in dim light, especially if the eye has recently been exposed to bright light.

Clinical Effects of Vitamin A Deficiency

The clinical effects of vitamin A deficiency are:
"night blindness";
drying and metaplasia of ectodermal tissues;
xerosis conjunctivae, xerophthalmia and keratomalacia;
? follicular hyperkeratosis.
"*Night blindness*".—In vitamin A deficiency the rate of light adapta-

tion can be shown to be reduced, although it is uncommon for the patient to complain of this.

Xerosis conjunctivae and xerophthalmia.—The conjunctiva and cornea become dry and wrinkled with squamous metaplasia of the epithelium and keratinisation of the tissue, resulting from deficiency of mucus secretion. *Bitot's spots,* seen in more advanced cases, are elevated white patches found in the conjunctivae and composed of keratin debris. If deficiency continues, *keratomalacia* occurs with ulceration and infection and consequent scarring of the cornea, causing blindness. Keratomalacia is an important cause of blindness in the world as a whole, but is rarely seen in affluent countries.

Squamous metaplasia occurs in other epithelial tissues. Skin secretion is diminished, and hyperkeratosis of hair follicles may be seen (*follicular hyperkeratosis*): these dry horny papules, varying in size from a pinhead to quarter-inch diameter, are found mainly on the extensor surfaces of the thighs and forearms. It is not certain if follicular hyperkeratosis is directly due to vitamin A deficiency.

Squamous metaplasia of the bronchial epithelium has also been reported and may be associated with a tendency to chest infection.

Causes of Vitamin A Deficiency

Hepatic stores of vitamin A are so large that clinical signs only develop after a year or more of dietary deficiency.

Deficiency is very rare in affluent communities. In steatorrhoea overt clinical evidence is rarely seen, but blood levels have been shown to be low. By contrast, deficiency is relatively common in underdeveloped countries, especially in children, and is a common cause of blindness.

Laboratory Diagnosis of Vitamin A Deficiency

This depends on the demonstration of low blood vitamin A levels. Specimens for this estimation should be kept in dark containers to prevent destruction of vitamin A by ultraviolet light.

Treatment of Vitamin A Deficiency

In treatment of xerophthalmia doses of 50 000 to 75 000 international units of vitamin A in fish liver oil should be given. Response to treatment of "night blindness" and of early retinal and corneal change is very rapid. Corneal scarring is irreversible.

Hypervitaminosis A

Vitamin A in large doses is toxic. Acute intoxication has been reported in Arctic regions as a result of eating polar bear liver, which has a very high vitamin A content, but a more common cause is overdosage with vitamin preparations: the symptoms of acute poisoning are nausea and vomiting, abdominal pain, drowsiness and headache. In chronic hyper-vitaminosis A there is fatigue, insomnia, bone pains, loss of hair and desquamation and pigmentation of the skin.

VITAMIN D (CALCIFEROL)

Source of Vitamin D

Vitamin D, like vitamin A, is stored in animal liver, and fish liver oils are rich in this factor.

The metabolism and functions of "vitamin D", and the effects and treatment of its deficiency are discussed in Chapter XI.

Causes of Vitamin D Deficiency

Liver stores of vitamin D are usually adequate for several months' needs and clinical deficiency is slow to develop. The commonest cause in affluent communities is steatorrhoea, and rickets is now rare in Britain, except in recent immigrants.

Hypervitaminosis D

Overdosage with vitamin D causes hypercalcaemia, with all its attendant dangers (p. 236). In chronic overdosage liver stores of cholecalciferol are large, and hypercalcaemia may persist for several weeks after the therapy is stopped.

VITAMIN K

Vitamin K cannot be synthesised by man but, like many of the B vitamins, it can be manufactured by the bacterial flora of the colon: unlike them it can probably be absorbed from this site and dietary deficiency is therefore not seen. In steatorrhoea the vitamin, whether taken in the diet or produced by bacteria, cannot be absorbed normally and deficiency may occur (p. 269).

Vitamin K is necessary for the synthesis of prothrombin and other coagulation factors in the liver, and deficiency is accompanied by a bleeding tendency with a prolonged prothrombin time. If these findings are due to deficiency of the vitamin they can be cured by parenteral administration (p. 292).

VITAMIN E (TOCOPHEROLS)

Vitamin E is a fat-soluble vitamin. In experimental animals deficiency has been reported to cause fetal death and sterility in both sexes, and also to be related to muscular dystrophy. Deficiency has not been shown to produce clinical effects in man, and therapeutic trials for various conditions have produced disappointing results.

WATER-SOLUBLE VITAMINS

The water-soluble vitamins are:
the B complex
thiamine (aneurine: B_1)
riboflavin (B_2)
nicotinamide (Pellagra Preventive (PP) factor: niacin)
pyridoxine (B_6)
(biotin and pantothenate)
folate (pteroylglutamate)
the vitamin B_{12} complex (cobalamins)
ascorbate (vitamin C)

THE B COMPLEX

This group of food factors was originally classified together in the B group (with the exception of vitamin B_{12} and folate, which were later discoveries). Most of them act as coenzymes, but it is not easy to relate the clinical findings to the underlying biochemical lesion.

Many of these vitamins are synthesised by colonic bacteria. Opinions vary as to the importance of this source in man, but since the absorption of water-soluble vitamins from the large intestine is poor, most of it is probably unavailable.

Clinical deficiency is rare in affluent communities. When deficiency does occur it is usually multiple, involving most of the B group and protein: for this reason it may be difficult to decide which signs and symptoms are specific for an individual vitamin and which are part of a general malnutrition syndrome.

THIAMINE

Source of Thiamine and Cause of Deficiency

Thiamine cannot be synthesised by animals: it is found in most dietary components, and wheat germ, oatmeal and yeast are particularly rich in the vitamin. Adequate amounts are present in a normal diet, but the deficiency syndrome is still prevalent in rice-eating areas: polished rice has the husk removed and this is the only source of thiamine in this food. In other areas thiamine deficiency occurs most commonly in alcoholics.

Function of Thiamine

Thiamine is a component of thiamine pyrophosphate, which is an essential coenzyme for *decarboxylation of α-oxoacids* (cocarboxylase): one of these important reactions is the conversion of pyruvate to acetyl coenzyme A (acetyl CoA). In thiamine deficiency pyruvate cannot be metabolised and accumulates in the blood. Thiamine pyrophosphate is also an essential cofactor for transketolation reactions. One such reaction is that catalysed by *transketolase* in the pentose-phosphate pathway; the keto (oxo) group is transferred from xylulose-5-phosphate to ribose-5-phosphate to produce glyceraldehyde-3-phosphate and sedoheptulose-7-phosphate.

Clinical Effects of Thiamine Deficiency

Deficiency of thiamine causes the syndrome known as *beri-beri*, which includes anorexia and emaciation, neurological lesions (motor and sensory polyneuropathy, Wernicke's encephalopathy), and cardiac arrhythmias. In the so-called "wet" form of the disease there is oedema, sometimes with cardiac failure. Some of these findings may be due to associated protein deficiency rather than to that of thiamine.

Beri-beri can be aggravated by a high carbohydrate diet, possibly because this leads to an increased rate of glycolysis and therefore of pyruvate production.

Laboratory Diagnosis of Thiamine Deficiency

The most reliable test is probably the estimation of *erythrocyte transketolase activity*, with and without added thiamine pyrophosphate. Blood pyruvate levels have been used, but are too variable and non-specific to be reliable.

Treatment of Thiamine Deficiency

True beri-beri responds to 5–10 mg of thiamine daily, although occasionally higher dosages may be required. In cases where multiple deficiency is suspected a mixture of the vitamins in the "B complex" should be given.

RIBOFLAVIN

Source of Riboflavin

Riboflavin is found in large amounts in yeasts and germinating plants such as peas and beans.

Function of Riboflavin

There are about 15 flavoproteins, mostly enzymes incorporating riboflavin in the form of flavin mononucleotide (FMN) and flavin adenine dinucleotide (FAD). FMN and FAD are reversible *electron carriers* in biological oxidation systems and are, in turn, oxidised by cytochromes.

Clinical Effects of Riboflavin Deficiency

Ariboflavinosis causes a rough scaly skin, especially on the face, cheilosis (red, swollen, cracked lips), angular stomatitis and similar lesions at the mucocutaneous junctions of anus and vagina, and a swollen tender, red tongue, which is described as magenta coloured. Congestion of conjunctival blood vessels may be visible clinically on microscopic examination of the eye with a slit lamp.

Laboratory Diagnosis of Riboflavin Deficiency

Riboflavin acts as a coenzyme for *glutathione reductase*, increasing its activity. Assay of the erythrocyte enzyme has been used to detect riboflavin deficiency.

NICOTINAMIDE

Source of Nicotinamide

Nicotinamide can be formed in the body from nicotinic acid. Both substances are plentiful in animal and plant foods, although much of that in plants is bound in an unabsorbable form. Some nicotinic acid can also be synthesised in the mammalian body from tryptophan. Probably both dietary and endogenous sources are necessary to provide sufficient nicotinamide for normal metabolism.

Function of Nicotinamide

Nicotinamide is the active constituent of the important coenzymes in *oxidation-reduction reactions*, nicotinamide adenine dinucleotide (NAD), and its phosphate (NADP). Reduced NAD and NADP are, in turn, re-oxidised by flavoproteins, and the functions of riboflavin and nicotinamide are closely linked. NAD and NADP are essential, for amongst other things, glycolysis and oxidative phosphorylation.

Clinical Effects of Nicotinamide Deficiency

Nicotinamide deficiency produces a clinical syndrome often remembered by the mnemonic "three Ds"—diarrhoea, dermatitis and dementia. The dermatitis is a sunburn-like erythema, especially marked in areas exposed to the sun, and later leading to pigmentation and thickening of the dermis. The "dementia" takes the form of irritability, depression and anorexia, with loss of weight. As in riboflavin deficiency, there may be glossitis and stomatitis. Many of the symptoms of the multi-deficiency disease *pellagra* may be due to nicotinamide deficiency, and nicotinic acid has been called "pellagra preventive factor" (PP factor).

Causes of Nicotinamide Deficiency

Dietary deficiency of nicotinamide, like that of the other B vitamins, is rare in affluent communities.

Hartnup disease is due to a rare inborn error of renal, intestinal and other cellular transport mechanisms for the monoamino monocarboxylic acids, including tryptophan (p. 358). Subjects with the disease may present with a pellagra-type rash, which can be cured by nicotinamide therapy of between 40 and 200 mg daily. Probably, if the supply of tryptophan for synthesis in the body is reduced, dietary nicotinic acid is insufficient to supply the body's needs over long periods of time: under these circumstances only a slight reduction of intake may precipitate pellagra. A similar clinical picture has been reported in the *carcinoid syndrome*, when tryptophan is diverted to the synthesis of large amounts of 5-hydroxytryptamine (p. 432).

Laboratory assessment of nicotinamide deficiency is unreliable.

PYRIDOXINE

Source of Pyridoxine and Cause of Deficiency

Pyridoxine is widely distributed in food and dietary deficiency is very rare. The antituberculous drug *isoniazid* (isonicotinic hydrazide) has been reported to produce the picture of pyridoxine deficiency, probably by competition with it in metabolic pathways.

Function of Pyridoxine

Pyridoxal phosphate is a coenzyme for the *transaminases*, and for *decarboxylation of amino acids*.

Clinical Effects of Pyridoxine Deficiency

Deficiency may produce roughening of the skin. A very rare hypochromic, microcytic anaemia responds to large doses of pyridoxine when there is no evidence of deficiency of the vitamin.

Laboratory Diagnosis of Pyridoxine Deficiency

Pyridoxal phosphate is required for conversion of tryptophan to nicotinic acid. In pyridoxine deficiency this pathway is impaired. *Xanthurenic acid* is the excretion product of 3-hydroxykynurenic acid, the metabolite before the "block", and in pyridoxine deficiency is found in abnormal amounts in the urine after an oral *tryptophan load*.

The increase in *erythrocyte aspartate transaminase* activity after addition of pyridoxine may be measured. The more severe the pyridoxine deficiency the greater the increase in activity.

BIOTIN AND PANTOTHENATE

Lack of these two vitamins of the B group probably almost never produces clinical deficiency syndromes.

Biotin is present in eggs, but large amounts of raw egg white in experimental diets have caused loss of hair and dermatitis thought to be due to biotin deficiency. Probably the protein avidin, present in the egg white, combines with biotin and prevents its absorption. Biotin is a coenzyme in carboxylation reactions.

Pantothenate is a component of coenzyme A (CoA), which is essential for fat and carbohydrate metabolism. It is very widely distributed in foodstuffs.

Figure 40 summarises some of the biochemical interrelationships of

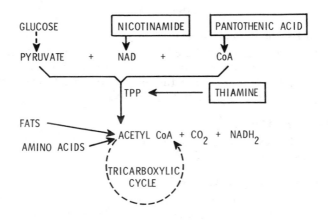

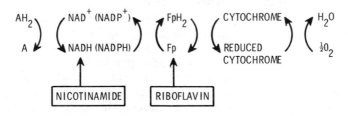

FIG. 40.—Some biochemical interrelationships of B vitamins.

the B vitamins discussed so far. Note that, as a general rule, deficiency of this group results in lesions of skin, mucous membranes and the nervous system.

FOLATE AND VITAMIN B_{12}

These two vitamins are included in the B group and are essential for the normal maturation of the erythrocyte; deficiency of either causes *megaloblastic anaemia*. Their effects are so closely interrelated that they are usually discussed together. A fuller discussion of diagnosis and treatment will be found in haematology textbooks.

Folate is present in green vegetables and some meats. It is easily destroyed in cooking and *dietary deficiency* may rarely occur. It is absorbed throughout the small intestine and in contrast to most of the other B vitamins (except B_{12}) clinical deficiency occurs relatively commonly in intestinal *malabsorption syndromes*. During *pregnancy* and lactation low blood folate levels are common and may be associated with megaloblastic anaemia. The active form of the vitamin is tetrahydrofolate, and this is essential for transfer of "one carbon" units: it is particularly important in *purine and pyrimidine* (and therefore DNA and RNA) synthesis. Competitive *folate antagonists* such as aminopterin have been used as cytotoxic drugs in the treatment of malignancy because they inhibit DNA synthesis.

The vitamin B_{12} group includes several cobalamins. Deoxyadenosyl and methyl cobalamin have, like folate, coenzyme activity in nucleic acid synthesis. Cyanocobalamin and hydroxycobalamin (the latter is the form most commonly used in therapy) are converted in the body to the coenzyme forms. All forms of vitamin B_{12} are absorbed mainly in the terminal ileum and in *malabsorption syndromes* affecting this region deficiency can occur: it also occurs in the *blind-loop syndrome* because of bacterial competition for the vitamin.

Absorption of the vitamin depends on its combination with intrinsic factor, secreted by the stomach. In true *pernicious anaemia* intrinsic factor antibodies result in malabsorption of B_{12}.

Dietary deficiency of vitamin B_{12} is rare.

Deficiency of vitamin B_{12}, like that of folate, causes megaloblastic anaemia: unlike that of folate it can result in *subacute combined degeneration* of the cord. Although the megaloblastic anaemia of B_{12} deficiency can be reversed by folate, this therapy should never be given in pernicious anaemia because it does not improve the neurological lesions and may even aggravate them.

Table XXXI summarises the synonyms of the B vitamins, their known biochemical actions, and the clinical syndromes associated with deficiencies of each.

TABLE XXXI

THE "B COMPLEX" VITAMINS

Name	Synonyms	Biochemical function	Clinical deficiency syndrome
Thiamine	Aneurine Vitamin B$_1$	Cocarboxylase (as thiamine pyrophosphate)	Beri-beri (neuropathy) Wernicke's encephalopathy
Riboflavin	Vitamin B$_2$	In flavoproteins (electron carriers as FAD and FMN)	Ariboflavinosis (affecting skin and eyes)
Nicotinamide	PP factor Niacin	In NAD and NADP (electron carriers)	Pellagra (dermatitis, diarrhoea, dementia)
Pyridoxine	Adermin Vitamin B$_6$	Coenzyme in decarboxylation and deamination (as phosphate)	? Pyridoxine responsive anaemia ? Dermatitis
Biotin		Carboxylation coenzyme	Probably unimportant clinically
Pantothenate		In coenzyme A	
Folate	Pteroyl glutamate	Metabolism of purines and pyrimidines	Megaloblastic anaemia
Vitamin B$_{12}$ group	Cobalamins	Coenzyme in synthesis of nucleic acid	Megaloblastic anaemia Subacute combined degeneration of the cord

ASCORBATE

Source of Ascorbate

Ascorbate is found in fruit and vegetables and is especially plentiful in citrus fruits. It cannot be synthesised by man and other primates, nor by guinea-pigs.

Functions of Ascorbate

Ascorbate can be reversibly oxidised in biological systems to dehydroascorbate and, although its functions in man are not well worked out, it probably acts as a hydrogen carrier. It seems to be required for normal collagen formation.

Causes of Ascorbate Deficiency

Deficiency of ascorbate causes scurvy and was commonly seen on the long sea voyages of exploration in the 16th, 17th and 18th centuries. Although Hawkins, as early as 1593, realised that oranges and lemons could cure the disease, and although in 1601 Sir James Lancaster introduced the regular use of oranges and lemons in the ships of the East India Company, a good description of scurvy can be found in Anson's account of his voyage around the world in 1740–44 (*Anson's Voyage Round the World*, edited by S. W. C. Pack in Penguin Books, Chapter 10); this demonstrates the importance of reading the literature.

Dehydroascorbate is easily oxidised further and irreversibly in the presence of oxygen, and loses its biological activity; this reaction is catalysed by heat. Scurvy was at one time fairly common in bottle-fed infants, as the ascorbate was often destroyed in the preparation of the feeds. With our knowledge of the aetiology of the disease, and with dietary supplementation, it is now rarely seen in this age group. It is most commonly seen in old people (especially men) living on their own on poor incomes, who do not eat fresh fruit and vegetables and who tend to cook in frying pans, where the combination of heat and the large area of food in contact with air irreversibly oxidises the vitamin. Ascorbate deficiency can occur in iron overload (p. 391).

Clinical Effects of Ascorbate Deficiency

Anson described "large discoloured spots", "putrid gums" and "lassitude" as characteristic of scurvy.

Many of the signs and symptoms of scurvy can be related to poor collagen formation.

1. Deficiency in *vascular walls* leads to a bleeding tendency with a positive Hess test, petechiae and ecchymoses ("large discoloured spots"), swollen, tender, spongy bleeding gums ("putrid gums") and, occasionally, haematuria and epistaxis. In infants subperiosteal bleeding and haemarthroses are extremely painful and may lead to permanent joint deformities.

2. There is *poor healing* of all wounds.

3. Deficiency of *bone matrix* causes osteoporosis and poor healing of fractures. In children bone formation ceases at the epidiaphyseal junctions, which look "frayed" radiologically.

4. There may be *anaemia*, possibly partly due to impairment of erythropoiesis. This anaemia may sometimes be cured by ascorbate alone. Bleeding aggravates the anaemia.

This florid form of the disease is rarely seen nowadays and the patient most commonly presents complaining that bruising occurs with only minor trauma.

All the signs and symptoms are dramatically cured by the administration of ascorbate.

Diagnosis of Scurvy

The laboratory diagnosis of scurvy can only be made *before* therapy has started. Once ascorbate has been given it is difficult to prove that deficiency was previously present.

The *ascorbic acid saturation test* is based on the fact that if the tissues are "saturated" with ascorbate any further vitamin administered will be lost in the urine (see Appendix).

Plasma ascorbate can be measured, but levels vary widely in normal subjects. Leucocyte ascorbate is said to give a better indication of scurvy than that of plasma or whole blood. For practical purposes the saturation test is still probably the best one in the clinical diagnosis of scurvy.

SUMMARY

1. Vitamins have important biochemical functions, most of which are now well understood. Unfortunately the relationship of these to clinical syndromes is not always obvious.

2. The fat soluble vitamins, especially "vitamin D", may be deficient in steatorrhoea. Both vitamin A and "vitamin D" are stored in the liver and deficiency takes some time to develop.

3. **Vitamin A** is necessary for the formation of visual purple and for normal mucopolysaccharide synthesis. Deficiency is associated with poor vision in dim light and with drying and metaplasia of epithelial surfaces, especially those of the conjunctiva and cornea.

4. **Vitamin D** as 1,25 dihydroxycholecalciferol is necessary for normal calcium metabolism and deficiency causes rickets in children and osteomalacia in adults.

5. Both vitamin A and vitamin D are toxic in excess.

6. **Vitamin K** is necessary for prothrombin formation and deficiency is associated with a bleeding tendency.

7. **Thiamine** deficiency causes beri-beri.

8. **Riboflavin** deficiency causes ariboflavinosis.

9. **Nicotinamide** can be manufactured from tryptophan in the body, but dietary deficiency causes a pellagra-like syndrome, which may also be seen in Hartnup disease, when tryptophan absorption is deficient, and in the carcinoid syndrome, when tryptophan is used in excess for 5-hydroxytryptamine synthesis.

10. **Pyridoxine** responsive anaemia may occur.

11. **Folate and vitamin B_{12}** deficiency produce megaloblastic anaemia, and deficiency of vitamin B_{12} can also cause subacute combined degeneration of the cord. Compared with the other B vitamins, deficiency

of these is relatively common in malabsorption syndromes, and vitamin B_{12} deficiency can be a feature of the "blind-loop syndrome". Classical pernicious anaemia is due to intrinsic factor deficiency with consequent malabsorption of vitamin B_{12}.

12. **Ascorbate** deficiency causes scurvy.

FURTHER READING

MARKS, J. (1968). *The Vitamins in Health and Disease.* London: J. & A. Churchill.
GRAY, S. P. (1971). Laboratory studies in vitaminology. *Lab. Equipm. Dig.*, **9** (No. 11), 59.

APPENDIX

ASCORBIC ACID SATURATION TEST

One gram of ascorbic acid is administered orally at 8.00 hours.

The bladder is emptied at 11.00 hours and the urine discarded. Any urine passed between 11.00 and 14.00 hours is put in a collection bottle. At 14.00 hours the bladder is again emptied and the urine added to the collection.

Because of the ease with which ascorbate can be destroyed the specimen must be sent to the laboratory immediately for estimation. If "saturation" is complete (see below) the test can be terminated; if it is not the procedure is repeated on the following day.

Interpretation

"Saturation" is said to be complete when at least 50 mg of the 1 g dose of ascorbic acid is excreted in the 3-hour urine collection.

In normal subjects "saturation" is complete on the first or second day of the test. In scurvy it takes much longer, but not usually more than a week.

Warning.—It should be noted that in malabsorption syndromes "saturation" may appear not to occur, even after several weeks. This is not diagnostic of scurvy, but is probably due to the fact that the absorption of the ascorbic acid is slower than usual and that it does not reach the kidneys in normal amounts during the 3 hours. Scurvy is very rare in such syndromes.

Chapter XXI

PREGNANCY AND ORAL CONTRACEPTIVE THERAPY

ANTENATAL TESTING

THE reasons for testing during pregnancy fall into two main groups:

1. Testing *late* in pregnancy (after about 28 weeks gestation) so that action may be taken to *improve the chance of survival* of the infant:

(a) by detecting incipient *fetal damage* caused, for example, by placental insufficiency or by severe blood group incompatibility, so that labour may be induced before such damage is severe or irreversible;

(b) by assessing *fetal maturity*, so that induction may, if possible, be delayed until the newborn infant stands a good chance of survival. Maturity of the fetal lung is of especial importance.

2. Testing *early* in pregnancy (before about 20 weeks gestation) to detect serious *congenital abnormalities* with a view to *termination* if such abnormalities are found.

Most of these tests are carried out on patients who are known to be at risk. Changes in the composition of maternal urine or plasma often reflect changes in fetal and placental metabolism, and sampling of these is safe and simple. Occasionally there are indications for testing amniotic fluid obtained by amniocentesis.

Amniocentesis.—Amniotic fluid may be sampled through a needle inserted into the uterus through the maternal abdominal wall at any time after about 14 weeks gestation. The procedure carries a small risk to the fetus, even in the best hands, and *should only be performed when there are very strong clinical indications*. Analytical results may be *dangerously misleading* if for example, the specimen is *contaminated with maternal or fetal blood or maternal urine, or is not fresh and properly preserved*. Both the safety and reliability of the procedure are improved if it is performed by someone with considerable experience, preferably with the use of ultrasound to indicate the position of the fetus, placenta and maternal bladder. The laboratory analysing the sample should also have experience and, as always, close liaison between the clinician and laboratory helps to ensure the suitability of the specimen and the speed of assay.

Amniotic fluid is probably derived from both maternal and fetal sources, but its value in reflecting fetal abnormalities arises from its intimate contact with the fetus, and from the increasing contribution of fetal urine in later pregnancy.

TESTS FOR FETO-PLACENTAL IMPAIRMENT

The production of placental metabolites increases from early pregnancy until near term, and some of them may be detected in maternal urine and plasma. Oestriol or human placental lactogen (HPL) production are relatively easy to assess. The "normal" range of both these and of similar parameters is very wide at every stage of pregnancy, and sudden changes in levels are more significant than a single "abnormal" result; for this reason serial estimations are usually performed. Even during normal pregnancy there is some day-to-day variation, and a change in the same direction in two or more consecutive specimens increases the probability that it is clinically significant. Results should always be interpreted in conjunction with clinical and other findings. Sometimes the result of one type of assay is consistently abnormal for no apparent reason: in rare cases of serious doubt both oestriol and HPL production should be serially assessed.

Assessment of Feto-Placental Oestriol Production

Soon after fertilisation the ovum is implanted in the uterine wall: secretion of chorionic gonadotrophin by the developing placenta maintains the corpus luteum of the luteal phase and the continued secretion of oestrogen and progesterone from the corpus luteum prevents the onset of menstruation. Oestriol secretion at this stage is very low and it rises only very slowly during the first weeks of pregnancy.

Chorionic gonadotrophin secretion reaches a peak at about 13 weeks of pregnancy, and then falls. At this stage the feto-placental unit takes over hormone production, and secretion of both oestrogen and progesterone rises rapidly.

At this stage the fetus and the placenta are an integrated endocrine unit and both are required for the production of oestriol. The hormone passes into the maternal circulation and ultimately into the urine.

The production of oestriol over a 24-hour period is usually assessed by measuring *total oestrogen* output in the *urine*. After about the 28th week of gestation most of this oestrogen is oestriol: before about the 28th week oestriol concentration is low and less easy to measure.

Collection of a 24-hour urine specimen delays the production of a result by a day and, as always, is potentially inaccurate. It was hoped that both these problems might be overcome by measuring *plasma oestriol* by more specific methods. Despite the theoretical advantages the concentration varies from day-to-day and the results seem to have no better predictive value than the simpler and cheaper estimation of daily oestrogen output.

Assessment of Human Placental Lactogen (HPL) Production

HPL is a peptide hormone, of unknown physiological significance, synthesised by the placenta. It is detectable by rapid radioimmunoassay methods after about the eighth week of gestation, and has been used in the assessment of threatened abortion. Levels rise until the 36th week, after which they fall slightly. The concentration is a measure of *placental* function only, unlike urinary oestriol which indicates fetal function as well. HPL certainly has no better predictive value than urinary oestriol, but may sometimes, as already discussed, be useful as a supplementary estimation.

ASSESSMENT OF THE SEVERITY OF FETO-MATERNAL BLOOD GROUP INCOMPATIBILITY

Amniotic fluid levels of *bilirubin* are used in conjunction with maternal antibody titres, to assess the effects on the fetus of rhesus (or other blood group) incompatibility. Normally amniotic fluid bilirubin levels decrease during the last half of pregnancy. The level at any stage may be correlated with the severity of haemolysis: used with tests of maturity the optimum time for termination of pregnancy, or the need for intra-uterine transfusion, may be assessed.

ASSESSMENT OF FETAL MATURITY

Assessment of fetal age from the date of the last menstrual period is often inaccurate. Fetal maturity may be assessed by estimating the level of products of fetal metabolism in *amniotic fluid*. The "mature" ranges are wide, but values for each parameter have been chosen above which the fetus is almost certainly mature: lower values do not necessarily indicate immaturity. These tests should be interpreted in conjunction with clinical and other findings.

Pulmonary Maturity

The lungs will not expand normally at birth if they are immature and the infant may then suffer from the *respiratory distress syndrome* (hyaline membrane disease); he may need intensive care in a special baby unit until he is mature. It is therefore important to have evidence of pulmonary maturity before labour is induced.

In late pregnancy (after 32 to 34 weeks gestation) the cells lining fetal alveolar walls start synthesising a surface tension lowering complex ("surfactant"), 90% of which is lipid in nature; most of this lipid is *lecithin*, which contains *palmitic acid*. Surfactant is probably washed from, or is perhaps secreted by, the alveoli into the surrounding amniotic fluid, in which both lecithin and palmitic acid concentrations increase steadily. The concentration of another lipid, sphingomyelin, remains constant, and the *lecithin-sphingomyelin* (*L/S*) *ratio rises*: measurement of this ratio is the most commonly used estimate of pulmonary maturity. Other estimates have included total lecithin and total palmitic acid concentrations. The predictive value of all these parameters is similar.

Renal Maturity

As the kidneys develop, the composition of amniotic fluid increasingly reflects the composition of the fetal urine excreted into it. Fetal urine is hypo-osmolar, but has a relatively high concentration of the nitrogenous end-products of metabolism, urea, creatinine and urate. *Creatinine* concentration is the most commonly used of these estimations; however, estimates of pulmonary maturity are more important than those indicating maturity of the renal tract.

TESTING FOR CONGENITAL ABNORMALITIES

α-Fetoprotein

α-Fetoprotein is a low molecular weight glycoprotein synthesised mainly in the fetal liver and yolk sac. Its production is almost completely repressed in the normal adult. Because of its low molecular weight it can diffuse slowly through capillary membranes, and appears in fetal urine—and hence in amniotic fluid—and in maternal plasma. Severe fetal neural tube defects (such as open spina bifida and anencephaly) are associated with abnormally high concentrations in these fluids: the reason for this is not clear, but the protein may leak from the exposed neural tube blood-vessels. Many other fetal causes of raised α-fetoprotein concentration in amniotic fluid and maternal plasma have been reported; one of these is multiple pregnancy, but almost all the others are associated with serious fetal abnormalities.

Estimation of α-fetoprotein levels in *maternal serum* may be used as a screening test for neural tube defects. It is a relatively crude diagnostic tool, with an appreciable incidence of false-positive and false-negative results and should never be acted on in isolation. The blood must be tested at *between 16 and 18 weeks gestation*. Earlier the levels are not high enough to be detectable, and later the risk to the mother of terminating pregnancy increases. Positive results should be confirmed on a fresh specimen, and maternal causes sought. It has been suggested that all pregnant women should have serum α-fetoprotein levels estimated at 16 to 18 weeks of pregnancy, and trials are now being carried out to assess the feasibility and desirability of such a screening programme.

α-Fetoprotein estimation on *amniotic fluid* is a much more precise diagnostic tool. It should be reserved for subjects known to be at risk, either because of a family history of neural tube defects, or because of the finding of a high level in maternal plasma. Amniocentesis should not usually be carried out for this purpose unless the parents are willing to consider termination if the result is positive. It should be performed by experts, with experienced laboratory back-up. It is important to remember that α-fetoprotein concentration in normal fetal blood is high at 16 to 18 weeks, and that *blood-stained amniotic fluid can yield dangerously misleading results.*

Amniotic Fluid Cell Culture

Some *congenital abnormalities* may be detected by chemical or enzymatic assay on cultures of cells in the fluid. These tests are performed only in special centres, and only on subjects with a genetic history of the condition.

METABOLIC EFFECTS OF PREGNANCY AND ORAL CONTRACEPTIVE THERAPY

The level of many plasma constituents is affected by steroid hormones: for example, the "normal" ranges of plasma urate and iron differ in males and females after puberty. It is therefore not surprising to find that pregnancy, which is associated with very high circulating levels of oestrogens and progesterone, affects plasma concentrations of many substances. Oral contraceptive therapy is said to prevent ovulation by mimicking pregnancy, as the tablets contain synthetic oestrogens and "progestogens", and many of the metabolic changes during pregnancy are also found in some women taking the "pill". Most of these changes have been attributed to the oestrogen rather than the progestogen fraction, but it may be very difficult to be sure exactly which hormone is producing the change. Many of the findings are less extreme since the introduction of the "low dose pill". The mechanism of the changes is poorly understood, but steroids are known to affect protein synthesis.

EFFECT ON CARBOHYDRATE METABOLISM

Steroids are known to affect carbohydrate metabolism. Cushing's syndrome is associated with diabetes and patients receiving large doses of corticosteroids may become diabetic. Cortisone has been given to subjects suspected of being "prediabetic" in spite of a normal glucose tolerance curve to uncover "latent diabetes".

During pregnancy some women develop a *diabetic type of glucose tolerance curve*, which usually reverts to normal after parturition. Similarly, some subjects taking oral contraceptive preparations develop impaired glucose tolerance, very rarely, if ever, severe enough to be accompanied by a raised fasting blood glucose level. There is no good evidence that such tablets cause true diabetes in normal subjects, but in those with latent diabetes the disease may become overt.

There may also be *renal glycosuria* both in pregnancy and in subjects taking oral contraceptives. The glomerular filtration rate increases by about 50 per cent during pregnancy, and glycosuria may partly be due to an increased glucose load on normal tubules.

EFFECT ON SPECIFIC CARRIER PROTEINS AND ON SUBSTANCES BOUND TO THEM

The plasma level of many specific carrier proteins is increased in pregnant subjects and in those taking oral contraceptive preparations. If this fact is not recognised an erroneous diagnosis may be made. In most cases the rise in the level of carrier protein is accompanied by a proportional increase of the substance bound to it, without any change in the unbound fraction. As the protein bound fraction is a transport form and because, in most cases, it is the free substance that is physiologically active, this rise in concentration is only of importance in interpretation of tests.

Thyroxine-Binding Globulin (TBG)

The level of TBG is increased both in pregnant subjects and in those taking some oral contraceptive preparations. This causes a rise in *plasma thyroxine* (T_4) concentration, with an increase in free binding sites. The true cause of the rise in T_4 will be recognised if the binding sites are also assessed (p. 164); estimation of T_4 only may lead to a false diagnosis of hyperthyroidism. Since the free thyroxine level is unaffected thyroid function is normal.

Cortisol-Binding Globulin (CBG)

The level of cortisol-binding globulin, and therefore of plasma cortisol, increases as a result of pregnancy and of some oral contraceptives. The free cortisol level, and therefore adrenal function, is normal in subjects on the "pill", although slightly raised in pregnancy. If these facts are not recognised a false diagnosis of Cushing's syndrome may be made.

Transferrin

The plasma concentration of the iron-binding protein, transferrin, also increases within a day or two of starting some oral contraceptive preparations. This is reflected in a rise in total iron-binding capacity (TIBC). The rise in pregnancy occurs after about the 20th week. *Plasma iron* concentrations also increase but, unlike the substances already discussed, the rise is probably independent of that of the binding protein.

Caeruloplasmin

The copper-containing protein, caeruloplasmin, is increased in concentration in subjects taking some oral contraceptives and in pregnancy. It also rises during oestrogen therapy. There is an associated rise in *plasma copper* levels, but this is unlikely to lead to misdiagnosis. The increase in this bluish protein may impart a green colour to the plasma.

Lipoproteins

The plasma level of some lipoproteins is increased in subjects taking some oral contraceptive preparations and is associated with an increased plasma triglyceride, and sometimes cholesterol, concentration. The plasma triglyceride level appears to be related to oestrogen content, while cholesterol levels are higher when progestational preparations are taken. In late pregnancy triglycerides are also increased but, compared with subjects taking oral contraceptives, there is a more marked rise in plasma cholesterol.

Other Proteins

The reduction of serum protein concentration in pregnancy has been thought to be due to dilution following fluid retention. Albumin and total protein levels have been reported to fall slightly during oral contraceptive therapy, but this change is unlikely to cause diagnostic confusion. It seems that *only binding proteins* are *increased* in these subjects.

PLASMA FOLATE AND VITAMIN B_{12} LEVELS

Plasma folate levels have been reported to be reduced during oral contraceptive therapy. The relationship to the low folate levels and anaemia of pregnancy is not certain.

Total plasma vitamin B_{12} levels are also slightly reduced during pregnancy.

LIVER FUNCTION TESTS

Some subjects develop a cholestatic jaundice during pregnancy and are prone to a similar syndrome when taking some oral contraceptive preparations. This is probably an extremely rare complication in normal people.

ALKALINE PHOSPHATASE

The placenta produces one of the alkaline phosphatase isoenzymes and the level of this rises in late pregnancy. This enzyme is stable at a temperature at which alkaline phosphatase from other sources is inactivated. Hopes that estimation of this heat-stable enzyme might be of value as a test of placental function have been disappointed, as results are too variable to be useful. Pregnancy should be remembered as a cause of high plasma alkaline phosphatase activity.

TABLE XXXII

SOME METABOLIC EFFECTS OF PREGNANCY AND ORAL CONTRACEPTIVE THERAPY
WHICH MAY CONFUSE DIAGNOSIS

Test	Effect	Comments
Hormone Levels:		
Adrenal hormones		
Plasma cortisol	Raised	Secondary to rise in binding protein Adrenal function normal
Thyroid hormones		
Plasma T_4	Raised ⎱	Secondary to rise in binding protein
Free binding sites	Increased ⎰	Thyroid function normal
Pituitary hormones		
Plasma gonadotrophin	Low normal	Feedback suppression
Plasma prolactin	Raised	
Carbohydrate metabolism		
Glucose tolerance test	"Diabetic" ⎱	Reversible abnormalities
"Renal threshold"	Reduced ⎰	
Iron metabolism		
Plasma TIBC (transferrin)	Raised on oral contraceptives and in late pregnancy	
Plasma iron	Relatively raised	
Plasma alkaline phosphatase	Raised	Due to placental isoenzyme (heat stable)

HORMONE SECRETION

The effect of oral contraceptives on cortisol-binding globulin and therefore on plasma cortisol has already been mentioned. *Gonadotrophin levels* are very low in subjects taking oral contraceptives because of feed-back suppression. These revert to normal when the tablets are stopped.

The changes which may occur during pregnancy and oral contraceptive therapy are summarised in Table XXXII.

SUMMARY

Antenatal Testing

1. Testing is carried out after the 28th week of pregnancy:
   to predict possible fetal damage if pregnancy is allowed to continue;
   to assess the maturity of the fetus before induction of labour is undertaken.

2. Fetoplacental function may be monitored by assessing the production of oestriol or of HPL. Daily maternal urinary total oestrogen output usually reflects oestriol production after about the 28th week of pregnancy. HPL concentration is measured in maternal plasma.

3. Fetal pulmonary maturity may be assessed by measuring the lecithin-sphingomyelin (L/S) ratio, or total lecithin or palmitic acid concentrations in amniotic fluid.

4. Amniotic fluid bilirubin levels are used in assessing the effect on the fetus of blood group incompatibility.

5. α-Fetoprotein levels are high in maternal serum and in amniotic fluid if the fetus has severe neural tube defects.

Metabolic Effects of Pregnancy and Oral Contraceptive Therapy

Pregnancy and oral contraceptive therapy produce similar metabolic effects which, if not recognised, may lead to misdiagnosis. The more important of these are listed in Table XXXII.

FURTHER READING

WILDE, C. E., and OAKEY, R. E. (1975). Biochemical tests for the assessment of fetoplacental function. *Ann. clin. Biochem.*, **12**, 83.

HOLTON, J. B. (1977). Diagnostic tests on amniotic fluid. In *Essays in Medical Biochemistry* (Eds. Marks, V., and Hales, C. N.) **3**, 75. The Biochemical Society and the Association of Clinical Biochemists.

Chapter XXII

BIOCHEMICAL EFFECTS OF TUMOURS

SOME rare syndromes are associated with neoplasia of cells which, although normally producing hormones, are, because of their scattered nature, not usually thought of as endocrine organs. Many malignant tumours of non-endocrine tissues can elaborate hormones usually foreign to them. Pearse and others have explained both these types of syndrome on the basis of the so-called APUD system. Although this is by no means the only possible explanation, we will start with a brief outline of the APUD concept.

THE DIFFUSE ENDOCRINE (APUD) SYSTEM

A group of cells, scattered throughout the body, has common cytological characteristics, and Pearse believes that these cells have a common origin in embryonic ectoblast. Most have endocrine or neurotransmitter properties. The acronym APUD is derived from their ability for *A*mine *P*recursor *U*ptake and *D*ecarboxylation to produce amines. Tumours of APUD cells have been called APUDomas. Some of these cells secrete *amines* which are physiologically active. In many the amine production seems to be linked to synthesis and secretion of *peptide hormones*. Some are apparently normally *non-secretory*.

Many of the peptide hormone-producing cells form recognisable *endocrine glands* (pituitary, parathyroid, calcitonin-producing cells of the thyroid). Other APUD cells, both peptide and amine-secreting, are found in *specialised nerve tissue* (the trophic hormone and ADH secreting cells of the hypothalamus, and the adrenaline and noradrenaline secreting cells of the sympathetic nervous system—including the adrenal medulla). Abnormalities of most of these tissues have been discussed elsewhere in this book; in this chapter we shall describe the secreting tumours of the sympathetic nervous system—phaeochromocytoma and neuroblastoma.

Hormone-secreting tumours of the APUD cells scattered through tissues of a non-ectodermal origin are being increasingly recognised. Many occur in the *gastro-intestinal tract and pancreas*. Insulinomas are discussed in Chapter IX. The pancreas contains many other types of peptide hormone-secreting APUD cells which are common to it and the intestinal tract and the amine, or amine-precursor, secreting carcinoid tumours occur mainly in the intestine. In this chapter we will consider briefly some of these rare tumours.

Probably all APUD cells have the potential to secrete any of the APUD hormones. A common site for apparently non-secretory cells is the bronchial tree. Many of the malignant tumours, apparently of non-endocrine tissue origin, which can sometimes elaborate hormones and other peptides normally foreign to that tissue, are bronchial carcinomas. Some of these syndromes may be due to malignancy of APUD cells in the region, rather than to that of the host tissue; this is not the only possible explanation. The second part of this chapter is devoted to a brief discussion of these interesting, and not uncommon, syndromes.

CATECHOLAMINE-SECRETING TUMOURS

The adrenal medulla and the sympathetic ganglia consist of sympathetic nervous tissue: this is derived from the embryonic neural crest and is composed of two types of cells—the *chromaffin cells* and the *nerve cells* —both of which can elaborate the active catecholamines. Adrenaline (epinephrine) is almost exclusively a product of the adrenal medulla, while most noradrenaline (norepinephrine) is formed at sympathetic nerve endings.

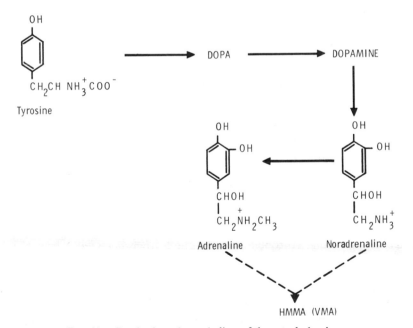

FIG. 41.—Synthesis and metabolism of the catecholamines.

METABOLISM OF THE CATECHOLAMINES

The amines adrenaline and noradrenaline are formed from tyrosine (the amine precursor) via dihydroxyphenylalanine (DOPA), and dihydroxyphenylethylamine (DOPamine). DOPA, DOPamine, adrenaline and noradrenaline are all catecholamines (dihydroxylated phenolic amines). The most important breakdown product of adrenaline and noradrenaline is 4-hydroxy 3-methoxymandelic acid (HMMA; vanillylmandelic acid, VMA) (Fig. 41): this is excreted in the urine.

ACTION OF THE CATECHOLAMINES

Both adrenaline and noradrenaline act on the cardiovascular system. Noradrenaline produces generalised vasoconstriction, and therefore hypertension and pallor, while adrenaline may cause dilatation of muscular vessels, with variable effects on blood pressure.

Adrenaline increases the rate of glycogenolysis (p. 179) and this, together with its other anti-insulin effects, may cause hyperglycaemia.

For a more detailed discussion of the action of these two hormones the student is referred to textbooks of pharmacology.

CATECHOLAMINE-SECRETING TUMOURS

Tumours of the sympathetic nervous system, whether adrenal or extra-adrenal, can produce catecholamines. Such tumours are of two main types:

tumours of chromaffin tissue. *Phaeochromocytoma.*
  in adrenal medulla 90 per cent;
  extra-adrenal 10 per cent.
tumours of nerve cells. *Neuroblastoma.*
  in adrenal medulla 40 per cent;
  extra-adrenal 60 per cent.

Although both these types of neoplasm secrete catecholamines their age incidences and the clinical pictures are quite different.

Phaeochromocytoma occurs mainly in *adult* life and is usually *benign*. The symptoms and signs can be related to excessive secretion of catecholamines and include *paroxysmal hypertension* accompanied by *anxiety, sweating,* a throbbing *headache* and either *facial pallor* or *flushing* during the attack. The attack may also be accompanied by hyperglycaemia and glycosuria. Occasionally hypertension may be persistent. This tumour accounts for only about half a per cent of all cases of hypertension, but if a young adult presents with high blood pressure for no obvious reason its presence should be sought, as the condition may be cured by surgery. Table XXXIII summarises some relatively rare causes

TABLE XXXIII

SOME METABOLIC CAUSES OF HYPERTENSION

Investigations

Renal disease	Plasma urea or creatinine.
	Examination of urine for casts and protein.
	Clearance tests if necessary (p. 20).
Primary aldosteronism	Plasma potassium and T_{CO_2} (p. 54).
Phaeochromocytoma	Urinary HMMA (VMA).

of hypertension which should be excluded in such patients before "essential" hypertension is diagnosed, and in which biochemical investigations aid the diagnosis.

Neuroblastoma is a very *malignant* tumour of sympathetic nervous tissue occurring in *children*. Secretion of catecholamines in these cases is often as high as, or higher than, that of a phaeochromocytoma, but the clinical syndrome described above is rare: the reason for this is not clear, but may be due to release of hormones into the blood stream in an inactive form. Some of the tumours secrete DOPamine in excess.

DIAGNOSIS OF CATECHOLAMINE-SECRETING TUMOURS

Because HMMA (VMA) is the major catabolic product of catecholamines, chemical diagnosis of the tumours can usually be made by estimating its excretion in a 24-hour specimen of urine. An excretion of more than twice "normal" is diagnostic, provided that all vanilla products, bananas and coffee are excluded from the diet for 24 hours before the test. Slightly raised excretion may be found in cases of essential hypertension.

Rarely HMMA excretion is normal, but that of total catecholamines is increased. Estimation of the latter should be reserved for cases clinically highly suggestive of phaeochromocytoma, but with a normal excretion of HMMA.

If hypertension is paroxysmal, urine should be collected during and immediately after the attack: this may be the only time at which increased excretion can be demonstrated.

THE CARCINOID SYNDROME

NORMAL METABOLISM OF 5-HYDROXYTRYPTAMINE

Small numbers of APUD *argentaffin cells* (that is, cells which will reduce, and therefore stain with, silver salts) are normally found in tissues derived from embryonic gut. The commonest sites for these cells are the ileum and appendix, but some are found in the pancreas, stomach and rectum. They synthesise the biologically active amine, *5-hydroxytryptamine* (5-HT: serotonin), from the amine precursor tryptophan, the intermediate product being *5-hydroxytryptophan* (5-HTP). 5-HT is inactivated by deamination, and oxidation by mono-amine oxidases to *5-hydroxyindole acetic acid* (5-HIAA) (Fig. 42): the latter enzymes, as well as the aromatic amino acid decarboxylase, are found in other tissues as well as in argentaffin cells. 5-HIAA is normally the main urinary excretion product of argentaffin cells.

Argentaffin cells may also secrete a peptide, *Substance P*, an excess of which causes flushing, tachycardia, increased bowel motility and hypotension.

CAUSES OF THE CARCINOID SYNDROME

The most usual sites for tumours of argentaffin cells, which may be benign or malignant, are the ileum or appendix: neoplasms at the latter site rarely metastasise and may be histochemically different from the other small intestinal argentaffin tumours. Less commonly the neoplasm is bronchial, pancreatic or gastric in origin, any other site being extremely rare.

The carcinoid syndrome is usually associated with an excess of circulating 5-HT. Ileal and appendiceal tumours do not produce the clinical syndrome until they have metastasised, usually to the liver, but primary tumours at other sites do cause symptoms. It is thought that at least some of the secretions of intestinal neoplasms are inactivated in the liver, whereas at other sites they are released in an active form into the systemic circulation.

CLINICAL PICTURE OF THE CARCINOID SYNDROME

The clinical syndrome includes *flushing, diarrhoea* and *bronchospasm*. There may be *fibrotic lesions of the heart*, typically *right-sided*, except in the case of bronchial carcinoid. The diarrhoea may be so severe as to cause a malabsorption syndrome. The signs and symptoms are more likely to be due to Substance P than to 5-HT. Histamine has also been suggested as a cause for the flushing, especially in tumours elaborating a preponderance of 5-HTP. A *pellagra*-type syndrome can rarely develop because of diversion of tryptophan from nicotinamide to 5-HT syn-thesis (Fig. 42).

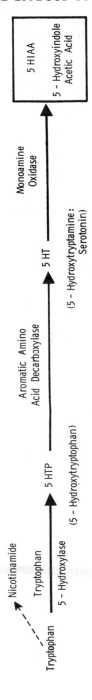

FIG. 42.—Metabolism of tryptophan.

DIAGNOSIS OF THE CARCINOID SYNDROME

In the carcinoid syndrome urinary 5-HIAA secretion, estimated on a 24-hour specimen of urine, is usually greatly increased. An excretion of more than 130 μmol (25 mg) in 24 hours is diagnostic, provided that walnuts and bananas are excluded from the diet for 24 hours before the collection is made.

In very rare cases, usually of bronchial or gastric tumours, the cells lack aromatic amino acid decarboxylase. Urinary 5-HIAA may not be obviously increased, and there is increased secretion of 5-HTP. If there is a strong clinical suspicion of the carcinoid syndrome and the urinary 5-HIAA excretion is normal, it may be more useful to estimate total 5-hydroxyindole excretion (which includes 5-HTP, 5-HT and 5-HIAA). This is very rarely necessary in practice.

PEPTIDE-SECRETING TUMOURS OF THE ENTEROPANCREATIC SYSTEM

All these tumours are rare. They usually form in the pancreatic islets. We have made no attempt to be comprehensive.

Gastrinomas may cause the *Zollinger-Ellison syndrome*. G cells, usually in the pancreatic islets (most commonly in the form of tumours), produce large amounts of gastrin: of these tumours about 60 per cent are malignant; of the remaining 40 per cent two-thirds are multiple and only one-third (13 per cent of the total) are single, resectable adenomas. More rarely the syndrome is due to hyperplasia of G cells in the gastric antrum. Acid secretion by the stomach in this condition is very high. The consequent ulceration of the stomach and upper small intestine may cause severe diarrhoea: the low pH, by inhibiting lipase activity, may cause steatorrhoea. Associated benign adenomas may be found in other endocrine glands, such as the parathyroid, pituitary, thyroid and adrenal (multiple endocrine adenopathy): these are only occasionally functional.

Radioimmunoassay of plasma gastrin is available in some centres. In the Zollinger-Ellison syndrome fasting concentrations are from 5 to 30 times the upper limit of normal. Gastrin levels may be high in a variety of conditions but the simultaneous findings of a very low gastric juice pH and a very high plasma gastrin level indicates autonomous hormone secretion not under feedback control. (Gastrin secretion is normally inhibited by gastric acidity, p. 277).

Glucagonomas usually arise from the glucagon-producing A cells of the pancreatic islets. The presenting feature is a bullous rash known as *necrolytic migratory erythema*. This is often accompanied by psychiatric disturbance, thrombo-embolism, glossitis, weight loss, impaired glucose tolerance, anaemia, and a raised ESR. The relationship of the rash to high glucagon levels is not clear. Diagnosis depends on the finding of very high plasma glucagon levels; this estimation is available in special centres. Resection of the tumour cures the syndrome. Although reports of glucagonomas are few, it has been suggested that, because of the rather non-specific clinical picture, some cases may not have been diagnosed.

VIPomas are tumours, usually of pancreatic islet cells, producing large amounts of vasoactive intestinal peptide (VIP), a hormone which increases intestinal motility. They are *extremely rare*, and are associated with very profuse and watery diarrhoea. This syndrome has been known as the *Verner-Morrison* or *WDHA syndrome* (WDHA stands for *W*atery *D*iarrhoea, *H*ypokalaemia, and *A*chlorhydria). It is not certain whether all cases of the Verner-Morrison syndrome are due to VIPomas.

MULTIPLE ENDOCRINE ADENOPATHY (MEA)

In the rare syndromes of MEA (pluriglandular syndromes), two or more endocrine glands secrete inappropriate amounts of hormone, usually from adenomas. There are two main groups of syndromes.

MEA I may involve two or more of the following (in order of decreasing incidence):

parathyroid gland (hyperplasia or adenoma)
pancreatic islet cells
    gastrinomas
    insulinomas
anterior pituitary gland
adrenal cortex
thyroid

MEA II includes:

medullary carcinoma of the thyroid
phaeochromocytoma
parathyroid adenoma or carcinoma

Other combinations are very rare. The reason for the grouping is not clear, but may be associated with lines of development of the APUD system.

HORMONAL EFFECTS OF TUMOURS OF NON-ENDOCRINE TISSUE

This section will best be understood after the student has read the chapters concerned with the hormones discussed.

Hormone secretion at sites other than that of the tissue normally producing it is known as *"ectopic"*; *inappropriate* secretion (p. 52) may or may not be ectopic in origin (see ADH), but ectopic secretion is always inappropriate since it is not under normal feedback control. Many tumours of non-endocrine tissue may secrete hormonal substances very similar to, and probably identical with, the natural hormone: some such tumours have been shown to secrete two or more hormones.

Mechanism of Ectopic Hormone Production

It is not yet certain why ectopic hormone secretion should occur, but although many theories have been propounded we feel that the one below is by far the most likely. For further discussion the interested student should consult the references at the end of the chapter.

DNA, through RNA formation, codes for peptide and protein synthesis, the sequence of bases in the DNA molecule determining the structure of the peptide synthesised. All cells in the body are derived from one cell, the fertilised ovum. The DNA complement of this cell is determined by that of the unfertilised ovum and that of the sperm, and, during the subsequent repeated cell division, the DNA is replicated so that every daughter cell has a genetic complement identical with that of the fertilised ovum. Unless mutation were to occur during differentiation of tissues (and there is evidence that it does not), every cell in the body must have the potential to produce any peptide coded for in the fertilised ovum.

It is thought that during functional cell differentiation (which, of course, proceeds at the same time as histological changes) various parts of the DNA molecule are consecutively "repressed" (stopped from functioning), and "derepressed": for instance in the early stages of blastula formation no RNA is produced and DNA replication is dominant, while at other times different proteins are manufactured. In the fully differentiated cell much of the genetic information is probably repressed and only the peptides essential for cell metabolism and those concerned with the special function of the organ are made. If the cell undergoes neoplastic change, the altered histological picture reflects altered chemical function probably resulting from the changed pattern of repression of the DNA molecule. It is now easy to see how a cell could revert to synthesising a peptide or protein which is foreign to it in its fully differentiated state.

If this were the correct explanation, the ectopic hormone should be chemically identical with that normally produced by endocrine tissue. Hormonal substances have been extracted from tumour tissue. In many cases the complete structure has been worked out, and evidence based on biological activity and chemical, physical and immunological properties also strongly suggests that many of these are identical with the naturally occurring hormones. As might also be expected if the above theory were correct, tumours have been described secreting two or more peptide hormones.

Another possible explanation is that some of these syndromes are due to malignancy of the APUD cells normally found in the tissue. In normal numbers these cells might secrete undetectable amounts of hormone, which, if the cells multiply, reach concentrations high enough to produce clinical effects; alternatively, derepression of normally non-secretory APUD cells might lead to hormone production.

It seems to us that derepression of tissue, rather than APUD, cells is the most likely explanation for this interesting set of syndromes. However, this derepression cannot be non-specific because some hormones are secreted much more commonly by one type of tumour than another. This may reflect a differing chemical background in different parts of the body.

The student should remember that neither theory is yet proven. We hope we may have interested him in the subject and that he will be stimulated to read further.

HORMONAL SYNDROMES

Many of these syndromes have been discussed in the relevant chapters. The following may be due to hormone secretion by the tumour:

hypercalcaemia (parathyroid hormone)
hyponatraemia (antidiuretic hormone)
hypokalaemia (adrenocorticotrophic hormone)
polycythaemia (erythropoietin)
hypoglycaemia ("insulin-like substance")
gynaecomastia (gonadotrophin)
hyperthyroidism (thyrotrophin)
hypertrophic pulmonary osteoarthropathy (? growth hormone)
carcinoid syndrome (5-HT and 5-HTP)

It should be noted that all the hormones mentioned above are peptides except 5-HT and 5-HTP: the syndrome associated with the latter may be the result of overproduction of a single enzyme (see below).

None of these hormones produced at ectopic sites is under normal feedback control. Secretion therefore continues under conditions in which it should be absent, and is inappropriate.

While absolute proof of hormone secretion depends on its isolation from the tumour tissue, in many cases strong presumptive evidence enables the clinician to diagnose the syndrome with a high degree of confidence.

Hypercalcaemia Due to Parathyroid Hormone (PTH) Secretion ("Pseudohyperparathyroidism")

Tumour types and incidence.—In our experience this is probably the most common of these syndromes. Parathyroid hormone (PTH), or a peptide very similar to it, has been extracted from a wide variety of tumour types, and presumptive evidence for its secretion is available in a great many cases.

Biochemical syndrome.—As discussed in Chapter XI, inappropriate PTH secretion, whether ectopic or parathyroid in origin, is associated with high plasma calcium and low plasma phosphate concentrations, unless accompanied by renal glomerular failure. If both these findings are present, and especially if there is no obvious evidence of bony metastases, it is probable that the hypercalcaemia is due to ectopic PTH secretion.

The hypercalcaemia of some cases of malignancy is said to be due to lysis of bone by secondary deposits, with release of calcium and phosphate into the blood stream. Prostaglandin or osteolytic sterol secretion by some tumours has been suggested as a cause of hypercalcaemia. As discussed on p. 244 we feel that these are very rare causes and that most cases are due to ectopic PTH secretion. The student should keep an open mind and study the cases he meets, noticing particularly the level of phosphate *in relation to that of plasma urea* (p. 244).

The hypercalcaemia of malignancy, unlike that of primary hyperparathyroidism, may often be suppressed by large doses of corticosteroids (p. 248).

With the development of improved therapy of malignant disease it is now important to control potentially lethal hypercalcaemia in these patients, who should have plasma calcium estimated at frequent intervals. The treatment is discussed on p. 252.

Hypokalaemia is a common accompaniment of hypercalcaemia (p. 258); it should always be sought. Its presence does not necessarily, or even usually, indicate simultaneous production of ACTH.

Hyponatraemia Due to Inappropriate Antidiuretic Hormone (ADH) Secretion

Tumour types and incidence.—Ectopic ADH production is most commonly associated with the relatively rare oat-cell carcinoma of the bronchus, although the syndrome has been reported in a wide variety of tumours. Mild hyponatraemia may be overlooked in a severely ill patient, in whom such a finding is common; it is our experience that if further evidence is sought the syndrome is found to be relatively common.

Inappropriate ADH secretion can occur with various non-malignant syndromes, (including hypothyroidism) but in many of these cases the hormone is thought to originate in the hypothalamic-posterior pituitary region and therefore not to be ectopic in origin. This may also be true of the inappropriate ADH secretion associated with non-bronchial tumours.

Biochemical and clinical syndrome.—ADH causes water retention. Water retention without sodium retention causes hyponatraemia, and therefore a reduction in plasma osmolality (p. 52), which normally cuts off ADH production by feedback control. If the low plasma osmolality is to be corrected the urinary osmolality must be even lower, so that the excess of water is excreted: this is the "appropriate" response. If, however, ADH is not under normal feedback control, water retention continues in spite of low plasma osmolality, and the urine passed is relatively concentrated. Thus the essential points in the diagnosis of inappropriate ADH secretion are:

a low plasma osmolality;

in spite of this low plasma osmolality, the urinary osmolality is relatively high.

It should be noted that both findings are essential. A low plasma osmolality occurs, for instance, in Addison's disease, but the urine is dilute. A high urinary osmolality is present in dehydration but the plasma osmolality is also high.

Continued expansion of the plasma volume reduces aldosterone secretion (p. 52), and the urine is of relatively high sodium concentration (note that plasma osmolality can be attributed almost entirely to sodium salts: urinary osmolality depends on a great many other substances). Retention of fluid dilutes other plasma constituents and these patients tend to have low plasma urea and protein concentrations.

The fall of plasma sodium concentration is usually gradual and allows time for equilibrium between cells and extracellular fluid: symptoms are therefore slight.

If the apparatus for measuring osmolality (an osmometer) is not available a presumptive diagnosis can be made if the following findings are present:

a *well-hydrated, normotensive*, relatively fit patient in spite of a low plasma sodium concentration;

a *normal or low plasma urea* concentration;

a *urine of relatively high sodium* concentration and a *relatively high specific gravity and urea* concentration.

In the hyponatraemia of Addison's disease the patient is usually dehydrated, hypotensive and severely ill by the time there is a marked fall in plasma sodium concentration. The plasma urea concentration is usually raised. If the patient has been overloaded with fluid of low sodium concentration in the presence of a normal feedback mechanism the urine, though containing sodium, will be of low specific gravity and urea concentration.

Treatment.—In most cases, which are relatively asymptomatic, restriction of fluids is adequate. However, if the sodium concentration has fallen rapidly, and if the symptoms of cerebral oedema are present, intravenous mannitol may relieve the latter: the osmotic diuresis causes predominant water loss by the kidneys (p. 9), and helps to increase plasma sodium concentration.

Hypokalaemia Due to Adrenocorticotrophic Hormone (ACTH) Secretion

Tumour types and incidence.—Inappropriate ACTH secretion is probably rarer than the two syndromes already discussed. It occurs most commonly with the relatively rare oat-cell carcinoma of the bronchus, but has been described less frequently with a variety of other tumours, especially those of thymic and pancreatic origin.

Clinical and biochemical syndrome.—The ACTH stimulates secretion of all adrenal hormones except aldosterone (p. 134). In spite of this the picture of ectopic ACTH secretion is more like that of primary aldosteronism than of Cushing's syndrome: the patient usually presents with hypokalaemic alkalosis and relatively rarely shows the Cushingoid clinical picture. The reason for this is not clear.

Treatment.—Potassium replacement is only effective if accompanied by administration of aldosterone antagonists or triamterene (p. 72).

Polycythaemia Due to Erythropoietin Secretion

The association of polycythaemia and renal carcinoma (hypernephroma) has been recognised for some years. It is now thought that the erythroid hyperplasia is often due to excessive production of an erythropoietin-like molecule by the tumour; erythropoietin stimulates marrow erythropoiesis. As this hormone is a normal product of the kidney this is not an example of ectopic hormone secretion: however, the syndrome has been reported rarely in association with other tumours, especially hepatocellular carcinoma (primary hepatoma).

Hypoglycaemia Due to an "Insulin-like" Hormone

Severe hypoglycaemia has been reported in association with many tumours: these are usually large neoplasms of retroperitoneal mesenchymal tissues resembling fibrosarcomata, or hepatocellular carcinoma. Various theories have been suggested for the aetiology of the hypoglycaemia, but there is evidence that at least some of the neoplasms produce a substance with insulin-like biological properties. This is said usually to be immunologically distinct from insulin, although immunoreactive insulin has sometimes been demonstrated. In our experience the syndrome is very rare.

Gynaecomastia Due to Gonadotrophin Production

Various types of tumour, including carcinoma of the bronchus, breast and liver, have been reported to cause gynaecomastia and be associated with high circulating levels of chorionic gonadotrophin or luteinising hormone; follicle-stimulating hormone activity is rare. In children with hepatoblastoma there may be precocious puberty. Although this syndrome is rarely recognised it is possible that mild degrees of gynaecomastia are overlooked.

Hyperthyroidism Due to "Thyrotrophin" Production

Some tumours of trophoblastic cells (choriocarcinoma, hydatidiform mole and one case with a testicular teratoma) have been shown to elaborate a thyrotrophin-like substance. In spite of markedly raised plasma levels of hormone, clinical signs of thyrotoxicosis are rare: tachycardia is the commonest finding. Laboratory findings are those of hyperthyroidism.

Growth Hormone (GH) Secretion

GH has been demonstrated in cells cultured from bronchial carcinoma. Although some of the patients have hypertrophic pulmonary osteoarthropathy, this clinical syndrome is not invariable: its relationship to the presence of GH is not proven.

Carcinoid Syndrome Due to 5-Hydroxytryptamine (5-HT) and 5-Hydroxytryptophan (5-HTP) Production

All the syndromes so far discussed have been due to peptide hormones and, as we have seen, derepression of DNA could lead to their production. In the carcinoid syndrome, a rare accompaniment of oat-cell tumours of the bronchus, there is usually excessive urinary excretion of 5-HTP and 5-HT, out of proportion to that of 5-HIAA. How can we explain this overproduction of a non-peptide hormone? One possible explanation is that the tumour is one of APUD rather than true bronchial cells.

Another, different, theory is speculative and is meant only as a working hypothesis. The student should study Fig. 42, p. 433: the production of 5-HTP from tryptophan requires only one enzyme, tryptophan-5-hydroxylase, while the decarboxylase and monoamine oxidase are normally found in non-argentaffin tissue. If derepression resulted in excessive production of this one enzyme there would be an excess of 5-HTP. 5-HT and 5-HIAA could be produced in other tissues in relatively smaller amounts than are usual in the carcinoid syndrome. It would, of course, be more difficult to explain overproduction of a substance synthesised by a pathway requiring several enzymes.

NON-HORMONAL PEPTIDES AS INDICATORS OF MALIGNANCY

The production of non-hormonal peptides by derepression of DNA will be less clinically obvious than the production of hormones. Recently such circulating peptides have been identified by immunological techniques. Two such proteins are discussed briefly here: both are normally present in fetal life, but production appears to be largely repressed in normal adults.

α-**Fetoprotein** may be found in the serum of many patients with *hepatocellular carcinoma* (primary hepatoma) and *teratoma* (p. 314). False positives may occur in hepatitis.

Carcinoembryonic antigen (CEA) appears to be produced by *malignant tumours, especially of the gastro-intestinal tract*. As it may also be detected in non-malignant disease of the gastro-intestinal tract, its diagnostic value remains to be established. Serial CEA estimations have been used to monitor the progress of known gastro-intestinal tumours after treatment.

SUMMARY

Catecholamine-Secreting Tumours

1. Tumours of sympathetic nervous tissue are associated with increased urinary excretion of the catecholamines, adrenaline and noradrenaline, and their breakdown product, hydroxymethoxymandelic acid (HMMA: VMA).

2. Phaeochromocytoma is a rare tumour, usually of adult life, occurring most commonly in the adrenal medulla. It is associated with hypertension and other symptoms of increased catecholamine secretion.

3. Neuroblastoma is a tumour of childhood, occurring in the adrenal medulla or in extra-adrenal sympathetic nervous tissue. Catecholamine secretion is increased, but symptoms are rarely referable to this.

The Carcinoid Syndrome

1. Argentaffin cells manufacture 5-hydroxytryptamine (5-HT) which is converted to 5-hydroxyindole acetic acid (5-HIAA) and excreted in the urine.

2. Tumours of argentaffin tissue are usually found in the intestine, and these do not produce typical symptoms of the carcinoid syndrome until they have metastasised.

3. The carcinoid syndrome is usually associated with an increased secretion of 5-HIAA in the urine.

Peptide-Secreting Tumours of the Enteropancreatic System

1. Gastrinomas cause the Zollinger-Ellison syndrome. Gastric hyperacidity causes peptic ulceration, diarrhoea and sometimes steatorrhoea.

2. Glucagonomas are associated with necrolytic migratory erythema and non-specific systemic signs and symptoms.

3. VIPomas (tumours secreting vasoactive intestinal peptide) may cause the Verner-Morrison syndrome—very severe watery diarrhoea which often causes hypokalaemia (WDHA).

All these tumours are very rare.

Multiple Endocrine Adenopathy

In these syndromes two or more endocrine glands secrete excessive amounts of hormones.

Hormonal Effects of Tumours of Non-Endocrine Tissue

1. Many malignant tumours produce hormonal substances normally foreign to them, which may be identical with the hormones produced in endocrine glands.

2. The commonest of these is parathyroid hormone. Hypercalcaemia should be sought and treated in cases in which the malignancy is thought to be responsive to therapy.

Non-Hormonal Peptides as Indicators of Malignancy

Some non-hormonal peptides may be detected by immunological techniques and have been used to diagnose malignancy of specific tissues.

FURTHER READING

The APUD System and Gut Hormones

PEARSE, A. G. E. (1974). The APUD Cell Concept and its Implications in Pathology. *Pathology Annual* (Ed. Sommers, S. C.) **27**. New York: Appleton Century Crofts.

PEARSE, A. G. E., POLAK, J. M., and BLOOM, S. R. (1977). The newer gut hormones, *Gastroenterology*, **72**, 746.

BLOOM, S. R. (Ed.) (1978). Articles in *Gut Hormones*. Edinburgh: Churchill-Livingstone.

Phaeochromocytoma

WOLF, R. L. (1974). Phaeochromocytoma. *Clinics in Endocr. & Metab.*, **3**, 609.

The Carcinoid Syndrome

GRAHAME-SMITH, D. G. (1968). The carcinoid syndrome. *Hosp. Med.*, **2**, 558.

Ectopic Hormone Production

ANDERSON, G. (1973). Paramalignant Syndromes. In: *Recent Advances in Medicine*, 16th edit. Chap. 1, p. 1. Eds. Baron, D. N., Compston, N., and Dawson, A. M. London: J. & A. Churchill.

RATCLIFFE, J. G., and REES, L. H. (1974). Clinical manifestations of ectopic hormone production. *Brit. J. hosp. Med.*, **11**, 685.

Lancet (1967). Editorial: Hormones and Histones? **1**, 86.

Tumour Markers

BAGSHAWE, K. D., and SEARLE, F. (1977). Tumour Markers. In: *Essays in Medical Biochemistry* (Eds. Marks, V., and Hales, C. N.) **3**, 75. The Biochemical Society and the Association of Clinical Biochemists.

Chapter XXIII

THE CEREBROSPINAL FLUID

CEREBROSPINAL fluid (CSF) is formed from plasma by the filtering and secretory activity of the choroid plexus, and is reabsorbed into the blood stream by the arachnoid villi. The mechanism of its production is not fully understood, but it seems predominantly to be an ultrafiltrate of plasma, since it contains very little protein. Some active secretion of, for example, chloride may occur.

Circulation of CSF is very slow, allowing long contact with cerebral cells: their uptake of glucose may account for its relatively low concentration in the CSF.

Concentrations in the CSF should always be compared with those of plasma, as alterations in the latter are reflected in the CSF even when cerebral metabolism is normal.

EXAMINATION OF THE CSF

Biochemical investigation of the CSF is usually of relatively little value compared with simple inspection and bacteriological and cytological examination of the fluid. Textbooks of microbiology should be consulted for further details.

TAKING THE SAMPLE

CSF should be taken into sterile containers and sent for *bacteriological examination first*. Any remaining specimen can be used for relevant chemical investigations, but if the tests are performed in the opposite order bacteriological contamination may occur. If possible, a few millilitres of CSF should be taken into two or three separate containers (see below—Appearance).

Specimens for glucose estimation, like those for blood glucose, should be taken into a tube containing fluoride to minimise glycolysis by any cells present.

APPEARANCE

Normal CSF is completely clear and colourless and should be compared with water: slight turbidity is most easily detected by this method.

Colour

Bright red blood may be due to:
a recent haemorrhage involving the subarachnoid space;
damage to a blood vessel during puncture.

If CSF is collected in three separate aliquots, all three will be equally blood-stained in the first case but progressively less so in the second.

Xanthochromia (yellow coloration).—This may be due to the presence of:

altered haemoglobin several days after a cerebral haemorrhage;

large amounts of *pus*. The cause will be obvious from the gross turbidity of the fluid, and from the presence of pus cells on microscopy;

cerebral tumours near the surface of the brain or spinal cord, impairing circulation of the CSF. Specimens from these cases have a very high protein content and tend to clot spontaneously after withdrawal, due to the presence of fibrinogen;

jaundice due to a rise in unconjugated bilirubin, which may impart a yellow colour to the CSF.

Turbidity

This is due to an *excess of white cells* (*pus*). Slight turbidity will, of course, occur after haemorrhage, but the cause of this will be apparent from the colour of the specimen.

Spontaneous Clotting

This occurs when there is an excess of fibrinogen in the specimen, usually associated with a high total protein content.

The following are the most frequently requested chemical estimations.

PROTEIN CONTENT

Normal CSF is mainly plasma ultrafiltrate and contains very little protein. The normal figures are usually quoted as 0·2 to 0·4 g/l. Quantitative estimation at this level is relatively inaccurate and little significance can be attached to concentrations up to 0·6 g/l.

Total Protein

CSF will have a high protein content under the following conditions:

in the presence of blood (due to haemoglobin and plasma proteins);

in the presence of pus (due to cell protein and exudation from inflamed surfaces).

These two causes will be obvious from inspection and microscopic examination of the specimen, and nothing further is to be gained by estimating protein.

in non-purulent inflammation of the cerebral tissues, when there is a moderate rise of protein concentration. It is when such conditions are suspected that the estimation is most useful:

tuberculous meningitis;
syphilitic meningitis;
multiple sclerosis;
encephalitis;
polyneuritis.

in blockage of the spinal canal, when stasis results in fluid re-absorption. There is xanthochromia and protein concentrations are very high (usually 5 g/l or more).

Tests for Abnormal Ratios of Proteins in the CSF

These tests are not useful in the presence of blood or pus. They may, however, detect abnormalities when the total protein is *only equivocally increased or normal.*

In the inflammatory conditions mentioned above CSF immuno-globulin, especially IgG, concentration is increased. Various methods have been used to detect this increase. These tests are most valuable if multiple sclerosis is suspected—in this condition they are positive in about 50–60 per cent of cases.

Qualitative tests.—Excess of immunoglobulin may be detected by qualitative tests of limited value.

Pandy reaction.—The abnormal CSF becomes turbid when added to phenol.

Nonne-Appelt reaction.—The abnormal CSF becomes turbid when added to ammonium sulphate.

Lange colloidal gold reaction.—Gold sols are precipitated when the ratio of protein fractions is abnormal. The CSF is serially diluted but no reliance should be placed on Lange "curves", in which precipitation at different dilutions was said to indicate specific clinical conditions.

Zinc sulphate precipitates γ-globulins. The resultant turbidity can be compared with that of standards.

Electrophoresis.—Electrophoresis may be carried out on CSF. Interpretation requires experience, and, because an abnormal plasma pattern may be reflected in the CSF, *should not be attempted without comparison with the patient's serum.*

The electrophoretic pattern of normal ventricular CSF is mainly that of a plasma ultrafiltrate. When it is compared with normal lumbar fluid there is evidence that both the concentration and type of some proteins have been modified during passage down the spinal canal. In diseases such as multiple sclerosis there is sometimes an increase in the level of

γ-globulin; this increased fraction is often composed of discrete bands of IgG *which are not present in the serum.*

Immunological techniques.—The level of CSF immunoglobulins, especially of *IgG*, may be quantitated. The diagnostic precision of this estimation is improved if the ratio of the concentrations of CSF IgG: serum IgG is expressed as a percentage of the ratio of the concentrations of CSF albumin:serum albumin (a *"quotient"*). Albumin is a smaller molecule than IgG. If raised CSF IgG levels were due to the increased vascular permeability of non-specific inflammation the quotient would be low or normal; if permeability were only slightly increased more albumin than IgG would diffuse from the plasma into the CSF and the quotient would be low, but more commonly the permeability is such that albumin and IgG diffuse at almost the same rate and the quotient is normal. However, in conditions such as *multiple sclerosis,* the *quotient is high* because CSF IgG has been synthesised locally.

GLUCOSE CONTENT

Provided that CSF for glucose estimation has been mixed with fluoride a low glucose concentration occurs in:
1. infection;
2. hypoglycaemia.

1. CSF glucose is normally metabolised only by cerebral cells. If many leucocytes and bacteria are present these also utilise glucose and abnormally low levels are obtained. If obvious pus is present the estimation of CSF glucose adds nothing to diagnostic precision. It is most useful when the CSF is clear and *tuberculous meningitis* is suspected, although levels are not as low in this condition as in pyogenic meningitis.

2. CSF glucose concentration parallels that of blood, although there is a lag before changes in blood glucose are reflected in the CSF. In the presence of hypoglycaemia (which may cause coma) CSF glucose levels may be low although there is no primary cerebral abnormality. *Both blood and CSF concentrations should be measured.*

In hyperglycaemia CSF glucose levels will be high.

CHLORIDE CONTENT

The chloride concentration in the CSF is higher than that in plasma by about 20 mmol/l. This is mostly accounted for by the Donnan effect, the additional difference probably being due to secretion of the ion by the choroid plexus.

Changes in *CSF chloride concentration parallel those in plasma.* The level has been said to be lowered specifically in tuberculous meningitis, but there is evidence that this is merely a reflection of plasma chloride levels: tuberculous meningitis is of insidious onset and the patient often presents after some weeks of vomiting with chloride loss. *CSF chloride estimation has no value as a diagnostic aid.*

The recommended procedure for examination of CSF is outlined on the next page.

SUMMARY

1. Biochemical analysis of the CSF is less important than simple inspection and bacteriological examination.

2. Estimation of CSF protein concentration is most useful if non-purulent inflammation of cerebral tissues is suspected.

3. Tests indicating increased CSF immunoglobulin concentration are of most value when multiple sclerosis or neurological syphilis are suspected.

4. Estimation of CSF glucose concentration is most useful in cases of suspected tuberculous meningitis.

5. CSF chloride estimation has no value as a diagnostic aid.

FURTHER READING

GANROTT, K., and LAURELL, C. B. (1974). The measurement of IgG and albumin content of the cerebrospinal fluid and its interpretation. *Clin. Chem.*, **20**, 571.

PROCEDURE FOR EXAMINATION OF CSF

CSF very cloudy.—Send for bacteriological examination. Chemical estimations unnecessary.

CSF heavily bloodstained in three consecutive specimens. Cerebral haemorrhage. Chemical estimation unnecessary.

CSF clear or only slightly turbid.—Send for bacteriological examination and for estimation of *glucose* and *protein* concentration. Always send blood for glucose and protein estimation at the same time. If multiple sclerosis or neurological syphilis is suspected, tests indicating increased CSF immunoglobulin levels may be of value.

CSF xanthochromic.—Send specimen, with blood, for protein estimation. Examine under microscope for erythrocytes.

Chapter XXIV

THE CLINICIAN'S CONTRIBUTION TO VALID RESULTS

CHEMICAL pathology is the study of physiology and biochemistry applied to medical practice and overlaps clinical medicine and surgery in particular. An understanding of the basis of chemical pathology, as we have seen, is essential for much diagnosis and treatment.

The medical student or the clinician need not know technical details of laboratory estimations. However, correct interpretation of results requires some understanding both of the acceptable analytical reproducibility and of physiological variations in normal individuals, and this subject will be discussed in the next chapter. It is also desirable that the clinician should be aware of the speed with which tests can be completed, and of laboratory organisation in so far as it affects this; above all he should realise that *the technique of collecting specimens can affect results drastically*, and should co-operate with the laboratory in its attempt to produce rapid, accurate answers, quickly identifiable with the relevant patient. To this end he should understand the importance of *accurately completed forms, correctly labelled specimens, taken at the right time by the right technique, and of speedy delivery to the laboratory.* In an emergency *therapy based on correctly estimated results from a wrongly labelled or collected specimen may be as lethal as faulty surgical technique*: moreover, even if the error is recognised, *precious time could have been saved by a few minutes' thought and care in the first place. An emergency warrants more, not less, than the usual accuracy in collection and identification of specimens.*

REQUEST FORMS

CLINICAL INFORMATION

Control of accuracy of results is largely the concern of the laboratory and most departments take stringent precautions to this end. However, when very large numbers of estimations are being handled it is impossible to be sure that every result is correct; fortunately the error rate is very low. The clinician can play his part by co-operating with the pathologist in minimising the chance of error, not only by taking suitable specimens, but also by giving *relevant* clinical information. "Unlikely" results are checked in most laboratories and, for instance, if the plasma urea concentration of a patient had fallen by 33 mmol/l (200 mg/dl) in 24 hours, both estimations would be repeated before the latest was reported: on the other hand, if it were known that haemodialysis had been performed in the interval the result would have been expected and time, money and worry would have been saved. In this example "post-haemodialysis" would be more informative, and take no longer to write in the "Clinical Details" space than "chronic renal failure".

PATIENT IDENTIFICATION

Accurate information about the patient, including *surname* and *first names* correctly and consistently spelt, and legibly written, *age or date of birth* and *hospital case number* are essential for comparing current with previous results on the same patient. If the laboratory uses "cumulative" reporting, the results on each patient on successive days are entered on one form (Fig. 43): this type of form enables the clinician to follow the progress of the patient more easily than by looking at a single result and, in addition, the laboratory can detect sudden changes, so that the cause can be sought. For the system to work successfully, accurate patient identification is essential: a surprising number of patients have the same names, even when these are apparently uncommon; it is less likely that they will also have the same age in years, and even less probable, the same date of birth; they should not have the same hospital number. Any of these items may be written inaccurately on the form and, unless there is complete agreement with previous details, results can be entered on the wrong patient's record: this can lead to confusion, and even danger to the patient. Many departments use computers to report results: computers cannot think, telephone the ward, or make intelligent guesses, and if the information fed into them is inaccurate it may either be rejected or, worse still, results may be reported as belonging to another patient. *It is at least as important to provide adequate identification to a computerised department as to one depending on filing clerks.*

LOCATION OF THE PATIENT AND CLINICIAN

It is obvious that if the *ward or department* is not stated it may take time and trouble to find out where results should be sent. The consultant's *name*, and the *signature of the doctor requesting the test* are desirable if urgent or alarming results are to be notified rapidly, and advice given about treatment.

The forms designed by pathology departments ask only for information essential to ensure the most efficient possible service to the clinician

DEPT. OF CHEMICAL PATHOLOGY, W.H. Tel. 2440

Please Use a Ball Point Pen 2-120a

Ward A Z   Consultant M^A W   *Hospital:

(Use addressograph label on both copies)

Hosp. No. 12 3 4 5
Surname P....
First Name G.....
D of B 13.7.53

(W.H.)
Gordon
All Saints
W. C. H.
*Circle hosp. of origin

Mr./Mrs./Miss

Clinical Details:—

Renal calculus
Parathyroid adenoma removed
17/7/78

Lab No	Date	Total CO_2 mmol/l	K mmol/l N.R. 3.3-4.6	Na mmol/l	Urea mmol/l	T Protein g/l	Corrected Calcium mmol/l	Crea-tinine µmol/l	Ca mmol/l	P mmol/l	Alk P'ase Units N.R. 90-330	Serum Acid P'ase T.L. Units N.R. 0.4-3.3	Urate mmol/l	Mg mmol/l	Signed
K1D	1/7/77	29	4.7	140	4.1	74	2.69	—	2.72	0.83	240	—	—	—	R
F32	2/9/77	—	—	—	4.4	78	2.49	—	2.59	0.79	—	—	—	—	R
D69	8/11/77	26	4.4	140	4.6	71	2.78	—	2.76	0.89	—	—	—	—	R
P86	4/1/78	—	—	—	3.9	77	2.74	—	2.82	0.85	192	—	—	—	R
C54	6/2/78	28	4.1	137	3.9	78	2.74	—	2.84	0.88	—	—	—	—	R
U84	17/7/78	28	3.9	138	4.6	70	2.66	—	2.65	0.60	187	—	—	—	R
1/7/78 PARATHYROID ADENOMA REMOVED (RIGHT UPPER GLAND)															
R57	18/7/78	27	3.8	137	5.3	67	2.29	—	2.21	0.92	—	—	—	—	R
F04	16/8/78	27	4.0	136	4.1	72	2.43	—	2.43	1.12	—	—	—	—	O

See Path. Lab. Services Guide for information on type of specimen and interpretation of results.

Fig. 43.—A sample of cumulative reporting of the calcium group of tests on a patient with hypercalcaemia due to primary hyperparathyroidism.
In some hospitals all types of result are reported on the same sheet.

and the patient. All pathologists have met the form containing as the only information "Smith"; sometimes not even the investigations requested are stated. Unless the pathologist is endowed with psychical powers it is difficult for him to help the clinician under these circumstances.

ADDRESSOGRAPH SYSTEMS

Some hospitals use "Addressograph" labels. All the required information (except clinical details) can be printed on a set of these labels when the patient is admitted, and one label can be attached to each request form and another to its accompanying specimen. With this system the possibility of error is much reduced.

COLLECTION OF SPECIMENS

COLLECTION OF BLOOD

If the laboratory obtains a clinically improbable result on a specimen, it will usually check this on that specimen. If the second result agrees well with the first, a fresh specimen should be obtained. Before doing this it is essential to try to find out why the first one gave a false answer (if it was false). Contamination of the syringe, needle or tube into which the specimen was collected, although an obvious possibility, is relatively rare. It should not be accepted as the cause until other more common ones have been excluded.

The errors to be discussed below arise outside the laboratory and are, in our experience, relatively common. Any examples given are genuine ones.

Effect on Results of Procedures Prior to Venepuncture

Oral medication.—Specimens should not be taken to measure a substance just after a large oral dose of the same substance has been given. For example, in the presence of iron deficiency, plasma iron concentration may rise to normal, or even high, levels for a few hours after a large dose of oral iron.

Significant hypokalaemia may occur for a few hours after taking potassium-losing diuretics. This is due to rapid clearance from the extracellular fluid, and plasma potassium returns to its "true" level as equilibration occurs between cells and extracellular fluid.

Interfering substances.—Previous administration of a substance may affect plasma levels for some time. For instance, paracetamol reacts in many urate methods, and falsely high urate levels are obtained in the presence of this drug. The effect of interfering substances on analytical methods may be more widespread than generally recognised. A whole issue of a clinical chemistry journal has been devoted to the topic.

Palpation of the prostate.—The prostate contains tartrate-labile acid phosphatase and the plasma concentration of this enzyme is used as an index of spread of carcinoma of the gland (p. 343). However, palpation of a non-malignant prostate may release relatively large amounts of this enzyme into the blood stream. These falsely elevated levels may persist for several days after rectal examination, passage of a catheter, or even after straining at stool. For example, a specimen was received by the department from a patient who had had a rectal examination a few hours previously. The tartrate-labile acid phosphatase level was three times the upper limit of "normal". Three days later a further specimen gave a level which was still twice the upper limit of "normal". Eight days after the examination the concentration was at the lower end of the "normal" range.

Any marginally raised tartrate-labile acid phosphatase level should be checked on a specimen taken a few days later. If possible, blood for this estimation should be withdrawn before performing a rectal examination. If, as is often the case, the result of such an examination suggests the need for the estimation it is best, if possible, to wait a week before taking blood. To save time, the specimen may be taken immediately and *a note made on the request form that rectal examination has been performed.* If the result is normal no further action is required and time has been saved: if it is not, another specimen will be requested.

Effect on Results of the Technique of Venepuncture

Venous stasis.—When blood is taken a tourniquet is usually applied proximally to the site of puncture to ensure that the vein "stands out", and is easier to enter with the needle. If this occlusion is maintained for more than a short time the combined effect of raised intravenous pressure and hypoxia of the vein wall results in passage of water and small molecules from the lumen into the surrounding extracellular fluid. Large molecules, such as protein, and erythrocytes and other cells, cannot pass through the vein wall: their concentration therefore rises. It should be remembered that there will not only be a rise in total protein levels but also in all protein fractions, including immunoglobulins, and day-to-day variations in these can often be attributed to this factor, as well as to changes in posture (p. 310).

Many plasma constituents are, at least partially, bound to protein in the blood stream. Prolonged stasis can falsely raise calcium concentrations (Fig. 44), possibly to high or equivocal levels. If such levels are found it is important to take another specimen, without stasis, for analysis. Other important protein-bound substances include lipids, and thyroxine.

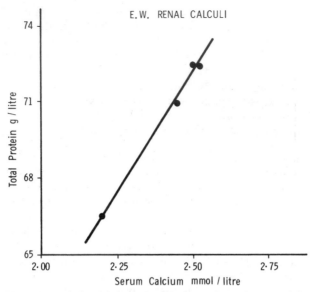

FIG. 44.—Relationship between plasma total protein and calcium concentrations.

Examples of the effects of stasis in a normal subject are given below.

Stasis for	0	2	4	6 minutes
Calcium (mmol/l)	2·38	2·45	2·52	2·58
Total protein (g/l)	72	74	77	80
Albumin (g/l)	39	40	42	43
Haemoglobin (g/dl)	14·7	14·8	15·1	15·5

Prolonged stasis, with associated local hypoxia, may also cause such intracellular constituents as potassium to leak from the cells into the plasma, causing falsely high potassium results.

Some patients have "bad veins", difficult to enter without stasis. Under such circumstances a tourniquet may be applied until the needle is in the vein lumen; if it is now released, and a few seconds allowed before removal of blood, a suitable specimen will be obtained.

Site of venepuncture.—Many patients requiring chemical pathological investigations are receiving intravenous infusions. In veins in the same limb, whether proximal or distal to the infusion site, the drip fluid has not mixed with the whole of the blood volume; local concentrations will therefore be unrepresentative of those circulating in the rest of the patient. Blood taken from the opposite arm will, however, give valid results.

The following example illustrates the point well. The clinical details were given as "post-op". On the previous day the electrolytes had been normal, the plasma urea 16·7 mmol/l (100 mg/dl), and the plasma total protein 67 g/l. The relevant results on the day in question were as follows:

plasma sodium	65 mmol/l
plasma potassium	2·2 mmol/l
plasma bicarbonate	11 mmol/l
plasma protein	52 g/l
plasma urea	8·7 mmol/l (52 mg/dl)

The patient was "doing well". As all measured constituents were diluted it was assumed that dextrose was being infused. A plasma glucose of 50 mmol/l (900 mg/dl) supported this view. A call to the ward confirmed that the specimen had been taken from the arm into which a dextrose infusion was flowing. Analysis of blood taken from the opposite arm gave results almost identical with those of the previous day. It should be noted that during dextrose infusion there will be some hyperglycaemia, even in blood from the opposite arm; if the level is above 10 mmol/l there is likely to be glycosuria. Only if the hyperglycaemia persists after stopping the glucose infusion is the patient likely to have diabetes mellitus.

Such an extreme example is easily detected and, although time has been wasted and the patient subjected to two venepunctures, no serious harm has been done. If, for example, isosmolar saline were being infused, electrolyte results might *appear* to be correct. Under such circumstances the wrong therapy might be applied.

If no veins are available in another limb a suitable sample can be obtained by stopping the infusion, disconnecting the tubing from the needle, aspirating 20–30 ml of blood through this needle and discarding before withdrawing the specimen for analysis.

Containers for Blood

Many hospital laboratories issue a list of the types of container required for each specimen, and this will vary slightly from hospital to hospital. For instance, most departments require that fluoride be added to blood taken for glucose estimation: this inhibits glycolysis, which would otherwise continue in the presence of erythrocytes.

To ensure accuracy of results, laboratories will only accept blood in the correct containers. However, errors can arise if blood is decanted from one container to another. Oxalate and sequestrene (ethylene-diamine tetracetate, EDTA) act as anticoagulants by removing or chelating calcium. Estimation of the latter is therefore invalidated by the presence of these chemicals. The potassium salt of sequestrene is usually used and this will invalidate potassium estimation: sodium oxalate (and sodium heparin instead of lithium heparin) would, of course, upset sodium estimations.

As an example blood was received from the out-patient department apparently in the correct tube. Clinical details were "uretero-sigmoidostomy". A calcium value of 0·4 mmol/l (1·6 mg/dl) was obtained. The plasma potassium was 7·5 mmol/l in spite of the fact that the patient felt very well. Enquiry confirmed that blood had been taken, at the same time, into a sequestrene bottle (for haematological investigations), and that to bring the blood level in the tube "to the mark" some had been tipped into that for chemical pathology.

Effects of Storage and Haemolysis of Blood

Erythrocytes contain very different concentrations of many substances from those of the surrounding plasma (for instance the potassium concentrations are about 25 times as high). If haemolysis occurs the contents will be released and false answers will be obtained. Plasma from haemolysed blood is red and this will be detected by the laboratory. To minimise the chance of haemolysis blood should be treated gently. The plunger of the syringe should not be drawn back too fast, and there should be an easy flow of blood. The needle should be removed from the syringe before the specimen is expelled *gently* into the correct container.

The maintenance of differential concentrations across the red cell wall requires energy, which is supplied by glycolysis. In whole blood outside the body the erythrocytes will soon use up available glucose (hence the need for fluoride in blood glucose specimens), after which no energy source remains: concentrations in erythrocytes and plasma will tend to equalise by passive diffusion across cell membranes. If blood is left unseparated for more than an hour or two the effect on plasma levels will, therefore, be the same as that of haemolysis, with the important difference that, to the naked eye, the plasma looks normal. If the container is undated, or wrongly dated, the error may not be detected. Low temperatures *slow* erythrocyte metabolism so that differential concentrations cannot be maintained, and refrigerating whole blood has the same effect, *in a shorter time*, as allowing it to stand at room temperature. It is therefore important to separate plasma from red cells before storing, *even in the refrigerator*, overnight.

An example of the effect of allowing whole blood to stand at room temperature on plasma potassium and blood glucose levels is given below.

Blood separated after	0	4	8	24 hours
Potassium (mmol/l)	4·0	4·3	4·8	6·4
Glucose (mmol/l)	4·8	3·9	3·0	1·9

Bilirubin is one of the many plasma constituents which deteriorate even in properly stored plasma. This can be minimised by ensuring, when possible, that blood reaches the laboratory early in the working day, at a time when assays are running.

COLLECTION OF URINE

Many urine estimations are carried out on timed specimens. Because of large circadian variations in excretion only qualitative tests, or those of tubular concentrating ability, are performed on random collections. Results are expressed as units/time (for example, mmol/24 h), and to calculate this figure the concentration (for example, mmol/l) is multiplied by the total volume collected. Clearance estimations, too, depend on comparing the amount of, for example, urea excreted per 24 hours with its concentration in the blood (p. 21). In both these examples the accuracy of the final answer depends largely on that of the urine collection: this is surprisingly difficult to ensure. In some cases the difficulty is insurmountable unless a catheter is inserted, and this is undesirable because of the risk of urinary infection: for instance, the patient may be incontinent or, because of prostatic hypertrophy or neurological lesions, be incapable of complete bladder emptying. However, more often errors arise because of a misunderstanding on the part of the nurse, doctor or patient collecting the specimen.

Let us suppose that a 24-hour collection is required between 8 a.m. on Monday and 8 a.m. on Tuesday. The volume *secreted by the kidneys* during this time is the crucial one: urine already in the bladder at the start of the test and secreted some time before should not be included; that in the bladder at the end of the test and secreted between the relevant times *should* be included. The procedure is therefore as follows:

8 a.m. on Monday—Empty bladder completely. *Discard specimen.* Collect all urine passed until:

8 a.m. on Tuesday—Empty bladder completely. *Add this to the collection.*

The error of not carrying out this procedure is very great for short (for example, hourly) collections of urine.

A preservative must usually be added to the urine to prevent bacterial growth and breakdown of the substance being estimated. Before starting the collection the bottle containing the correct preservative should be obtained from the laboratory.

COLLECTION OF FAECES

Rectal emptying is much more erratic than bladder emptying, and cannot usually be performed to order. Estimations of 24-hourly faecal content of, say, fat may vary by several hundred per cent from day to day. If the collection were continued for long enough the *mean* 24-hourly output would be very close to the true daily loss from the body (which includes that in faeces in the rectum at any time). There must be a reasonable compromise on time, and most laboratories collect for between 3 and 5 days. Provided that the patient is not constipated the mean daily loss is usually reasonably representative: if no stool is passed during this period, excess faecal loss of any kind is most unlikely, and the test almost certainly unnecessary. To render the collection more accurate many departments use coloured "markers" (see Appendix to Chapter XII).

Faecal estimations and collections are time-consuming and unpleasant for all concerned. It is important that *every* specimen passed during the time of collection is sent to the laboratory if a worthwhile answer is to be obtained. Administration of purgatives or enemas during the test alters conditions and invalidates the answer.

LABELLING SPECIMENS

It is important to label a specimen accurately to correspond with the accompanying form in all particulars. The date, and sometimes the time of taking the specimen should be included, and should be written *at the time* of collection. If it is done in advance the clinician may change his mind, and the information will be incorrect; there is also the danger of using a container with one patient's name on it for another patient's specimen.

Blood Specimens

Specimens in wrongly labelled tubes may cause danger to one or more patients. The date of the specimen is important, both from the clinical point of view and, as discussed on p. 458, to assess the suitability of the specimen for the estimation requested. It is important to state the time when a specimen was taken, particularly if the concentration of the substance being measured varies during the day: for instance, blood glucose varies according to the time since the last meal. If more than one specimen is sent for the same estimation on one day *each must be timed* so that it is known in which order they were taken.

Urine and Faecal Specimens

Timed urine specimens should be labelled with the date and time of starting and completing the collection, so that the volume per unit time is known. Faecal collections are best labelled, not only with date and time, but with the specimen number in the series of 5-day collections, so that the absence of a specimen is immediately obvious.

SENDING THE SPECIMEN TO THE LABORATORY

If, in an emergency, a result is needed quickly, many types of estimations can be carried out in a short time. However, it is more economical in staff, reagents and time, as well as easier to organise, if estimations are batched as far as possible. For this reason most laboratories like to receive non-urgent specimens early in the morning. A constant "trickle" of specimens may cause delay in reporting the whole batch, and possibly in noticing a result requiring urgent treatment.

If a patient is seen for the first time late in the day and the results are not required urgently, the specimen should be sent to the laboratory with a note to that effect. Plasma can then be separated from cells and stored overnight.

In cases of true clinical emergency, the department should be notified.

TABLE XXXIV

SOME EXTRA-LABORATORY FACTORS LEADING TO FALSE RESULTS

Cause of error	Consequence
Keeping blood overnight before sending to laboratory.	High plasma K, total acid phosphatase, LD, HBD, AST.
Haemolysis of blood.	As above.
Prolonged venous stasis during venesection.	High plasma Ca, total protein and all protein fractions, lipids, T_4.
Taking blood from arm with infusion running into it.	Electrolyte and glucose concentrations approaching composition of drip fluid. Dilution of everything else.
Putting blood into "wrong" bottle or tipping it from this into Chemical Pathology tube.	e.g. EDTA or oxalate cause low Ca, with high Na or K.
Blood for glucose not put into fluoride tube.	Low glucose (fluoride inhibits glycolysis by erythrocytes).
Palpation of prostate by rectal examination, passage of catheter, enema, etc., in last few days.	High tartrate-labile acid phosphatase.
Inaccurately timed urine collection.	False timed urinary excretion values (e.g. per 24 hours). False and erratic renal clearance values.
Loss of stools during faecal fat collection. Failure to collect for long period between markers.	False faecal fat result.

The clinical details warranting urgency may be given on the form, but preferably the laboratory should be warned before the specimen is taken, so that they may be prepared to deal with it quickly. Usually a specimen not known to require urgent attention, and certainly one accompanied by no information about clinical details, will be assumed to be non-urgent. *It is the clinician's responsibility to indicate the degree of urgency.*

Table XXXIV summarises some of the errors which have been discussed in this chapter.

SUMMARY

The clinician's responsibility for maintaining accuracy and speed of reporting of results includes:

taking a suitable specimen of blood
  (*a*) at a time when a previous procedure will not interfere with the result;
  (*b*) from a suitable vein;
  (*c*) with as little stasis as possible;
  (*d*) with precautions to avoid haemolysis.

putting the specimen into the correct container;

labelling the specimen accurately;

completing the form accurately, including *relevant* clinical details;

ensuring that the specimen reaches the laboratory without delay, and that plasma is separated from cells before storing the specimen;

in the case of urine and faeces, collecting accurate and complete timed specimens.

FURTHER READING

CHRISTIAN, D. G. (1970). Drug interference with laboratory blood chemistry determinations. *Amer. J. clin. Path.*, **54**, 118.

Clinical Chemistry (1975). Effects of drugs on clinical laboratory tests. **21**, No. 5 (whole issue).

PANNALL, P. (1971). Pitfalls in the interpretation of blood chemistry results. *S. Afr. med. J.*, **45**, 1184.

ZILVA, J. F. (1970). Collection and preservation of specimens for chemical pathology. *Brit. J. hosp. Med.*, **4**, 845.

Chapter XXV

REQUESTING TESTS
AND INTERPRETING RESULTS

REQUESTING TESTS

WHY INVESTIGATE?

THE clinician now has a great many tests at his disposal. These can often provide helpful information if used critically: if used without thought the results are at best useless, and at worst misleading and dangerous. Writing a request form should not be considered as casting a magic spell, which will benefit the patient merely by doing it.

Investigation should be used to improve the management of the patient: it should not be used to show how "clever" the doctor is. This may seem obvious, but is often forgotten. It is more truly intelligent and satisfying to know what one hopes to gain from investigation and to achieve this aim as economically as possible in time and money. One test is not necessarily better than another because it is newer, more expensive or more difficult to perform: if it *is* better it should be used instead of (not as well as) the other one.

Far from helping the patient, over-investigation may harm him by delaying treatment, causing him unnecessary discomfort or danger, or more insidiously, by using money that may be more usefully spent on other aspects of his care. Of course, under-investigation is just as undesirable as over-investigation: the cost to the patient of omitting a necessary test is just as high as carrying out unnecessary ones.

Before requesting an investigation the doctor should ask himself:

will the answer, whether it be high, low, or normal, *affect my diagnosis?*

will the answer affect the treatment?

will the answer affect my estimate of the patient's *prognosis?*

can the abnormality I am seeking *exist without clinical evidence* of it? If so, *is such an abnormality dangerous*, and *can it be treated?*

If, after careful thought, the answer to *all* these questions is a clear "no", there is no need for the test. If the answer to *any* of them is "yes", the test should be performed.

Sometimes, even if the clinical diagnosis is obvious, the test may still be necessary. For example, a patient may have overt myxoedema. Once treatment is started both the clinical and biochemical features are obscured. Later, another doctor seeing the patient for the first time may not be sure that the patient was ever hypothyroid, and may need to stop the therapy to verify the diagnosis. *One* unequivocally abnormal result before treatment is started (such as a very low free thyroxine index) provides objective documentation: more than one in these circumstances is unnecessary. To perform *all* available tests routinely, such as estimation of neck uptake with TSH stimulation, *and* serum T_4 *and* TSH is unnecessary; such a "battery" should be reserved for genuine diagnostic problems.

WHY NOT INVESTIGATE?

The student is warned against the following unqualified statements.

"It would be nice to know." Ask yourself if it will help the patient.

"We would like to document it fully." Will this extra documentation make any difference to your management of the patient?

"Everyone else does it." They may be right, but do you know their reasons? Perhaps they too do it because everyone else does it. Do not accept anything from anyone (not even this book) uncritically. Re-assess dogma continually. If you lack experience, at least look at the reasoning behind what you read or are taught; use the statements of "experts" as working hypotheses until you are in a position to make up your own mind.

HOW OFTEN SHOULD I INVESTIGATE?

This depends on:

How quickly numerically significant changes are likely to occur. For instance, serum protein fractions are most unlikely to change significantly in less than a week, and the plasma urea concentration will not become significantly abnormal after 12 hours "anuria".

Whether a change, even if numerically significant, will alter treatment. For instance, transaminase levels may alter over 24 hours during acute hepatitis. Once the diagnosis is made this is unlikely to affect treatment. On the other hand, potassium concentrations may alter rapidly in patients on large doses of diuretics, and these *may* indicate the need for treatment.

Unless the patient is receiving intensive therapy of some kind, investigations are very rarely required more than once every 24 hours.

WHEN IS AN INVESTIGATION "URGENT"?

The only reason for asking for an investigation to be carried out urgently is that an earlier answer will alter treatment. This situation is very rare. For example, the doctor should ask himself how often treatment would really be different within the next 12 hours if the urea was 10 mmol/l (60 mg/dl) or 50 mmol/l (300 mg/dl).

INTERPRETING RESULTS

Before considering diagnosis or therapy based on a result received from the laboratory the clinician should ask himself three questions:

1. If it is the first estimation performed on this patient, *is it normal or abnormal*?
2. If it is abnormal, *is the abnormality of diagnostic value* or is it a non-specific finding?
3. If it is one of a series of results, *has there been a change*, and if so *is this change clinically significant*?

IS THE RESULT NORMAL?

Normal Ranges

The normal range of, for example, plasma urea is often quoted as between 3·3 and 6·7 mmol/l (20 and 40 mg/dl). It is clearly ridiculous to assume that a result of 6·5 mmol/l (39 mg/dl) is normal, while one of 6·9 mmol/l (41 mg/dl) is not. Just as there is no clear-cut demarcation between "normal" and "abnormal" for body weight and height, the same applies to any other measurement which may be made.

The majority of a normal population will have a value for any constituent near the mean value for the population as a whole, and all the values will be distributed around this mean, the frequency with which any one occurs decreasing as the distance from the mean increases. There will be a range of values where "normals" and "abnormals" overlap (Fig. 45): all that can be said with certainty is that the *probability* that a value is abnormal increases the further it is from the mean until, eventually, this probability approaches 100 per cent. For instance, there is no reasonable doubt that a plasma urea value of 50 mmol/l (300 mg/dl) is abnormal, whereas it is possible that a urea of 7·5 mmol/l (45 mg/dl) is normal for the individual concerned. It should also be noted that a "normal" result does not necessarily exclude the disease sought: because of intersubject variation, a value within the "normal" range may, nevertheless, be abnormal for the individual. It is perhaps more acceptable to talk of "reference", rather than "normal" ranges.

To stress this uncertainty on the borders of the normal values it is better to quote limits between which values for 90 and 95 per cent of the "normal" population fall, than to give a "normal range". Statistically the 95 per cent limits are two standard deviations from the mean. Although the probability is high that a value outside these limits is

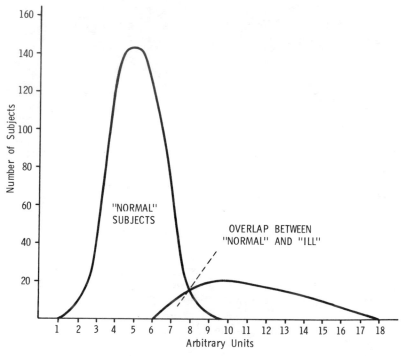

FIG. 45.—Theoretical distributions of values for "normal" and "ill" subjects, showing overlap at upper end of "normal range".

abnormal, obviously 2·5 per cent of the "normal" population may have such a value at either end. The range of variation for a single subject is usually less than that for the population as a whole.

In assessing a result one can only take all factors, including the clinical picture, into account, and reach some estimate of the probability of its being normal.

Physiological Differences

Certain physiological factors affect interpretation of results. For instance, the "normal" levels of plasma urate or iron vary with *sex*, being higher in males than in females (pp. 373 and 383); the plasma urea concentration tends to rise with *age* especially in male subjects, and normal values in children are often different from those in adults; in different parts of the world mean values of many parameters are different, because of either *racial* or *environmental* factors.

It is clear, therefore, that we are not so much expressing "normal" values as the most usual ones for a given population. A plasma urea level persisting at 7·5 mmol/l (45 mg/dl) at the age of 20 suggests mild renal impairment, which may progress to clinically severe damage in later life: the same value at the age of 70 suggests the same degree of renal impairment, but usually the subject will die of some other disease before this becomes severe. In other words, this rising mean value of urea with age is not strictly normal and probably does reflect disease. Note that we are talking of *mean* values for a population of a certain age: in an individual there may be no change with advancing years.

Differences Between Laboratories

From the above discussion it will be seen that, even if the same method is used in the same laboratory, it is difficult to define a normal range clearly. With some constituents interpretation becomes even more difficult if results obtained in different laboratories are compared, because different analytical methods may be used. For most estimations agreement between reliable laboratories is close. However, with certain constituents, such as serum proteins and especially albumin, different methods, even in the best hands, give different results: this reflects the fact that different techniques measure different properties of protein, and that different methods of fractionation do not necessarily separate exactly the same fractions. It should be noted that the definition of "international units", in which the results of enzyme assays are expressed, does not include the temperature at which the assay is performed. Different laboratories, for various technical reasons, may use temperatures varying from 25°C to 37°C: results at these temperatures will be very different, although apparently expressed in the same units. If reproducibility is acceptable, one method is often no better than another for clinical purposes, provided that the results are compared with the "normals" for the laboratory in which the estimation was performed and provided that serial estimations are carried out by the same method.

Is the Abnormality of Diagnostic Value?

Serum or plasma values express only extracellular concentrations. Moreover, some abnormalities are non-specific and of no diagnostic or therapeutic import.

Relationship Between Plasma and Cellular Levels

Intracellular constituents are not easily estimated, and plasma levels do not always reflect the situation in the body as a whole; this is particularly true for such constituents as potassium, which have very high intracellular concentrations compared with those in the surrounding fluid. A normal, or even a high, plasma potassium concentration may be associated with cellular depletion, if conditions are such that the equilibrium across cell membranes is disturbed (for instance, in acidosis).

Relationship Between Extracellular Concentrations and Total Body Content

The numerical value of a concentration depends not only on what is measured, but also on the amount of water in which it is distributed (for instance, mmol/l). A low plasma sodium is not necessarily (or even usually) due to sodium depletion: it is more often due to water excess. Under such circumstances there may even be excess of sodium in the body (p. 59). Conversely, hypernatraemia is more often due to water deficit than sodium excess (p. 59). It is very important to recognise this fact and adapt therapy accordingly. It has already been noted that protein concentrations can be affected by stasis during venesection, but if this factor is eliminated, significant day-to-day variations of protein concentration over a short period of time can be used to assess changes of hydration of the patient (in other words, the amount of protein is not, but that of water is, changing significantly).

Another cause of falsely low plasma sodium concentration is gross lipaemia or hyperproteinaemia (p. 35).

Non-specific Abnormalities

Circulating levels of, for example, albumin, calcium and iron, vary considerably in diseases unrelated to the primary defect in metabolism.

The concentration of albumin, as well as of all other protein fractions (including immunoglobulins) and of protein-bound substances, may fall by as much as 15 per cent after as little as 30 minutes recumbency, possibly due to fluid redistribution in the body. This effect may, partly at least, account for the non-specific low albumin concentration found in quite minor illnesses. In-patients usually have blood taken early in the morning, while recumbent, and tend to have lower values for these parameters than out-patients.

Routine laboratory methods for estimating calcium measure the total protein-bound plus ionised concentrations: changes in albumin levels are associated with changes in those of the calcium bound to it, without an alteration of the physiologically important ionised fraction, and this can occur either artefactually, as discussed on p. 455, or because of true changes in albumin. It is most important not to attempt to raise the total calcium level to normal in the presence of significant hypoalbuminaemia.

Plasma iron is very labile, and levels fall in the presence of anaemia other than that of iron deficiency: giving iron, especially by parenteral routes, to patients with anaemia and a low plasma iron can be dangerous unless other, more reliable, evidence of iron deficiency is present (p. 393).

Note that different methods may give different answers for albumin concentration on the *same specimen*. Even when the same method is used on different specimens from the *same patient* the value obtained depends upon the amount of stasis used during venepuncture, and on whether the patient is ambulant or not. Moreover, a very low albumin concentration may be the *cause* of oedema, or may be the *result* of overhydration. A decision on management based primarily on whether the result is above or below an arbitrary figure of, say, 20 g/l, is to misunderstand the difficulties of interpreting such a figure.

HAS THERE BEEN A CLINICALLY SIGNIFICANT CHANGE?

To interpret day-to-day changes in results, and to decide whether the patient's biochemical state has altered, one must know the degree of variation to be expected in results from a normal population.

Reproducibility of Laboratory Estimations

In reliable laboratories most estimations should give results reproducible to well within 5 per cent: some (such as calcium) should be even more reproducible but the variability of, for example, many hormone assays is much greater. Changes of less than the reproducibility of the method are probably clinically insignificant. The approximate precision of some assays is given in Table XXXV, at the end of this Chapter.

Physiological Variations

Physiological variations occur in both plasma levels and urinary excretion rates of many substances and false impressions may be gained from results of several types of investigation if this fact is not taken into account.

Physiological variations may be regular or random.

Regular variations.—Regular changes occur throughout the 24-hour period (circadian or diurnal rhythms, like that of body temperature), or the month: there may be seasonal variations. There is a marked circadian variation in the urinary excretion of, for example, electrolytes, steroids, phosphate and water. For this reason estimates of excretion carried out on accurate 24-hour collections are more valuable than measurements of concentration of random specimens. In interpreting such results the effect of diet and of fluid intake should be remembered: because of these effects it is difficult to give "normal" values for many urinary constituents (for example, of electrolytes).

Blood glucose concentration varies with the time after a meal, and the concentration of plasma protein and of protein-bound substances varies with posture. Plasma iron shows very marked circadian variation, apparently unrelated to meals or other activity: it may fall by 50 per cent between morning and evening. The circadian variation of plasma cortisol is of diagnostic importance (p. 137), and it should be remembered that, superimposed on this regular variation, "stress" will cause acute rises. To eliminate the unwanted effect of circadian variation, blood should, ideally, always be taken at the same time of day (preferably in the early morning, with the patient fasting). This is not always possible, so that these variations should be borne in mind when interpreting results. Correct interpretation of blood glucose levels requires an estimation on blood taken with the patient fasting, or at a set time after a known dose of glucose (p. 191).

Some constituents show monthly cycles, especially in women (again, compare body temperature). These can be very marked in the case of plasma iron, which may fall to very low levels just before the onset of menstruation (p. 383). There are also, probably, seasonal variations in some constituents.

Although some of these changes, such as those of blood glucose related to meals, have obvious causes, many of them appear to be regulated by a so-called "biological clock", which may be, but often is not, affected by the alternation of light and dark.

Random variations.—Day-to-day variations in, for instance, plasma iron levels are very large, and may swamp regular changes (p. 383). The causes of these are not clear, but they should be allowed for when interpreting serial results. The effect of "stress" on plasma cortisol, and other hormone levels, and the many factors affecting serum protein concentration have already been mentioned.

CONSULTATION WITH THE LABORATORY STAFF

The object of citing the examples given in this and in the preceding chapter, is not to confuse the clinician, but to stress the pitfalls of interpretation of a figure taken in isolation. On most occasions, if care has been exercised while taking the specimen, a diagnosis can be made and therapy instituted *by relating the result to the clinical state of the patient.* However, if there is any doubt about the correct type of specimen required, or about the interpretation of a result, consultation between the clinician and chemical pathologist or biochemist can be helpful to both sides.

Laboratory errors do, inevitably, occur, even in the best-regulated departments. However, a discrepant result should not be assumed to be due to this. Consultation may help to find the cause. The estimation may already have been checked, and, if it has not, the laboratory is usually willing to do so in case of doubt. If it has already been checked, a fresh specimen should be sent to the laboratory after consultation to determine why the first specimen was unsuitable. If the result is still the same every effort must be made to find the cause.

Laboratory staff of all grades often take an active interest in patients whom they are investigating. On their side they often take trouble to keep the clinician informed of changes requiring urgent action, and they may suggest further useful tests: the clinician should reciprocate by giving the pathologist information relevant to the interpretation of a result, and of the clinical outcome of an "interesting" problem. Such exchange of ideas and information is in the best interests of the patient.

SUMMARY

The clinician should use the laboratory intelligently and selectively, in the best interests of the patient. In interpreting results the following facts should be borne in mind:

1. The "normal range" only indicates the *probability* of a result being normal or abnormal.

2. There are physiological differences in normal ranges and physiological variations from day to day.

3. There are small day-to-day variations in results due to technical factors and "normal ranges" may vary with the laboratory technique employed.

4. Using plasma or serum, extracellular concentrations are being measured. These depend on the relative amount of water in the extracellular compartment to that of the constituent measured, and may also be a very poor reflection of intracellular levels.

5. Changes in a given constituent may be non-specific, and unrelated to a primary defect in the metabolism of that constituent.

Finally, when in doubt, two heads are better than one. Pathologists and clinicians tend to see things from slightly different angles and full consultation between the two is in the patient's best interest.

FURTHER READING

RUSSE, H. P. (1969). The use and abuse of laboratory tests. *Med. Clin. N. Amer.* **53**, 223.

BOLD, A. M., and WILDING, P. (1975). *Clinical Chemistry Conversion Scales for S.I. Units, with Adult Normal (Reference) Values.* Oxford: Blackwell Scientific Publications.

A very salutary collection of essays on the general subject of common sense in medicine is:—

ASHER, R. (1972). *Richard Asher Talking Sense.* London: Pitman Medical. (The short section entitled "Logic and the Laboratory" on pp. 166–167 is excellent).

TABLE XXXV
"NORMAL" VALUES

The authors feel strongly that each laboratory should issue its own list of "normal" values, and that clinicians should consult that list, rather than a textbook, when interpreting results. However, so that students should have some idea of the order of magnitude of "normal" values, this list gives MEAN normals for the authors' laboratories. The student should fill in the blank column with the values of his own laboratory. Investigations marked with an asterisk indicate those most likely to have different values in different laboratories: this especially applies to enzyme values.

CV = Coefficient of variation in authors' laboratories at upper end of normal range. (This is a measure of the analytical variability if the estimation is repeated on the same specimen several times.) The percentage variation may differ with level, usually being higher at very low levels. It will be nearer 15–20 per cent for some radioimmunoassays (e.g. prolactin). NOTE THAT THIS IS ANALYTICAL VARIATION ONLY. IT DOES NOT INCLUDE VARIATIONS DUE TO PHYSIOLOGICAL FACTORS OR SPECIMEN COLLECTION, NOR THOSE DUE TO DRUG THERAPY AND INTERFERING SUBSTANCES. These are often greater than analytical variation.

Investigation PLASMA, SERUM or BLOOD (Plasma unless otherwise stated)	Approximate MEAN adult normal values for authors' laboratories		CV ±%	"Normal Range" for student's laboratory (fill in)
	SI or other "New" Units	"Old" Units		
*Amylase	Less than 200 U/l at 37°C	Less than	4	
Bilirubin	17 μmol/l	1 mg/dl	5	
Calcium	2·25 mmol/l	9 mg/dl	1	
"Cortisol" 9 a.m.	440 nmol/l	16 μg/dl	8	
Midnight	110 nmol/l	4 μg/dl		
*Creatine kinase	50 U/l at 35°C	—	4	
Creatinine	88 μmol/l	1·0 mg/dl	6	
Electrolytes Sodium	140 mmol/l	140 mEq/l	1	
Potassium	4 mmol/l	4 mEq/l	2	
Bicarbonate (standard and actual)	24 mmol/l	24 mEq/l	5	
Chloride	100 mmol/l	100 mEq/l		
Gases and pH (whole blood)				
Po_2	12·6 kPa	95 mmHg	1	
pH	7·4	—		
Pco_2	5·3 kPa	40 mmHg		

Glucose (whole blood)					
Fasting	5·3	mmol/l	95	mg/dl	3
*γ-glutamyltransferase (GGT)	30	U/l at 37°C	—		5
*HBD	125	U/l at 35°C	—		6
	90	U/l at 25°C			
Iron male	21·5	μmol/l	120	μg/dl	3
female	14·3	μmol/l	80	μg/dl	
Iron-Binding Capacity (Total)	54	μmol/l	300	μg/dl	4
Lipids					
Cholesterol (Total)	5·2	mmol/l	200	mg/dl	5
*Phospholipids	2·5	mmol/l	8	mg/dl (as P)	
*Triglycerides (fasting)	1·0	mmol/l	88	mg/dl (as triolein)	
Magnesium	0·8	mmol/l	1·6	mEq/l	5
*Phosphatases Acid (Tartrate-labile)	Up to 1·6	U/l at 37°C	Up to 0·9	KA Units	7
Alkaline	71	U/l at 37°C	10	KA Units	3
*Phosphate	1·3	mmol/l	4	mg/dl (as P)	2
*Proteins (Serum)					
Total	60	g/l (about 3 g/l higher for plasma)	6·0	g/dl (about 0·3 g/dl higher for plasma)	2
Albumin	40	g/l	4·0	g/dl	2
Thyroxine (T₄) and indices					
T₄	103	nmol/l	8	μg/dl	8
Free thyroxine index	103		8		10
*Transaminases					
ALT (SGPT)	12	U/l at 35°C	—		6
	9	U/l at 25°C	—		
AST (SGOT)	15	U/l at 35°C	—		5
	12	U/l at 25°C	—		
Urate male	0·33	mmol/l	5·5	mg/dl	3
female	0·27	mmol/l	4·5	mg/dl	
Urea	3·3	mmol/l	20	mg/dl	2

TABLE XXXV (continued)

Investigation URINE or FAECES	Approximate MEAN adult normal values for authors' laboratories			CV ±%	"Normal Range" for student's laboratory (fill in)
	SI or other "New" Units		"Old" Units		
URINE					
5-HIAA	31	μmol/24 h	6 mg/24 h	11	
HMMA (VMA)	20	μmol/24 h	4 mg/24 h	13	
"Cortisol" male	690	nmol/24 h	250 μg/24 h	10	
female	550	nmol/24 h	200 μg/24 h		
FAECES					
Fat (collected over 5-day period)	Up to 18 mmol/24 h (as fatty acid)		Up to 5 g/24 h (as stearic acid)		

THE INHERENT IMPRECISION OF COLLECTING A TIMED SPECIMEN USUALLY OUTWEIGHS ANALYTICAL VARIATION IN THESE URINARY AND FAECAL ESTIMATIONS (AND IN CLEARANCE ESTIMATIONS).

INDEX

The main page references are in **bold** type

486   INDEX